The Mathematical

Approach to

Physiological

Problems

A CRITICAL PRIMER

The Mathematical Approach to Physiological Problems

DOUGLAS SHEPARD RIGGS

THE M.I.T. PRESS
Cambridge, Massachusetts, and London, England

First MIT Press paperback edition, March 1970
Second printing, October 1972
Third printing, August 1976

ISBN 262 18046 4 (hardcover)
ISBN 262 68018 1 (paperback)

Library of Congress Catalog Card Number 63-13822
Printed in the United States of America

To

ROBIN PALMER RIGGS

whose giants and piskies
(though perhaps not quite correct dimensionally)
are in some ways just as interesting

as

$$(\dot{V}\text{cl}) = \dot{Q}_v/C_P$$

or

$$Q_{\text{tot}} = \dot{V}_{\text{tot}} \int_0^\infty C_{\text{art.}t} \, dt$$

"The hurried scribbles in our laboratory notebooks are intelligible only to ourselves, and the seminar and lecture have only a temporary and narrow influence. The published record, however, is permanent; it is there for all time as a source of pride or shame as the case may be."

HUBERT BRADFORD VICKERY

from his tribute to Rudolph J. Anderson,
for twenty-one years Editor-in-Chief of
The Journal of Biological Chemistry.

(*Journal of Biological Chemistry*, *233*, 1249–1250, 1958)

PREFACE

Before settling down in the village of Shepreth, in Cambridgeshire, England, to start working in earnest upon this book, I had the unusual pleasure of taking my family on a month-long youth hosteling trip through Northern England and the Scottish Highlands. For the most part, we bicycled, but occasionally we would make an excursion on foot up the steep hillsides and along the rocky ridges where no bicycle could go. We soon learned that the British discriminate carefully between "hill walking" and "mountain climbing." To be a mountain climber, you must coil 100 feet of nylon rope around you slantwise from shoulder to waist, and have a few pitons dangling somewhere about. You are then entitled to adopt an ever-so-faintly condescending attitude toward any hill-walkers whom you may encounter along the trail, even if you meet them where the trail is practically level. Hill-walkers, on the other hand, remain hill-walkers even when the "walk" turns into a hands-and-knees job up a 40° slope of talus which is barely anchored to the mountain by a few wisps of grass and a clump or two of scraggly heather.

Mathematically speaking, this is a hill-walking book. It is necessarily so, since I myself have never learned the ropes of higher mathematics. But I do believe that the amount of wandering I have done on the lower slopes, the number of sorry hours I have spent lost in a mathematical fog, and the miles I have stumbled down false trails have made me a kind of backwoods expert on mathematical pitfalls, and have given me some practical knowledge of how to plan a safe mathematical ascent of the more accessible physiological hills.

The Mathematical Approach to Physiological Problems is a *primer* in the sense that I hope it will be a *first* guide for the student who is just starting to apply mathematical reasoning to the solution of physiological problems. But the student who thinks that a primer, *i.e.*, an elementary textbook, is by definition an *easy* book will be disappointed. The elements of any new subject are likely to prove difficult at first, and the author confidently expects that the reader will come upon a few 40° slopes herein. If all difficulties were glossed over, the text would be both dull and dishonest.

The Mathematical Approach to Physiological Problems is a *critical* primer in the perfectly straightforward sense that I have found fault with work published by over 100 investigators, including myself. This immediately prompts the question—what right have I to be so critical? Well, it is certainly not a right derived from unusual scientific productivity or superior scholarly attainments. In fact, *the majority of the persons I criticize have made far more extensive, and far more important, contributions to the advancement of science than I have.* Rather do I claim the right, nay, the obligation, of a teacher to choose for his pupils whatever material most vividly and most practically illuminates the subject under discussion. If we want our students to excel us (as we must, else why teach at all?), we should be willing to display to them our falterings and our failings so that at least they will not fail and falter in the same way that we did.

It would be grossly unjust for a book so critical of others to be itself free of error. No such injustice is to be feared. I was told long ago—and I now believe it—that nobody has ever published a mathematical treatise without mistakes. I can only hope that the errors herein contained are neither so numerous nor so obscure as to impair the value of the book seriously. I shall very much appreciate having all such errors brought to my attention. And if the student finds that I myself commit the very sins I preach against, I shall have to fall back upon that hoariest of authoritarian dicta: "Don't do as I do; do as I say!"

The major purpose of a preface is to give the author an opportunity to thank those who have helped him complete his work. By far my heaviest debt of gratitude is to Dr. Joseph Benforado, who set aside his own interests to take charge of my department while I was on sabbatical leave during the academic year 1960 to 1961. To him in particular, and to my staff in general, go my profound thanks for making my sojourn abroad possible. Next, it is a pleasure to thank Dr. E. Basil Verney, then Professor of Pharmacology at the University of Cambridge, for his willingness to accept me as a somewhat shadowy Fellow in his department, for arranging library privileges, and for other courtesies too numerous to mention. I am happy to acknowledge the financial assistance of a Special Research Fellowship from the Division of General Medical Sciences, The National Institutes of Health, United States Public Health Service, during my year abroad, and I am indebted to Professors Otto Krayer, Ernest Witebsky, Louis S. Goodman, and Avram Goldstein for writing letters of recommendation in support of my fellowship application.

Several people have had a direct hand in this book. Of these, the most important has been Dr. J. E. Dowd, who critically reviewed the first eleven chapters and made many valuable suggestions for revision. I am indebted to Dr. Robert A. Spangler and Professor Fred M. Snell for their help with a section of the book which (through no fault of theirs!) was deleted from the final copy. I am grateful to many other colleagues, and to my students, for drawing illustrative material to my attention, for discussing with me portions of the text, and for working several of the problems. The excellence of the figures is due to our Director of Medical Illustration, Mr. Melford D. Diedrick, and to his two sons, Douglas and Curtis. I am grateful to the University Typing Service in Cambridge, and to Diane Denardo, Jane Mitchell, Clare Silverman, and Mary Wolney for typing, correcting, and assembling the manuscript, and I am indebted to my wife for help with the reading and correction of proof. Finally, I would like to express my gratitude to the staff of The Williams & Wilkins Company for their meticulous editing of a difficult manuscript. If *The Mathematical Approach to Physiological Problems* does not measure up to their motto, *Sans Tache*, it is my fault, not theirs.

DOUGLAS S. RIGGS
Eden, New York
March 1963

INTRODUCTORY NOTE

The reprinting of this book in paperback has given me a welcome opportunity to make certain minor corrections in the text. I particularly hope that the lower cost of the present edition will favor its purchase by graduate students, for whom it was primarily intended.

Writing *The Mathematical Approach to Physiological Problems* had a curious effect upon me. I became so fascinated by the behavior of biological feedback systems and (because of my inadequate background in mathematics and engineering) so frustrated by my inability to understand what was being published about them that I quit my chairmanship, begged two years' leave of absence, and went back to school. As a result, I am now completing a new monograph entitled *Control Theory and Physiological Feedback Mechanisms*. But in the process, I have lost some of the mathematical innocence of which I boasted in the Preface to *The Mathematical Approach to Physiological Problems*. If I were to revise it and bring it up to date as some of my more enthusiastic friends have suggested, I might be tempted to rewrite it altogether, thereby running the risk of making it more remote and less appealing. On balance, it seems better to let it stand; its faults being, as I hope, more of omission than of commission. If a thoroughgoing revision is eventually needed, let it be an entirely new book, by a new and younger author, for a new and more sophisticated audience.

Douglas S. Riggs, M.D.
Professor of Pharmacology
Eden, New York
February, 1969
The School of Medicine
State University of New York
at Buffalo

CONTENTS

xi

1

INTRODUCTION: DEFINITIONS AND SYMBOLS

The complexity of biological systems is notorious. Even the simplest particle of virus—incapable of independent life and growth—must consist of at least one macromolecule which is not only intricately patterned itself, but is also able to direct the precise synthesis of like particles from the much simpler substances supplied by the host cell. Biology begins where chemistry has already become immoderately complicated. It is therefore hardly surprising that hitherto so many of the great discoveries in biology have been triumphs of logical deduction from essentially *qualitative* observations assisted only by simple counts or measurements.

Although major break-throughs may still originate from astute qualitative observations, biological investigators are becoming more and more concerned with the *quantitative* behavior of living systems, and with the physical and chemical laws which govern that behavior. The development of accurate methods of microanalysis, the introduction of radioactive and stable isotopes for use as tracers in biological processes, and the design of elaborate electronic apparatus for the rapid and precise measurement of tensions, pressures, flows, voltages, and the like, have given biologists unprecedented tools for the quantitative investigation of biological phenomena. Equipped with microelectrode, pressure transducer, and scintillation counter, the physiological scientist can now make measurements which are astonishingly delicate and precise, and he can make them without disrupting the normal function of the cell, the organ, or the intact animal under observation. Never before have reliable quantitative data been so easy to come by. But unfortunately many biologists, because of disinclination or ill advice, are miserably prepared in mathematics. They find it far easier to gather than to interpret quantitative data.

The primary purpose of this book is to show how, even with quite limited mathematical training, a biologist can use simple mathematical methods to describe living systems and to advance biological theory. These methods are

1

"biophysical" in the sense that they necessarily treat highly complex biological systems as if they were much simpler physical systems, but it is quite wrong to regard them as belonging exclusively to an independent discipline of "biophysics." The elementary mathematical techniques described here should, in fact, lie well within the competence of every biological scientist interested in the function of living things. These are not esoteric mathematical tricks; they are (or ought to be!) everyday tools of the physiologist's trade.

1-1. The Inescapable Need for Making Assumptions and Devising Models

Precisely because living systems are so very complex, one can never expect to achieve anything like a complete mathematical description of their behavior. Before the mathematical analysis itself is begun, it is therefore invariably necessary to reduce the complexity of the real system by making various *simplifying assumptions* about how it behaves. In effect, these assumptions allow us to replace the actual biological system by an imaginary *model* system which is simple enough to be described mathematically. The results of our mathematical analysis will then be rigorously applicable to the model. But they will be applicable to the original biological system only to the extent that our underlying assumptions are reasonable. Hence, the ultimate value of our mathematical labors will be determined in large part by our choice of simplifying assumptions.

In the last three chapters of this book, detailed consideration will be given to making assumptions and devising model systems, but it is important from the very outset for the reader to recognize the *inescapable* need for this approach. Every chapter will deal more or less explicitly with model systems. There is no alternative.

1-2. The Mathematical Preparation Required for Using This Book

It is assumed that the reader of this book will have had algebra and an introductory course in calculus. Although steady states can often be described by algebraic equations, much of the real fascination of quantitative biology lies in the study of transient states, in which, for example, the amount of a drug in the body is decreasing, or the quantity of radioactive iodine in the thyroid gland is increasing, or the concentration of a diffusing substance is changing with the passage of time. The methods of differential and integral calculus are indispensable for dealing with such rate processes, and students who have not already studied calculus are most strongly advised to do so. For the problems in this book, a first course in calculus will suffice. If you cannot arrange to take a formal course, you can acquire the necessary background by working carefully through the text and problems in Daniels' *Mathematical Preparation for Physical Chemistry* (29). Additional training in higher mathematics is all to the good, and indeed is essential for anyone planning to specialize in a field such as biophysics.

Students who have made little use of algebra since secondary school may find that once-familiar rules for manipulating algebraic quantities have grown hazy. Usually, however, a little practice will restore these to mind. You should check any procedure about which you are dubious by substituting sensible numerical values for the algebraic terms and making sure that the procedure is correct for these numbers. To guard against the possibility that a wrong procedure may by chance give a correct result with the particular numbers chosen, you should check by this method of numerical substitution twice with two different sets of values. Infinity, which is not a number, and the unique numbers, zero and unity, should be avoided. Students whose knowledge of logarithms is rusty should review the relations between arithmetic scales and logarithmic scales, the use of negative logarithms, and the difference between common logarithms (to the base 10) and natural logarithms (to the base e) which are summarized in Appendix B.

1-3. Two Comments on Mathematical Work

There are two general aspects of mathematical work which may not be obvious to the student of biology. The first is that mathematical derivations of any complexity take far more time and thought than might be supposed from the final result.* A new mathematical analysis of a biological system cannot be tossed off in the odd hour between afternoon tea and cocktails. You must learn to allow time for false starts and blind alleys as well as for checking and rechecking the mathematical work. Furthermore, if you cannot seem to push a problem beyond a certain stage, it is often wise to break off the attack for a day or two and let the thing simmer gently in your mind. It is likely to be much clearer the next time you have a chance to tackle it.

The second feature which deserves emphasis is that mathematical work is either correct or incorrect, right or wrong; there is no middle ground.

---

* The story is told of a brilliant young instructor in mathematics who, by the regulations of his university, could not be promoted to the rank of Assistant Professor until he had written an acceptable thesis for the Ph.D. degree. By way of encouragement he was granted a year's leave of absence, at the end of which he submitted, duly bound and titled in gold, a thesis consisting of a single typewritten page of mathematical work. After careful scrutiny of this work the Ph.D. committee unanimously recommended that he be awarded the degree of Doctor of Philosophy *summa cum laude*.

From a given set of premises two independent investigators should be led inevitably (though perhaps by different routes) to exactly the same mathematical conclusion. This is quite different from a good deal of biological work. Two equally competent observers may not agree at all about whether a particular cell in a bone marrow smear is or is not a megaloblast, or about what determines the rate of secretion of aldosterone from the adrenal cortex, or about whether to erect a new genus for a new species of beetle. But if two independent investigators decide to use, let us say, the model of inert gas absorption depicted in Figure 13-5, and from this model each derives an equation for the time needed to attain a given proportion of the equilibrium quantity of inert gas in the body, both equations *must* be the equivalent of Equation 13-30. This is the great strength of mathematical analysis.

A necessary corollary of the "right or wrongness" of mathematical work is that an investigator who publishes an erroneous equation has no place to hide! It is therefore prudent to check each calculation, each algebraic manipulation, and each transcription from a table of figures before going on to the next step. Above all, whenever you are engaged in mathematical work you should keep asking yourself over and over and over again, *"Does this make sense?"* and *"Is this of the correct magnitude?"*

1-4. The Cardinal Need for Clear and Precise Definition of Terms

One of the commonest, most frustrating, and least pardonable mistakes made in presenting mathematical work is failure to define terms clearly and precisely. This obscures what may otherwise be a perfectly correct mathematical formulation and tempts the reader to conclude that the author is either discourteous or befuddled. If the author *does* know clearly what he is talking about, he should make it equally clear to the reader. If the author *does not* know clearly what he is talking about, he has no business publishing the mathematical analysis at all. An example will illustrate the difficulty caused by failure to give lucid definitions:

A paper entitled "Affinity, Intrinsic Activity, and Drug Interactions" was recently published as an important part of a symposium on drug antagonism (6). Let us see how the authors define *intrinsic activity*:

> In order to produce an effect, the drug has to satisfy at least two conditions. There must be an *affinity* between the drug and the specific receptors, in other words a pharmacon-receptor complex has to be formed and this complex must have the properties necessary to intervene with the biochemical or biophysical processes in such a way that an effect results. The contribution to the effect per unit of pharmacon-receptor complex is called *intrinsic activity*.

"... contribution to the effect...." The effect of a "pharmacon" (*i.e.*, a drug) is some change in a biological system which is caused by the drug.

For example, in Table I of the paper, the effect produced by the drugs listed is contracture of the isolated rectus abdominis muscle of the frog. This effect might be measured as the force of contracture in units of gram weight. ". . . per unit of pharmacon-receptor complex. . . ." The subsequent text makes it clear that this means per unit of concentration of drug-receptor complex expressed as millimoles per liter. We may suppose, therefore, that the intrinsic activity of a drug producing contracture in the frog rectus might be expressed as force in grams per (millimole per liter) of drug-receptor complex. This supposition is at first strengthened when the authors use intrinsic activity in an equation. (For clarity, the authors' symbol, α, for the intrinsic activity of drug A has been changed to α_A.)

I. DRUG-RECEPTOR INTERACTION

The drug A interacts in a reversible way with the receptor system R and produces an effect by means of the effector system E_R. This is represented by

$$[A] + [R] \xrightarrow[k_2]{k_1} [RA]$$

The authors then present the following equation for the effect, E_A, produced by the formation of the drug-receptor complex:

$$E_A = [RA]\alpha_A = \frac{\alpha_A[r]}{\frac{K_A}{[A]} + 1} \tag{1-1}$$

in which $[A]$ is the concentration of the pharmacon A, $[R]$ is the concentration of free receptors, $[r]$ is the total concentration of receptors (free and occupied), $[RA]$ is the concentration of pharmacon-receptor complex, $K_A = k_2/k_1$ is the dissociation constant of RA, α_A is the intrinsic activity of A.

We shall confine our attention to the first part of Equation 1-1: $E_A = [RA]\alpha_A$. From this, $\alpha_A = E_A/[RA]$, which is in complete agreement with the definition discussed above. It is therefore disturbing to find that the actual values for intrinsic activity listed in Table I of the paper are dimensionless ratios, ranging from zero to one, and having no units at all. Accordingly, we must suppose that intrinsic activity is a ratio of two concentrations because $[RA]$ is unquestionably a concentration. But if this be so, we must also regard E_A as a concentration, even though it was defined as the effect of drug A. To make matters worse, intrinsic activity is apparently measured experimentally as the ratio of two effects. So intrinsic activity is *defined* as one thing, *used* as a second thing, and *measured* as a third thing. The *effect* of a drug seems to be a *concentration*. It is all very confusing.

Let us see whether we can define intrinsic activity rigorously so as to dispel the confusion without doing violence to the really important basic concepts which the authors are trying to put before us:

>Let A be a drug which produces a contracture of the frog rectus by combining reversibly with a particular kind of receptor, R, to give a drug-receptor complex, RA.
>
>Let $[A]$ be the concentration of the drug, in millimoles per liter, to which the muscle is exposed.
>
>Let $[RA]$ be the concentration of drug-receptor complex, in millimoles per liter.
>
>Let E_A be the effect of A measured as the force of contracture in grams.

In general, an increase in $[A]$ will cause an increase in E_A. However, as $[A]$ is made larger and larger, E_A will eventually approach a maximum value asymptotically.

>Let $E_{A.\text{asymp}}$ be the value of this asymptote for drug A.

But the asymptotic value approached is not necessarily the same for all drugs which act by combining with R.

>Let M be a drug whose asymptotic effect is the greatest which can possibly be produced by *any* drug which acts by combining with R.
>
>Let $E_{M.\text{asymp}}$ be the value of this maximum asymptote.

Now we may define the intrinsic activity of A, α_A, as:

$$\alpha_A = E_{A.\text{asymp}}/E_{M.\text{asymp}} \tag{1-2}$$

Note that by this definition the intrinsic activity of M is unity.

Equation 1-2 is an explicit definition of intrinsic activity as a ratio of the asymptotic effects of two drugs. This ratio can actually be measured experimentally. But we cannot use it in any such mass-action equation as Equation 1-1 unless we make certain assumptions about the relation between the concentration of drug-receptor complex, which is not measurable, and the effect of the drug, which is measurable. These assumptions will now be stated. For complete clarity *all* of the equations which follow directly from a given assumption will be listed beneath the statement of that assumption. Some of these equations will then be combined with Equation 1-2 in an attempt to derive $E_A = [RA]\alpha_A$, which is the first part of Equation 1-1.

Assumption I

Assume that, in general, only a fraction, F, of the drug-receptor complexes actually contribute to the effect of a drug. Assume also that for a given drug, acting on a given system, this fraction remains constant. Accordingly, define the *effective* concentration of drug-receptor complex, $[RA]_E$, as

$$[RA]_E = F_A[RA] \tag{1-3}$$

similarly,

$$[RA]_{E.\text{asymp}} = F_A[RA]_{\text{asymp}} \tag{1-4}$$

$$[RM]_E = F_M[RM] \tag{1-5}$$

$$[RM]_{E.\text{asymp}} = F_M[RM]_{\text{asymp}} \tag{1-6}$$

Assumption II

Assume that the effect of a drug is directly proportional to the effective concentration of drug-receptor complex:

$$E_A = J[RA]_E \tag{1-7}$$

$$E_{A.\text{asymp}} = J[RA]_{E.\text{asymp}} \tag{1-8}$$

$$E_M = J[RM]_E \tag{1-9}$$

$$E_{M.\text{asymp}} = J[RM]_{E.\text{asymp}} \tag{1-10}$$

where J, a constant of proportionality, is the force of contracture in grams per (millimole per liter) of effective drug-receptor complex.

Assumption III

Assume that when the effect of a drug has reached its asymptotic value, all of the receptors are combined with the drug:

$$[RA]_{\text{asymp}} = [r] \tag{1-11}$$

$$[RM]_{\text{asymp}} = [r] \tag{1-12}$$

where $[r]$, as before, is the total concentration of receptors.

Assumption IV

Assume that for drug M whose asymptotic effect is the largest possible, every drug-receptor complex contributes to the effect of the drug:

$$F_M = 1 \tag{1-13}$$

Combine Equations 1-2, 1-8, and 1-10:

$$\alpha_A = [RA]_{E.\text{asymp}}/[RM]_{E.\text{asymp}} \tag{1-14}$$

Combine Equations 1-14, 1-4, and 1-6:

$$\alpha_A = F_A[RA]_{\text{asymp}}/F_M[RM]_{\text{asymp}} \tag{1-15}$$

Combine Equations 1-15, 1-11, and 1-12:

$$\alpha_A = F_A/F_M \tag{1-16}$$

Combine Equations 1-16 and 1-13:

$$\alpha_A = F_A \tag{1-17}$$

Notice that in spite of the simplicity of Equation 1-17, we have used all four assumptions in deriving it from Equation 1-2.

Combine Equations 1-17 and 1-3:

$$\alpha_A = [RA]_E/[RA] \tag{1-18}$$

Combine Equations 1-18 and 1-7:

$$E_A/J = [RA]\alpha_A \tag{1-19}$$

Thus, by making appropriate assumptions, we have been able to express intrinsic activity both as a ratio of effects (Equation 1-2) and as a ratio of concentrations (Equation 1-18). But it cannot *also* be expressed, as Equation 1-1 would have us believe, as an effect divided by a concentration. In Equation 1-19, which is the correct equation corresponding to the first part of Equation 1-1, E_A has been divided by J to preserve its original meaning. This is a matter of necessity, not choice.

Clear and precise definition does not mean giving a tediously minute description of the thing defined. It does mean giving enough details so that the reader (and the author!) knows beyond question exactly what each term signifies. It is not sufficient to define P_B as "blood pressure." Is it arterial or venous, left ventricular or capillary? If arterial, is it systolic or diastolic, mean or pulse? If mean, is it measured directly or calculated? If calculated, how? By graphical integration of a curve of pressure plotted against time?

$\bar{P}_{B.art}$ = mean arterial blood pressure, in mm. Hg, calculated by adding to the diastolic pressure one third of the difference between the systolic and diastolic pressures.

This may or may not be a satisfactory way to express mean arterial blood pressure, but at least the reader has no doubt what $\bar{P}_{B.art}$ stands for.

With each definition it is desirable to indicate in what units the quantity will be expressed. If the symbols are at all numerous, it is a real courtesy to the reader to list the symbols, definitions, and units together in a table. This avoids the annoyance of having to search through paragraphs or even pages of text to find the definition of a particular term.

1-5. The Choice of Symbols

A judicious selection of symbols is particularly important for biological scientists who are not so accustomed to dealing with abstract mathematical expressions as are physical scientists. For sheer mathematical correctness, any set of symbols, properly defined and consistently used, would be satisfactory. There would be nothing algebraically wrong in calling pressure V and volume P. But for ease of comprehension, this choice would be disastrous. Hence, though in principle symbols are entirely arbitrary, in practice they should be thoughtfully chosen with three desiderata in mind: ease of recognition of what each symbol stands for, brevity, and conformity with established usage.

To some extent, ease of recognition and brevity are incompatible. Immediate and effortless comprehension can be achieved only by the liberal use of words or familiar abbreviations, while maximum brevity can be achieved only by using a different single letter for each different quantity. For most purposes neither extreme is satisfactory. As an example, consider the following equivalent statements:

1. The renal plasma clearance of a substance, x, is equal to the concentration of x in the urine multiplied by the volume of urine formed per unit time and divided by the concentration of x in the plasma.

2. $$\begin{pmatrix} \text{Renal plasma} \\ \text{clearance of } x \end{pmatrix} = \frac{\begin{pmatrix} \text{concentration} \\ \text{of } x \text{ in urine} \end{pmatrix} \begin{pmatrix} \text{volume of urine} \\ \text{per unit time} \end{pmatrix}}{(\text{concentration of } x \text{ in plasma})} \qquad (1\text{-}20)$$

3. $C = UV/P$ $\qquad\qquad\qquad\qquad\qquad\qquad\qquad\qquad\qquad (1\text{-}21)$

   where C = renal plasma clearance of a substance, x

   U = concentration of x in the urine

   P = concentration of x in plasma

   V = volume of urine formed per unit time

Statement 1 is a clumsy but accurate sentence which simply illustrates the need for a more succinct way of expressing quantitative relationships. Statement 2 is an equation in which brief definitions have been used instead of symbols. This expression wants nothing for clarity, and in fact might be quite suitable for an elementary or nonmathematical presentation. But to carry such wordy items through a long series of algebraic manipulations would be wretchedly inconvenient. Statement 3 wants nothing for brevity, and the single letters are easily treated as algebraic quantities. But C, U, V, and P, once defined, let us say on page 19, will probably be used on subsequent pages without definition. The farther C, U, V, and P wander away from page 19, the harder it is to remember that C is *not* a concentration but a volume per unit time, that V is *not* a volume but a flow, that U does *not* mean urine but is a concentration of some substance in urine, and that P is *not* a pressure but a concentration of the same substance in plasma. May not part of the difficulty which students so often have in understanding the term "clearance" stem from an unfortunate choice of symbols?

A reasonable compromise between cumbersome clarity and abstruse brevity can be achieved by selecting a limited number of primary quantities to be symbolized by single letters. Each primary symbol can then be modified by appropriate accessory symbols (chiefly subscripts) to give it a specific meaning without obscuring its more general meaning. For example, V, however modified, will always be a volume; $\dot{V}$, a flow; $(\dot{V}cl)$, a clearance; C, a concentration; Q, a quantity; $\dot{Q}$, a quantity per unit time, etc. When several modifying subscripts are needed, they will be separated by periods, and will usually be arranged to designate "of what," "where," and "when," in that order. For example, $Q_{z.fat.eq}$ would mean "the quantity of z in fat at equilibrium."

In this system of notation, the expression for renal plasma clearance will be written

$$(\dot{V}_P cl)_x = \frac{C_{x.U}\dot{V}_U}{C_{x.P}} \tag{1-22}$$

where

$(\dot{V}_P cl)_x$ = volume of plasma cleared of x by the kidney per unit of time
$C_{x.U}$ = concentration of x in the urine
$\dot{V}_U$ = volume of urine formed per unit of time
$C_{x.P}$ = concentration of x in plasma

Note that each symbol reminds the reader of the essential information contained in its definition. For example, "volume (V) of plasma (P) cleared (cl) of x (x) per unit of time (dot over the V)." For maximum specificity all of these modifiers are needed; clearance is not a volume, but a volume per unit of time $(\dot{V})$. But by itself $\dot{V}$ would mean an actual flow, hence the need for cl as part of the primary symbol. The clearance might be of whole blood, hence the need for P. It is not clearance in general, but clearance of a particular substance, x. Finally, it might occasionally be necessary to insert "kid" after x to show that we mean clearance by the kidney, not the liver, or lungs, or some other organ. We do not need "kid" in Equation 1-22, because it is clear from the other symbols that we are dealing with the excretion of x in urine, a process which can occur only in the kidney. Now it must be admitted that such a long series of modifiers for a single symbol is unwieldy. Usually, however, at least one of the subscripts may be omitted because the restriction which it imposes is clear from the context. For example, if only *plasma* clearances are being dealt with, the subscript P is superfluous and may be dispensed with. The symbol for clearance of x then becomes $(\dot{V}cl)_x$.

A special problem arises with fractions or ratios for which the primary symbols F and R will be used. Usually a fraction or ratio is the quotient of two variables which are alike in all but one respect. Accordingly, the letter F or R might be followed by the usual symbol for the variables, written as a subscript, but with the distinction between the numerator and the denominator of the fraction shown as a parenthetical ratio. For example, in the previous section we defined α_A, the intrinsic activity of drug A, as $E_{A.\text{asymp}}/E_{M.\text{asymp}}$. If we were to use $F_{E_{(A/M)\text{asymp}}}$ instead of α_A, the reader could hardly forget that the intrinsic activity is the asymptotic effect of A expressed as a fraction of the asymptotic effect of M. But symbols with double subscripts are far too awkward for general use, and it will commonly suffice to append to the general symbol a subscript ratio which will remind the reader of the two things being compared. For example, we might use $F_{A/M}$ as the symbol for the intrinsic activity of A.

Symbols of this sort are so easily understood, so specific, and so readily adapted to various needs that their occasional unwieldiness may be forgiven. They have the further substantial advantage that they often help to show how simple is the relationship between one set of variables and another. Consider again Equation 1-21. By a slight rearrangement, it becomes

$$CP = UV \qquad (1\text{-}21a)$$

which suggests nothing.* But the same rearrangement of Equation 1-22,

$$(\dot{V}_P\text{cl})_x C_{x.P} = \dot{V}_U C_{x.U} \qquad (1\text{-}22a)$$

makes it clear that we are equating the product of a *plasma* concentration and volume-per-unit-time with the product of a *urine* concentration and volume-per-unit-time. Moreover, since the product of concentration and volume-per-unit-time is quantity-per-unit-time, $\dot{Q}$,

$$(\dot{V}_P\text{cl})_x C_{x.P} = \dot{Q}_{x.U} \qquad (1\text{-}23)$$

This makes it easy to see that the renal plasma clearance of x is the volume of plasma per minute from which x would have to be completely removed ("cleared") by the kidney to supply the quantity of x excreted in the urine per minute.

A few symbols are firmly established by long usage. It would be folly to call the ratio of the circumference of a circle to its diameter anything but π, or the base of natural logarithms anything but e. On the whole, however, symbology in the biological sciences has suffered far more from a chaotic lack of uniformity than from any bonds of tradition. A notable exception

---

* Except the ridiculous notion that "chemically pure" equals "ultraviolet." In these days of alphabetical jargon one cannot be too careful. A medical student, asked to define LD_{50} (lethal dose for 50 per cent of animals) replied, "An insecticide used to kill aphids." Presumably he had in mind "Black Leaf 40!"

is the field of respiratory physiology. Recognizing that "students trained with one set of symbols are bewildered by the multitude of different terminologies in the current literature" and that "even experts in the field find difficulty in deciphering the equations published by their colleagues from other laboratories," a group of eminent respiratory physiologists has recently adopted a set of standardized definitions and symbols for use in research and teaching. The symbols described above and to be used hereafter throughout this book have been patterned largely upon their recommendations (82). However, in extending their system to other fields, certain modifications seem desirable. In particular, "volume flow of blood" will be designated $\dot{V}_B$ rather than $\dot{Q}$ so that the symbols Q and $\dot{Q}$ can be used for "quantity" and "quantity per unit of time," respectively. Additional differences will be noted on comparing Appendix D with the list of symbols suggested by the respiratory physiologists.

For the present book, the author has adopted one or two "local ground rules" which he hopes will help the reader to make certain important distinctions. *When the first derivative of a quantity with respect to time remains constant*, it will be symbolized by placing a dot above the primary symbol for the quantity. For example, $\dot{V}_B$ will mean blood flow (volume per unit time) regarded as a constant. $\dot{Q}_{X .\text{in} \rightarrow A}$ will mean the constant quantity of X being infused into A per unit time. *But when the first derivative of a quantity with respect to time is not constant*, but is itself a time-dependent variable, it will be symbolized by a ratio of differentials. Thus, $-dQ_{S .W}/dt$ will mean the rate at which the quantity of S in W is decreasing *at a particular instant of time*, t. $(dQ_S/dt)_{W \rightarrow Z}$ will mean the rate at which S is being transferred from W to Z *at a particular instant of time*, t. For simplicity the subscript, t, meaning "at a particular instant of time t," will *not* be appended to these derivatives, since by the above convention it is clear enough that the derivative is itself a time-dependent variable. For all other quantities whose values change with time, the subscript, t, will be used as a way of reminding the reader that they are time-dependent variables.

Other symbols used in the present text have been drawn chiefly from the extensive list, compiled by the Committee on Letter Symbols and Abbreviations of the American Association of Physics Teachers, which is published in *The Handbook of Chemistry and Physics* (58).

1-6. Substitution of a Single Symbol for a Recurrent Group of Symbols

If two or more symbols appear repeatedly together in a particular grouping, it is often expedient to replace the entire group by a single symbol. This avoids the tedium of copying the individual symbols in the group over and over again. While the expedient symbol may be used freely throughout all intermediate steps of a derivation, in any final equation it should either be replaced by the original symbols, or should be redefined immediately below the final equation.

1-7. The Need for Consistency in the Use of Symbols

Once symbols have been chosen, meticulous care should be taken to use them consistently, at least throughout any single derivation or group of closely related equations. This is not mere pedantry. The language of mathematics is so terse and precise that carelessness in the use of symbols is almost sure to obscure the meaning. If both p and P are used in a series of equations, they must have different meanings. Yet consider the following example taken from a recent elementary textbook of biophysics (106):*

Boothby (1944) presents a derivation of the basic alveolar equation which contains many of the features included in other similar approaches. Thus, it is here considered that Boothby's equation is typical. His derivation is based on the relation:

$$RQ = \frac{F'CO_2}{FO_2 \left(\dfrac{F'N_2}{FN_2}\right) - F'O_2} \qquad \text{(Eq. 28-5)}$$

where RQ is the respiratory quotient, $F'CO_2$ is the fraction of expired air which is CO_2, FO_2 is the fraction of inspired air which is O_2, $F'N_2$ is the fraction of expired air which is N_2, FN_2 is the fraction of inspired air which is N_2, and $F'O_2$ is the fraction of expired air which is O_2.

This equation is the familiar one which includes the definition of *respiratory quotient*, i.e.,

$$RQ = \frac{CO_2 \text{ produced}}{O_2 \text{ absorbed}}$$

it contains a correction based on the nitrogen content of inspired and expired air, for the reduction in volume which follows when the CO_2 produced does not equal the O_2 removed from the air. Assuming the relation $F_x = P_x/P_b$ where F_x is the fraction of the gas in the mixture, P_x is the partial pressure of the gas, and P_b is the barometric pressure, by simple transposition in the above equation:

$$RQ = \frac{P'CO_2}{PO_2 \dfrac{P'N_2}{PN_2} - P'O_2} \qquad \text{(Eq. 28-5a)}$$

---

* From *Essentials of Biological and Medical Physics* by R. W. Stacy *et al.* McGraw-Hill Book Company, 1955. Used by permission. I am particularly grateful to Professor Stacy for graciously allowing me to quote this passage even though he knew that I was intending only to criticize it.

Using Equation 28-5, and assuming $PN_2 = P_b - PO_2$,

$$RQ = \frac{PN_2 \cdot P'CO_2}{PO_2 \cdot P'N_2 - PN_2 \cdot P'O_2}$$

and

$$RQ = \frac{(PB - PO_2) \cdot PCO_2}{PO_2 \cdot P'N_2 - (P_b - PO_2)P'O_2}$$

If we divide through by P_b and assume $p'N_2 = P_b - p'H_2O - p'O_2 - p'CO_2$:

$$RQ = \frac{(1 - FO_2)P'CO_2}{FO_2(P_b - PH_2O - PCO_2) - P'O_2} \qquad \text{(Eq. 28-6)}$$

If we solve this equation for $p'O_2$:

$$P'O_2 = FO_2(P_b - P'H_2O) - P'CO_2 \left[\frac{1 - FO_2(1 - RQ)}{RQ} \right] \qquad \text{(Eq. 28-7)}$$

Equation 28-6 is known as the *basic alveolar equation*; Equation 28-7 is an application of the basic equation.

In this single page, symbols are used altogether 67 times. Twenty of these uses are inconsistent. Although the authors needed only 15 symbols, they have actually used 22. In 2 instances they have used 3 different symbols for the same quantity: $p'CO_2$ and PCO_2 should both be $P'CO_2$; and $p'H_2O$ and PH_2O should both be $P'H_2O$. The omission of the primes is a particularly serious error for it changes the meaning from "partial pressure in the expired air" to "partial pressure in the inspired air." Finally, the reader who has the patience to struggle through this maze of mismatched symbols is ill-paid for his work. For although the authors call Equation 28-6 "the basic alveolar equation," all of its terms refer to inspired or expired air, not alveolar air. Furthermore, Equation 28-7 for the partial pressure of oxygen in the *expired* air seems a ridiculously complicated way of calculating a quantity which can be measured directly, and the modifications needed for applying this equation to *alveolar* air are by no means self-evident to the reader.

1-8. Abstract Symbols, Concrete Meanings

Too many biological scientists shy away from any kind of mathematical analysis because they have an unreasoning fear of replacing everyday terms by abstract symbols. For such biologists, "rate of excretion of sodium in the urine" somehow seems to lose its meaning, and to become foreign and forbidding when called "$\dot{Q}_{Na.u}$." Now in fact it *is* a great convenience to be able to manipulate symbols mathematically as if they had no more meaning than the x's, y's, and z's of an algebra text. But mathematical manipulation cannot rob our symbols of the carefully defined concrete meanings which we ourselves originally gave them. It is therefore a mistake to let symbols float about in your mind like featureless wisps of mist above a marsh. Name them! Think of them as what they really are—commonplace quantities drawn from the laboratory bench. Read them aloud with their units attached:

> "P_{bar} millimeters of mercury"
> "N beats per minute"
> "$C_{x.art}$ millimoles of x per liter of arterial blood"
> "$\dot{V}_A$ liters of alveolar ventilation per minute"

Finally, train yourself to translate simple equations into plain English which anyone can understand. "$-dQ_S/dt = kQ_{S.t}$" simply means "at any instant of time, t, the rate at which the quantity of S is decreasing is directly proportional to the quantity of S present at that time." This kind of translation of equation to statement (and the reverse translation of statement to equation) will soon strip the mystery away from mathematical formulations and will encourage you to seek the concrete meaning which lies at the heart of even the most abstract expression.

1-9. The Need for Skepticism

All too frequently, students are willing to accept on faith whatever mathematical formulations they encounter in their reading. And why not? After all, mathematics is *the* exact science, and presumably an author would not express his theories or his conclusions mathematically without due regard for mathematical rigor and precision. It is only by bitter experience that we learn never to trust a published mathematical statement or equation, particularly in a biological publication, unless we ourselves have checked it to see whether or not it makes sense (see Chap. 14). Misprints are common. Copying errors are common. Blunders are common. Editors rarely have the time or the training to check mathematical derivations. The author may be ignorant of mathematical laws, or he may use ambiguous notation. His basic premises may be fallacious even though he uses impressive mathematical expressions to formulate his conclusions. The present book—in text and in exercise—points again and again at published errors. *But there are bound to be similar errors in this very book. Caveat lector!* Let the reader beware!

EXERCISES. CHAPTER I

Exercise 1

In discussing how the pain-relieving effect of certain analgesic drugs might be measured in patients with postoperative pain, Denton and Beecher (32) offer the following definitions of "AD 90 per cent":

Page 1052: " ... the AD 90 per cent range ... is defined as the analgesic dose giving moderate to complete relief of pain in 90 per cent of the subjects ..."

Page 1053: " ... the establishment of the AD 90 of each new drug and of morphine ... was done in the following manner. First ... the percentage of moderate to complete pain relief was calculated on the basis of the effect of the total number of doses given. ..."

Page 1147: " ... the AD 90 per cent ... is defined as the analgesic dose giving moderate to complete relief of pain in 90 per cent of the trials."

Moreover, in Table 6, page 1147, we are told that a dose of 21 to 25 mg. of *dl*-isomethadone given to 27 patients for a total of 64 doses provided 87.5 per cent relief.

A. Under what circumstances would per cent of doses and per cent of subjects be the same?

B. What is meant by "per cent of the trials"?

C. Judging from Table 6, which of the definitions was actually used?

Exercise 2

Loewe (72) has published an interesting account of the relationships between the dose, *D*, of a drug, the magnitude of the effect, *E*, of the drug, and the tolerance of the individuals receiving the drug. The following are the only statements in the paper which bear upon the definition of *tolerance*.

Page 693: "The individuals in any population differ in tolerance, *T*. When large, randomly selected populations are tested in single-dose groups, the percentage of responders at any selected *E* level increases with the dose and hence can be taken as a measure of *T*."

Page 693: "The value *n* of *T* for any particular *D* can be regarded as the tolerance of the *n*th individual in a population of 100 individuals arrayed in series of increasing tolerance."

A. Is the author referring to the tolerance of an individual or of a population of individuals?

B. If you wanted to measure the tolerance of a single mouse to hexobarbital, how would you go about it on the basis of the definitions given?

C. Can you offer a more satisfactory and rigorous definition of tolerance?

Exercise 3

In a paper on the genesis of cerebrospinal fluid (CSF), Lloyd and Taylor (71) offer the following definitions of certain symbols:

Page 401: "At *A*, *a* represents an immediate reabsorption operating on the ultrafiltrate produced at a rate denoted by the constant *u* before it mixes with the bulk of ventricular CSF";

Pages 401–402: "Secretion of solute out of" the ventricular cerebrospinal fluid space "occurs at B, at a rate denoted by the transfer constant, s, and diffusion of solute between CSF and plasma takes place at D, the transfer constant being f."

A, B, and D represent locations in the cerebrospinal fluid system, and are properly identified on a diagram of the model system which the authors give in Figure 2 of their paper.

State in your own words what you think a, u, s, and f are. How would you define them if you were writing the paper?

Exercise 4

A triad of E's.

A. In a review of the role of calcium ions in neural processes, Brink (15), page 262, makes the following statement: "Experience shows that the stability of the membrane depends upon many variables:

$$E = E \ (\text{Pd}, \ L\text{-fraction}, \ CO_2, \ K^+, \ Ca^{++}, \ H^+, \ T, \ \text{etc.})"$$

What do the first and second E's mean?

B. Riker (90), discussing the excitation of effector cells by acetylcholine, states: "... it is essential that the activating or suprathreshold concentration of acetylcholine be presented to the effector cell within a minimal period of time. More generally, this may be represented as $E = k \ (dc/dt)$, where E is excitation and c is acetylcholine concentration."

In what units would you measure E, excitation?

Note about calculus textbooks: A more modern text than Daniels (29) is: S. Lang, *A First Course in Calculus*, Addison-Wesley Publishing Co., Reading, Mass., 1964. This and its companion volume, *A Second Course in Calculus* by the same author, are written in an exceptionally lucid style which is particularly welcome to the beginner. The material discussed has been so carefully selected that the student is in no danger of bogging down in minutiae or of being overwhelmed by technical terms.

A more comprehensive text of deservedly high reputation is: G. B. Thomas, *Calculus and Analytic Geometry*, Addison-Wesley Publishing Co., Reading, Mass., 4th Edition, 1968.

𝟤

DIMENSIONS AND UNITS

The variables which physiologists measure—pressures, flows, voltages, concentrations, and so on—are perfectly ordinary physical entities whose fundamental nature is in no way altered by their occurrence in a living organism. To deal mathematically with such variables, we must be able to describe them both qualitatively and quantitatively. In other words we must know exactly *what kind* of thing each variable is, as well as *how much* of it there is.

2-1. Dimensions

Any well-defined physical entity can be described qualitatively by specifying its *dimensions*. For example, distance, whether long or short, whether backwards or forwards, whether measured in miles or microns, has but one dimension, length (symbolized $[L]$). But area has two linear dimensions, length and breadth. Since these dimensions are qualitatively the same, being lengths in different directions, we say simply that area has the dimensions "length squared" or $[L^2]$. Likewise any volume, regardless of size or shape, has the dimensions $[L^3]$. In a precisely analogous way, all well-defined physical entities which are not electromagnetic quantities* can be described by simple products of the powers of four fundamental dimensions: mass, $[M]$, length, $[L]$, time, $[T]$, and temperature, $[\theta]$. For example, velocity is defined as distance per unit of time. Hence the dimensions of velocity are $[L/T]$, usually written $[LT^{-1}]$. Acceleration is a change in velocity per unit time; therefore, its dimensions are those of velocity/time, *i.e.*, $[LT^{-1}/T]$ or $[LT^{-2}]$. The dimensions of some common physical entities are listed in Table 2-1. Others are to be found in the

* For the description of electromagnetic quantities, either the dielectric constant of a vacuum, ϵ, or the magnetic permeability of a vacuum, μ, is commonly used in addition to mass, length, time, and temperature. Although the dimensions of electromagnetic quantities are not discussed here, the principles to be outlined are fully applicable to them.

TABLE 2-1

Dimensions of some common physical entities

Physical Entity	Description	Dimensions
Mass		M
Length		L
Time		T
Temperature		θ
Area	Length squared	L^2
Volume	Length cubed	L^3
Velocity	Distance per unit time	LT^{-1}
Acceleration	Rate of change of velocity	LT^{-2}
Flow	Volume per unit time	L^3T^{-1}
Resistance to fluid flow	Pressure difference per unit flow	$ML^{-4}T^{-1}$
Density	Mass per unit volume	ML^{-3}
Force	Mass $\times$ acceleration	MLT^{-2}
Momentum	Mass $\times$ velocity	MLT^{-1}
Pressure	Force per unit area	$ML^{-1}T^{-2}$
Work, energy	Force $\times$ distance	ML^2T^{-2}
Power	Work per unit time	ML^2T^{-3}
Viscosity		$ML^{-1}T^{-1}$
Fluidity	Inverse of viscosity	$M^{-1}LT$

TABLE 2-1—*Continued*

Physical Entity	Description	Dimension
Diffusivity	Coefficient of diffusion	L^2T^{-1}
Surface tension	Force per unit length	MT^{-2}
Thermal capacity	Heat per unit mass-degree	$L^2T^{-2}\theta^{-1}$
Gas law constant	Energy per mole-degree	$ML^2T^{-2}\theta^{-1}$

Dimensions of other physical entities may be found in *The Handbook of Chemistry and Physics* (58).

section entitled "Definitions and Formulas" in the *Handbook of Chemistry and Physics* (58).

The designation of mass, length, time, and temperature as the fundamental dimensions from which the dimensions of other physical entities are derived is reasonable, but by no means inescapable. By defining terms appropriately, it is perfectly possible to replace mass by force as a "fundamental" dimension, and this system is sometimes actually used. It is not even necessary to have four fundamental dimensions (68, 118). But although the choice of which dimensions to regard as fundamental is, to a considerable extent, arbitrary, the mass-length-time-temperature system is most widely used and will serve our needs perfectly well. There is no reason to confuse matters by discussing alternative systems in any detail.

Some physical entities are so poorly defined that they cannot be described dimensionally. For example, the "hardness" of a mineral is commonly measured by its ability to scratch, or to be scratched by certain standard minerals which have been assigned arbitrary degrees of hardness ranging from 1 to 10. This enables one to say, for example, that apatite has a hardness of 5 and diamond a hardness of 10. But there is no answer to the question, "5 what?" or "10 what?" For that matter, one cannot even say that diamond is twice as hard as apatite. Now this inability to define hardness in terms of mass, length, time, and temperature does *not* mean that hardness is unimportant or unscientific. It *does* mean that there is no way in which hardness can be related mathematically to other properties of a solid by means of a general theoretical equation. In the same way, many an important biological entity cannot be dealt with mathematically. Consider, for example, the terms "second-degree burn," "excitation of a nerve," "memory," and "unconsciousness." Each of these is important. Each can be given a reasonably precise working definition. Yet not one of them can be assigned dimensions, or be related to other variables by means of a theoretical equation. In contrast, the *area* of a second-degree burn, the electrical *current* needed to excite a nerve, the *time* between learning a fact and forgetting it, and the maximum *acceleration* which a man can tolerate under specified circumstances without losing consciousness are all dimensional entities which are legitimate grist for the mathematical mill.

The importance of knowing the dimensions of each variable is that there are certain rigid rules which specify how dimensional entities can be related to each other. To be valid, any equation which states a general or theoretical relationship between two or more variables must follow these rules for dimensional correctness just as it must follow arithmetical rules for numerical correctness. These two requirements are quite independent of each other. For example, the equation

$$(7 \text{ inches}) \ (10 \text{ inches}) \ = \ (700 \text{ square inches}) \tag{2-1}$$

is arithmetically wrong. But it is dimensionally correct because "inches" $[L]$ times "inches" $[L]$ yields an answer in "square inches" $[L^2]$ as it should. Most biologists check their arithmetic carefully, so that numerical errors of this kind are rare. In contrast, carelessness with dimensions, or plain ignorance of dimensional restrictions, is all too common. Equations are actually published which say something like

$$(7 \text{ peacocks}) \ (10 \text{ pencils}) \ = \ (70 \text{ square inches}) \tag{2-2}$$

Now 7 times 10 certainly equals 70, but the equation as a whole is utter nonsense.

The following paragraphs describe a simple technique for determining whether or not an equation is dimensionally correct. A much more extensive elementary discussion of dimensional analysis has been published by Abramson (1).

To be dimensionally correct, an equation must obey the following rules:
1. Quantities added to or subtracted from each other, and the resulting sum or difference must all have the same dimensions.
2. Quantities equal to each other must have the same dimensions.
3. Any quantity may be multiplied by or divided by any other quantity without regard to dimensions. However, the resulting product or quotient must have appropriate dimensions so that rule 2 is not violated.
4. The dimensions of a physical entity are entirely independent of its magnitude. Hence Δx and dx must have the same dimensions as x, even though the differential, dx, is "infinitesimally" small.
5. Pure numbers, such as e (the base of natural logarithms), Avogadro's number, number of moles, etc., have no dimensions. Two important general classes of pure numbers or dimensionless quantities are:

a. Exponents, including all logarithms.
b. Ratios of two quantities with the same dimensions, *e.g.*, π; all trigonometric functions such as sine, cosine, etc.; partition coefficient of a substance between oil and water; specific heat.
6. The dimensions of a quantity are not affected when it is multiplied or divided by a nondimensional quantity.

The easiest way to check an equation for conformity with these rules is to rewrite the equation with only dimensional symbols. Deal first with any sums or differences. Make sure that they follow rule 1, and then replace them by the appropriate dimensional symbols so that the dimensional equation will have no sums or differences. Next, replace all other symbols by their dimensional equivalents. Dimensionless quantities are simply disregarded or, more precisely, considered equal to unity.* Check any exponents to make sure that they are dimensionless as required by rule 5a. Finally, combine the dimensional symbols on each side of the equation just as though they were algebraic quantities. For example, $[ML^2T^{-2}L^{-2}T/M]$ reduces to $[T^{-1}]$, *i.e.*, to a rate of some sort; $[ML^{-1}T^{-1}/ML^{-1}T^{-1}]$ is a dimensionless ratio. The simplified dimensional equation should reduce to an identity as required by rule 2.

If the equation violates any of the rules either it contains a numerical constant with hidden dimensions (Section 2-6), or it is empirical (Section 2-9), or it is impossible. You should *never* neglect to check the dimensional correctness of any equation which you yourself derive, or which you find puzzling in published work.

Three simple examples of dimensional analysis follow:

Example 1

The volume, V, of the frustrum of a right cone with base radius, r_1, top radius, r_2, and altitude, h, is

$$V = \pi \frac{h}{3} (r_1^2 + r_1r_2 + r_2^2) \tag{2-3}$$

First, the three quantities in the parentheses, r_1^2, r_1r_2, and r_2^2 all have the dimensions $[L^2]$. Hence, by rule 1, they may be added, and their sum will also have the dimensions $[L^2]$.

---

* Pure numbers are often said to have the dimension "unity." For example, dimensionally the ratio of two sides of a triangle would be $[L/L]$, and if we treat this algebraically it does equal unity. But suppose we have an expression such as $F_A + F_B + F_C = 1$ where F_A is the fraction of A, F_B the fraction of B, and F_C the fraction of C in a mixture of A, B, and C. The F's are all dimensionless ratios. But if we were to replace each F by unity, we would have $1 + 1 + 1 = 1$, which is ridiculous. Actually this difficulty should not arise, for the entire sum should be treated dimensionally as a single unit.

Second, π is a dimensionless number and can be disregarded. The same is true of the $\frac{1}{3}$, the numerical coefficient of h.

Therefore, the dimensional transcription of Equation 2-3 is as follows:

$$V = \pi \frac{h}{3} (r_1{}^2 + r_1r_2 + r_2{}^2)$$

$$[L^3] = [L] \qquad [L^2]$$

(2-4)

which agrees with rule 1.

The equation is therefore dimensionally correct. However, it is not necessarily correct *in toto*, since we have not checked the dimensionless quantities. For example, had the equation been written

$$V = \pi \frac{h}{4} (r_1{}^2 + r_1r_2 + r_2{}^2)$$

(2-5)

it would still have been dimensionally correct though numerically wrong. But an error of the following sort:

$$V = \pi \frac{h}{3} (r_1 + r_1r_2 + r_2)$$

(2-6)

would have been revealed at once by the dimensional analysis.

Example 2

Consider the equation for a "decay" process where the quantity disappearing at a given time, t, ir proportional to the quantity, Q_t, present at that time:

$$\frac{-dQ}{dt} = kQ_t$$

(2-7)

In Equation 2-7, k is a proportionality constant. Does k have any dimensions? Assuming that Q_t is expressed as a mass, and letting $[k]$ stand for "the dimensions of k," the dimensional equation corresponding to Equation 2-7 is

$$[MT^{-1}] = [k] [M]$$

(2-8)

Solving for $[k]$,

$$[k] = [T^{-1}MM^{-1}] = [T^{-1}]$$

(2-8a)

Equation 2-8a tells us that k must have the dimension of reciprocal time, *i.e.*, k must be a rate. Indeed, this kind of "inverse time" proportionality constant is usually called a rate constant.

Incidentally, Equation 2-7 may be integrated to the form

$$Q_t = Q_0 e^{-kt} \tag{2-9}$$

where Q_0 is the quantity at time zero, and Q_t is the quantity at time t. Since $-kt$ is an exponent, it should have no dimensions (rule 5a). The dimensions of $-kt$ are $[T^{-1}T]$, or $[T/T]$ which is a dimensionless ratio, as it should be. Since e itself is dimensionless, the dimensional equation for Equation 2-9 is simply

$$[M] = [M] \tag{2-10}$$

Example 3

In Section 1-4, two equations for the effect of drug A were discussed, namely,

$$E_A = C_{RA}\alpha_A \tag{1-1}$$

and

$$E_A = C_{RA}\alpha_A J \tag{1-19}$$

In these equations (in which the symbol for concentration has been changed from square brackets to C),

E_A = effect of drug A; *e.g.*, force of contracture
Dimensions: force, $[MLT^{-2}]$

C_{RA} = concentration of drug-receptor complex; *e.g.*, millimoles per liter
Dimensions: number per volume, $[L^{-3}]$

α_A = the intrinsic activity of drug A. By Equation 1-18, $\alpha_A = C_{RA_E}/C_{RA}$
Dimensions: none

J = drug effect per unit of concentration of effective drug-receptor complex; *e.g.*, force of contracture per (millimole per liter) of C_{RA_E}
Dimensions: $[MLT^{-2}L^3] = [ML^4T^{-2}]$

The dimensional equation corresponding to Equation 1-1 is

$$[MLT^{-2}] = [L^{-3}] \tag{2-11}$$

This is impossible, and Equation 1-1 is therefore not dimensionally correct. But the dimensional equation corresponding to Equation 1-19 is an identity:

$$[MLT^{-2}] = [L^{-3}ML^4T^{-2}] = [MLT^{-2}] \tag{2-12}$$

Hence Equation 1-19 is dimensionally correct.

2-2. Relation between Dimensional Categories and Common Descriptive Terms

In scientific work, the term "area" can mean only something with the dimensions $[L^2]$, and anything with the dimensions $[L^2]$ can properly be described as an area. Similarly, the terms "volume," "acceleration," "force," "momentum," "power," and several others are each uniquely associated with a particular set of dimensions. But since dimensional categories are necessarily broad, it is hardly surprising that often two or more commonly used terms refer to different entities with the same dimensions. For example, work and energy both have the dimensions $[ML^2T^{-2}]$, and flow and clearance both have the dimensions $[L^3T^{-1}]$. Worse still, there are some terms in very common use whose meaning is broader than any single dimensional category. Consider, for example, the term "quantity." Basically, quantity means anything that can answer the question, "How much?", and it may therefore be used with equal propriety for things as different as matter, energy, and electricity. However, in the present text the symbol Q_x will be used to mean "quantity of x" only when x is some form of matter. Even when so restricted, quantity may be expressed as a number (dimensionless), e.g., number of moles; or as a mass $[M]$, e.g., grams; or as volume $[L^3]$, e.g., milliliters of a gas under standard conditions; or even as "counts per minute" $[T^{-1}]$, e.g., amount of a radioactive isotope. The term "concentration" is, if anything, even broader, being whatever answers the question, "How thickly scattered is x through y?" Now while x *may* be measured as a mass, and y is *usually* measured as a volume, the term concentration may also have a bewildering variety of other meanings (Table 2-2). (Note, by the way, that "mass of x per volume of x" is not a concentration, but is the density of x, even though it has the same dimensions, $[ML^{-3}]$, as one kind of concentration.) When dealing with such very general terms it is obviously important to know just what measures of quantity or concentration are being used in a particular situation.

2-3. Units of Measurement

Having characterized an entity qualitatively by specifying its dimensions, we must next characterize it quantitatively by specifying its magnitude. The magnitude of any dimensional entity is expressed as a multiple or fraction of some arbitrarily selected *unit of measurement*. For example, the accepted standard unit of length is the International Prototype Meter (the distance between two lines on a platinum-iridium bar), which has been determined to be 1,553,164.13 times as long as the wave-length of the red cadmium line in air at 760 mm. Hg pressure and 15°C. Similarly

TABLE 2-2

Various meanings of the word "concentration," as illustrated by three solutions which are commonly called "40 per cent ethanol," "0.9 per cent sodium chloride," and "epinephrine, 1:1000"

| Dimensions | Description | Concentrations of | | Epinephrine 1:1000 as the hydrochloride | |
		40% Ethanol	0.9% Sodium chloride	"Base"	"Salt"
L^{-3}	Molarity	6.890 moles/L.	0.1540 moles/L.	5.46 mM./L.	5.46 mM./L.
M^{-1}	Molality	10.86 moles/kg. H_2O	0.1546 moles/kg. H_2O		
L^{-3}	Normality		0.154 equiv./L.		
L^{-3}	Osmolarity		0.287 osmoles/L.		
ML^{-3}	gm./L. = mg./ml.	317.4	9.00	1.000	1.199
Unity	% (v/v)	40.0%			
ML^{-3}	% (w/v)	31.74%	0.900%	0.100%	
Unity	% (w/w)	33.35%	0.896%		
ML^{-3}	"mg. per cent" (= mg./100 ml.)		546 mg. % of Cl^- 354 mg. % of Na^+		
ML^{-3}	"1:n"*			1:1000	

"Epinephrine base" refers to the uncharged form of the amine with the empirical formula: $C_9H_{13}O_3N$. The hydrochloride "salt" of epinephrine is $(C_9H_{14}O_3N)^+Cl^-$.

* The expression "1:n" *ought* to mean "one part of the solute in n parts of solution." Thus 1:1000 ought to mean "one gram of solute per 1000 grams of solution." But in practice, at least among pharmacologists (who, by and large, can afford to be sloppy) and with aqueous solutions, 1:1000 is ordinarily taken to mean one *gram* of solute per 1000 *milliliters* of solution.

arbitrary, but precisely defined units have been specified for mass, time, and temperature. Many (too many!) other units, adopted for convenience or inherited from ancient systems of measurement, are defined as multiples or fractions of these fundamental units. For example, an inch (U. S. A.) is exactly 1/39.37 of a meter, or 100/39.37 = 2.540005 centimeters. The dimensionless number, 2.540005, is called a unit *conversion factor* and is simply "centimeters per inch." All of the common conversion factors are listed in the *Handbook of Chemistry and Physics* (58).

There is a very simple way to convert any magnitude from one set of units to another. Suppose you want to convert "15 pounds of weight per square inch" to "kilograms of weight per square centimeter":

$$15 \text{ lb./in.}^2 = 15\left[\frac{(1 \text{ lb.})}{(1 \text{ in.})(1 \text{ in.})}\right] \qquad (2\text{-}13)$$

$$1 \text{ lb.} = 1/2.2 \text{ kg. (approximately)} \qquad (2\text{-}14)$$

$$1 \text{ in.} = 2.54 \text{ cm. (approximately)} \qquad (2\text{-}15)$$

Substituting the values from Equations 2-14 and 2-15 into Equation 2-13,

$$15 \text{ lb./in.}^2 = 15\left[\frac{(1/2.2 \text{ kg.})}{(2.54 \text{ cm.})(2.54 \text{ cm.})}\right] \qquad (2\text{-}16)$$

or, performing the indicated arithmetic,

$$15\left[\frac{1}{(2.2)(2.54)(2.54)}\right]\left[\frac{\text{kg.}}{\text{cm.}^2}\right] = 1.056 \text{ kg./cm.}^2 \qquad (2\text{-}16\text{a})$$

2-4. The Importance of Consistency in the Use of Units

To avoid numerical chaos, one *must* adhere to the same units throughout any particular series of calculations. Consider, for example, the definition of renal plasma clearance of x which was discussed in Section 1-5:

$$(\dot{V}\text{cl})_x = C_{x.U}\dot{V}_U/C_{x.P} \qquad (1\text{-}22)$$

It is customary to express renal plasma clearance in units of milliliters per minute. Now suppose that $C_{x.P}$ were reported as 24 mg. of x per liter of plasma, and that $C_{x.U}\dot{V}_U$ were reported as 0.8 gm. per hour. If these numerical values were blindly substituted in Equation 1-22, we might conclude that

$$(\dot{V}\text{cl})_x = 0.8/24 = 0.0333 \text{ ml. of plasma cleared per min.}$$

which would be completely wrong. To get clearance in *milliliters per minute*, $C_{x.U}\dot{V}_U$ *must* be expressed in, say, milligrams *per minute*, and $C_{x.P}$ in milligrams *per milliliter*. (Grams, or any other unit of quantity, consistently used, would serve as well because clearance has no dimension corresponding to quantity.) Therefore, we must convert 0.8 gm. per hour to milligrams per minute:

$$0.8 \text{ gm./hr.} = 0.8 \ (1000/60) \text{ mg./min.} = 13.33 \text{ mg./min.}$$

and 24 mg. per liter to milligrams per milliliter:

$$24 \text{ mg./L.} = 24 \ (1/1000) \text{ mg./ml.} = 0.024 \text{ mg./ml.}$$

and substitute these values into Equation 1-22:

$$(\dot{V}\text{cl})_x = (13.33 \text{ mg./min.})/(0.024 \text{ mg./ml.})$$

$$= 556 \text{ ml. of plasma cleared/min.}$$

which is correct.

If $C_{x.U}\dot{V}_U$ is to be reported routinely in grams per hour and $C_{x.P}$ in milligrams per liter, it would be perfectly legitimate to combine the two conversion factors, $1000/60$ and $1/1000$, in such a way as to write Equation 1-22 in the *special* form:

$$(\dot{V}\text{cl})_x \text{ in ml./min.} = 16{,}667 \left(\frac{C_{x.U}\dot{V}_U \text{ in mg./hr.}}{C_{x.P} \text{ in mg./L.}} \right) \qquad (2\text{-}17)$$

If this is done, it will be clear that Equation 2-17 is valid only for a particular set of units, while the original *general* equation, 1-22, is valid for any units whatsoever provided that the units chosen are consistently used. If an equation of restricted validity is published, the restrictions should be stated, and each numerical conversion factor should be clearly identified.

2-5. The Uselessness of Units in Deriving Theoretical Equations

In the work-a-day world of balances, stopwatches, and voltmeters, units of measurement are indispensable. But when one retires to the privacy of one's study to derive a *theoretical* relationship between a set of variables, units become a needless encumbrance. For example, in sufficiently dilute solutions of an undissociated solute, the osmotic pressure, Π, is directly proportional to the number of moles, n, of solute per unit volume, V, of solution, and directly proportional to the absolute temperature, T. The constant of proportionality, R, is the ideal gas law constant. Hence,

$$\Pi = R(n/V)T \qquad (2\text{-}18)$$

(Notice that Equation 2-18 is identical in form to the equation of state for an ideal gas.) Now the *numerical* value of R will depend upon what units of measurement are chosen for V, T, and Π. But until we apply Equation 2-18 to actual data, it is quite unnecessary to specify any set of units, or to give R *any* numerical value. The validity of the theoretical relationship itself transcends all systems of measurement and requires only that the units be consistently used. Because of their universality and simplicity, the majority of the equations in the present book will be theoretical equations, independent of units of measurement, or, more correctly, true for any consistent set of units.

2-6. Numerical Constants and "Hidden" Dimensions

Although the use of a dimensionless unit-conversion factor restricts an equation to a particular set of units, it does not obscure the dimensional characteristics of the equation. Suppose, however, that one of the terms in an equation is a constant which has dimensions. If the numerical value of the constant, rather than its general symbol, is used in the equation, not only will the equation be restricted to a particular set of units, but it may also appear to be dimensionally incorrect because the dimensions of the constant are "hidden" in what may seem to be a pure number. Consider, for example, Einstein's famous equation relating mass, m, and energy, E,

$$E = mc^2 \qquad (2\text{-}19)$$

where c is the velocity of light in a vacuum. Since c is a constant whose value is very nearly $3 \cdot 10^{10}$ cm. per second, this equation *might* be written

$$E = 9 \cdot 10^{20} m \qquad (2\text{-}20)$$

which, when mass is expressed in grams, is correct if, and only if, energy is expressed in ergs. Now the constant in Equation 2-20, $9 \cdot 10^{20}$, actually has the units (cm. per sec.)^2 and therefore has the dimensions $[L^2 T^{-2}]$. If you were not aware of this, you might think that $9 \cdot 10^{20}$ was a pure number and conclude that the equation was dimensionally incorrect.

2-7. Hidden Dimensions: Weight and Mass

Occasionally the numerical constant with hidden dimensions is unity. Then it may not even appear in the equation! As an example, let us examine the relationship between *weight* and *mass*.

By definition, force is equal to the product of mass and acceleration:

$$F = ma \qquad (2\text{-}21)$$

where F = force $[MLT^{-2}]$
      m = mass $[M]$
      a = acceleration $[LT^{-2}]$

Like Equation 2-18 this equation is a general equation which is valid so long as consistent units are used. If mass is expressed in grams, length in centimeters, and time in seconds, then the force must be in dynes. If mass is expressed in pounds, length in feet, and time in seconds, force must be in poundals.

By definition also, weight, W, is the force resulting when the acceleration due to gravity, g, acts upon a mass, m:

$$W = mg \qquad (2\text{-}22)$$

This equation is a mere variant of Equation 2-21, and is likewise true regardless of the choice of units. As before, if m is in grams, and g is in centimeters per second-squared, weight will be in dynes.

Now suppose we define a new unit of weight, the gram weight, as equal to 980.665 dynes. Then

$$W \text{ in gm.} = W \text{ in dynes}/980.665$$
$$= (m \text{ in gm.})(g \text{ in cm./sec.}^2)/980.665 \qquad (2\text{-}23)$$

where $1/980.665$ is a dimensionless conversion factor. Note that Equation 2-23 is no longer valid for all consistent sets of units but, like Equation 2-17, is restricted to those specified.

On the surface of the earth, the standard value of g is 980.665 cm./sec.^{2} Hence (neglecting slight variations in g from place to place) Equation 2-23 may be rewritten *for weights on the earth's surface:*

$$W \text{ in gm.} = (m \text{ in gm})(1/980.665)(980.665 \text{ cm./sec.}^2) \qquad (2\text{-}24)$$

or,

$$W \text{ in gm.} = (m \text{ in gm.}) \qquad (2\text{-}25)$$

With the specified restrictions, Equation 2-25 is *numerically* correct. But dimensionally it is incorrect because weight, which has the dimensions of force, $[MLT^{-2}]$, cannot be equated to a mass, $[M]$. What went wrong? In essence, we defined our units so that a dimensional "constant" (the acceleration of gravity at the earth's surface) became equal to unity, and could too easily be dropped from Equation 2-24 because it made no numerical difference. The likelihood of confusion is increased by having the same names for certain units of weight and mass, so that whenever there is any doubt about which is meant we are forced to say, for example, "pound weight" or "pound mass." Thanks to man's invasion of outer space, the idea of weightlessness has become so familiar that the fundamental difference between weight and mass probably needs no further emphasis.

2-8. Hidden Dimensions: Solubility of a Gas in a Liquid

As a further (and more subtle) illustration of the confusion which may be caused by hidden dimensions, let us consider the two commonest ways in which the *solubility of a gas in a liquid* is expressed. The simpler definition is:

The *solubility*, S, of a gas in a liquid is the ratio

$$\frac{\text{(Concentration of the gas in the liquid phase)}}{\text{(Concentration of the gas in the gas phase)}}$$

at equilibrium; *i.e.*, when the partial pressure of the gas in solution equals the partial pressure of the gas above the solution.*

According to this definition, gas solubility is simply the equilibrium distribution ratio for the gas between the two phases. The numerical value of S depends upon what gas and what liquid are being used, and upon the temperature. The choice of units of concentration is entirely arbitrary, but the same units must, of course, be used for both numerator and denominator so that the ratio itself will be independent of the units chosen. Provided the gas obeys *Henry's law*, the ratio is also independent of pressure since the concentration of the gas in both phases is directly proportional to its partial pressure. As is true of any ratio of like quantities, S is dimensionless.

Physiologists usually prefer to express the solubility of a gas in a liquid by the Bunsen solubility coefficient, α:

The *Bunsen solubility coefficient*, or *absorption coefficient*, α, is the volume of gas, reduced to standard conditions (273°K., 760 mm. Hg), which will dissolve in one volume of liquid when the partial pressure of the gas is one atmosphere (760 mm. Hg).

From this definition it might appear at first glance that α is also a dimensionless ratio, being the volume of gas per volume of solution under specified conditions. But if this were true, α would not be very useful since volume of gas per volume of solution varies directly with the partial pressure of the gas. In fact, α is (*volume of gas/volume of solution*) *per unit of partial pressure*, and accordingly it has the dimensions of reciprocal pressure $[M^{-1}LT^{2}]$.

Thus far there seems to be no particular difficulty. Confusion about dimensions may, however, arise when we consider the relationship between S and α. This relationship may be derived as follows:

Let P_x = the actual partial pressure of a gas, x, both in the gas phase and in the liquid phase.

$V_{x(\text{act } T,P)\text{gas}}$ = the volume of x, measured at actual temperature and pressure, in the gas phase.

$V_{x(\text{std } T,P)\text{gas}}$ = the volume of x, measured at standard temperature and pressure, in the gas phase.

* This expression for gas solubility is the Ostwald solubility coefficient, often symbolized by λ.

$V_{x(\text{std } T,P)\text{liq}}$ = the volume of x, measured at standard temperature and pressure, in the liquid phase.

$V_{\text{gas(act } T,P)}$ = the volume of the gas phase, measured at actual temperature and pressure.

V_{liq} = the actual volume of the liquid phase.

T_{act} = the actual absolute temperature (held constant).

T_{std} = standard absolute temperature, *i.e.*, 273°K.

P_{act} = the actual pressure.

P_{std} = standard pressure, *i.e.*, 760 mm. Hg = 1 atmosphere.

By the definitions given above,

$$S = (V_{x(\text{std } T,P)\text{liq}}/V_{\text{liq}})/(V_{x(\text{std } T,P)\text{gas}}/V_{\text{gas(act } T,P)}) \qquad (2\text{-}26)$$

and

$$\alpha = (V_{x(\text{std } T,P)\text{ liq}}/V_{\text{liq}})/P_x \qquad (2\text{-}27)$$

Combine Equations 2-26 and 2-27 by eliminating the numerator which is the same in both:

$$S(V_{x(\text{std } T,P)\text{gas}}/V_{\text{gas(act } T,P)}) = \alpha P_x \qquad (2\text{-}28)$$

The partial pressure of x in the gas mixture at actual temperature and pressure is equal to the total pressure multiplied by the fraction of x in the mixture:

$$P_x = \left(\frac{V_{x(\text{act } T,P)\text{gas}}}{V_{\text{gas(act } T,P)}}\right) P_{\text{act}} \qquad (2\text{-}29)$$

Reduce $V_{x(\text{act } T,P)\text{gas}}$ to standard conditions:

$$V_{x(\text{act } T,P)\text{gas}} = V_{x(\text{std } T,P)\text{gas}} \left(\frac{T_{\text{act}}}{T_{\text{std}}}\right)\left(\frac{P_{\text{std}}}{P_{\text{act}}}\right) \qquad (2\text{-}30)$$

Substitute this value in Equation 2-29:

$$P_x = \left(\frac{V_{x(\text{std } T,P)\text{gas}}}{V_{\text{gas(act } T,P)}}\right)\left(\frac{T_{\text{act}}}{T_{\text{std}}}\right) P_{\text{std}} \qquad (2\text{-}31)$$

Substitute the value of P_x from Equation 2-31 into Equation 2-28:

$$S = \alpha\left(\frac{T_{\text{act}}}{T_{\text{std}}}\right) P_{\text{std}} \qquad (2\text{-}32)$$

Equation 2-32 is the *general* equation for the relationship between S and α.

Now the actual numerical value of α for a particular gas, a particular liquid, and a particular temperature will depend upon what units are chosen for pressure. Values of α are customarily given for pressures expressed in atmospheres. But when pressure is expressed in atmospheres, $P_{std} = 1$ atmosphere, and *numerically*, Equation 2-32 may be written

$$S = (T_{act}/T_{std})\alpha = (T_{act}/273°\text{K.})\alpha \qquad (2\text{-}33)$$

Here again, a unit of measurement has been so chosen that a dimensional constant, P_{std} equals unity, and may therefore be omitted from the equation without altering its *numerical* validity. Unfortunately this makes it easy to forget the dimensional constant entirely, and to argue wrongly from Equation 2-33 that since S is dimensionless, α must also be dimensionless.

2-9. Dimensions in Empirical Equations

An empirical equation is one which has been derived from experimental observations rather than from any underlying theory. It is chosen to represent, as well as possible, the trend of the relationship between two or more observed variables, and is therefore the equation of some curve which fits the experimental points. Now even when an experiment has been properly designed and carefully executed, the experimental results themselves (and hence any equation chosen to represent them) will inevitably be influenced by a large number of factors, immediate and remote, besides the independent variable which is under study. It is impossible even to list all of these factors, much less to specify their dimensions, and it would therefore seem *a priori* unreasonable to insist that every empirical equation be precisely correct dimensionally. In fact, however, the great majority of the empirical equations used to describe biological phenomena *must* be dimensionally correct as the following argument will show.

Suppose we have observed a relationship between a variable, x, whose dimensions are $[x]$, and a variable, y, whose dimensions are $[y]$. We want to find an empirical equation which will describe this relationship. The equation will contain not only x and y but also certain arbitrary constants, or parameters (see Section 4-1), a, b, c, k_1, k_2, k_3, etc., whose numerical values are to be determined from the observed points so that the equation will fit the points. Now with most empirical equations, the form of the equation itself unequivocally determines the dimensions of the constants in strict accordance with dimensional rules, because the equation not only specifies exactly how each constant is related to the variables, x and y, but it also specifies what exponents x and y may have. It is only the *numerical* values of the constants, and not their dimensions, which are determined by the outcome of the experiment. Take, for example, the polynomial

$$y = a + bx + cx^2 + dx^3 + \cdots \qquad (2\text{-}34)$$

of which the equation for a straight line (first two terms only),

$$y = a + bx \tag{2-35}$$

is a special case. We see at once that according to our dimensional rules, the dimensions of a must be $[y]$, of b, $[y]/[x]$, of c, $[y]/[x^2]$, etc., or, in general, if the nth term is kx^{n-1}, k must have the dimensions $[y]/[x^{n-1}]$. Similarly, in the equation for a sum of exponentials (see Section 6-12),

$$y = ae^{kx} + be^{k'x} + ce^{k''x} + \cdots \tag{2-36}$$

each k must have the dimensions $[x^{-1}]$, and each of the constants a, b, c, etc., must have the dimensions $[y]$.

Occasionally, however, one must use an empirical equation of such a form that the dimensions of the constants *cannot* be specified, in advance of the experiment, from the dimensions of x and y. Suppose we have the equation

$$y = ax^b \tag{2-37}$$

where, as before, x and y are experimental variables with the dimensions $[x]$ and $[y]$, and a and b are constants to be determined from the experimental data. Now even though the dimensions of x and y are precisely known, the dimensions of x^b will depend upon what numerical value of b is obtained from the experiment. Furthermore, the numerical value of b will be determined not only by the relation between x and y but also by whatever extraneous factors are responsible for the scatter of the experimental points. In other words, b, as estimated, will be subject to more or less experimental error. In this uncomfortable situation, we have but two alternatives. Either we must not insist upon dimensional correctness for empirical equations of this type, or we must allow the constant, a, to assume whatever dimensions are needed to make the equation dimensionally correct. But the second alternative would make a mockery of the whole idea of dimensions, for dimensions would then be subservient to numerical data; and what might be a dimensionless constant for one set of observations would become a constant with dimensions for another set. We must therefore accept the first alternative and exempt this sort of empirical equation from the ordinary requirements of dimensional correctness.

Consider the following example. Certain biological variables, notably basal metabolic rate, seem to vary directly with the *surface area of the body*. To facilitate comparison of the metabolic rates of animals of different sizes, it is important to have some way of calculating surface area, for to measure it directly, even in a dead animal, is exceedingly tedious. In contrast, both body length and body weight are relatively easy to measure. If we assume that the density, D, of the body (grams per cubic centimeter) and the acceleration due to gravity, g, are constant, body volume, V, can be estimated from body weight, W:

$$V = W/gD \qquad (2\text{-}38)$$

In order to avoid dimensional confusion in the following discussion we shall use volume, not weight, although in fact custom and convenience sanction the substitution of weight for volume.

Now let us examine specifically the calculation of surface area in the dog. If all dogs had the same shape, surface area, A, would be directly proportional both to the square of body length, l:

$$A = k_1 l^2 \qquad (2\text{-}39)$$

and to the $\frac{2}{3}$ power of body volume, V:

$$A = k_2 V^{2/3} \qquad (2\text{-}40)$$

where k_1 and k_2 are dimensionless constants of proportionality. Unfortunately, dogs vary considerably in shape so that if the constants are estimated for dogs of average proportions, equation 2-39 will tend to underestimate, and equation 2-40 will tend to overestimate the surface area of obese dogs. The opposite will be true for lean dogs. Recognizing this, Cowgill and Drabkin (24) assumed that surface area would be proportional to some power of body weight if a term were introduced which would correct for differences in the nutritional status (*i.e.*, the "shape") of different dogs. For each dog, d, they therefore calculated a "nutritive correction factor," $(Fcor)_d$, defined as

$$(Fcor)_d = (V^{1/3}/l)_{\max}/(V_d^{1/3}/l_d) = 0.34/(V_d^{1/3}/l_d) \qquad (2\text{-}41)$$

where $\quad V_d$ = the volume of dog d in cubic centimeters

$\qquad l_d$ = the length of dog d in centimeters (distance from nose to anus measured over the belly)

$(V^{1/3}/l)_{\max}$ = the ratio (cube root of volume/length) for the most obese dog which they could obtain

Using this correction factor, Cowgill and Drabkin proposed the following equation for the surface area of dog d, A_d, in square centimeters:

$$A_d = k(Fcor)_d V_d^b \qquad (2\text{-}42)$$

where k is a proportionality constant which, by analogy with k_2 in Equation 2-40, may be assumed to be dimensionless. Cowgill and Drabkin then measured surface area, volume, and length in seven dogs, and found the value of b which gave the smallest variation of k when k was calculated separately for each dog by means of Equation 2-42. The resulting equation,

$$A_d = 6.67(Fcor)_d V_d^{0.700} \qquad (2\text{-}43)$$

or its equivalent obtained by substituting in Equation 2-43 the value of $(Fcor)_d$ from Equation 2-41,

$$A_d = 2.268 l_d V_d^{0.367} \tag{2-44}$$

has been used more extensively than any other for calculating surface area in the dog.

Equations 2-43 and 2-44 are not quite correct dimensionally. But in view of the small number of observations available for calculating the values of k and b (actually, only five degrees of freedom) it is remarkable how close the calculated exponent of V came to its theoretical value. Cowgill and Drabkin themselves point out that a dimensionally correct equation,

$$A_d = 4.381 l_d^{0.725} V_d^{0.425} \tag{2-45}$$

agrees with their seven observations almost as well as does Equation 2-44. Furthermore, the exponents in Equation 2-45 are exactly the same as those in DuBois' equation for the surface area in man (35). So even in this example, in which we were prepared to accept a dimensionally incorrect equation if it best fitted the available data, there seems to be no compelling reason to prefer it to an alternative equation with correct dimensions. (For further consideration of this particular problem, see Exercise 10.)

EXERCISES. CHAPTER 2

Exercise 1

Of the following 20 terms, some are not definable mathematically, some are definable but dimensionless, and some have dimensions. Place each term in one of these three categories, and give dimensions for the dimensional quantities.

1. Absolute zero
2. Avogadro's number
3. Barometric pressure
4. Cardiac output
5. Clearance
6. "Counts per minute" as a measure of a radioactive isotope
7. $-dN/dt$ where N is number of bacteria; t is time
8. Excitation
9. Log_{10} of the injected dose
10. Neonatal period
11. Molarity
12. pH
13. Q_{10}, the temperature coefficient

14. R, the constant in the ideal gas law equation
15. Senility
16. Stroke volume
17. Turbulence
18. $\dot{V}_W \int_{t_1}^{t_2} C_{x.in}\, dt$

    where $\dot{V}_W$ is the flow of blood into region W, $C_{x.in}$ is the concentration of x in milligrams per liter in the inflowing blood, and t is time
19. White blood cell count
20. Zygote

Exercise 2

In *The Handbook of Chemistry and Physics*, Ed. 41, 1959–1960, (58) on page 3163 there is a list of units and conversion factors for quantities with the dimensions of acceleration $[LT^{-2}]$. One of the quantities listed is "radians per second per second." Is this properly listed?

Exercise 3

In a recent review, Wilbrandt and Rosenberg (116) (Equation 1, page 115) give the following equation for the rate of transport of a substance, S, which is a substrate for a carrier substance, C:

$$v = D' (CS_1 - CS_2)$$

where v = transport rate
D' = permeability constant
CS_1 = concentration of carrier-substrate complex on side 1 of the membrane
CS_2 = concentration of carrier-substrate complex on side 2 of the membrane

What are the dimensions of the "permeability constant," and what does it represent?

Exercise 4

Check the following equation, published by Osmond and Hoffer (81) for dimensional correctness:

"An equation for schizophrenia could be written

$$S = f(Ad + p + c + Sp + d + t)$$

where Ad = adrenolutin
p = personality and factors which have led to it
c = culture
Sp = specific perceptual functions of the brain which are altered
d = duration of illness
t = treatment given"

Exercise 5

In considering the analysis of data obtained after administration of an N^{15}-labeled amino acid, Wu *et al.* (123) discuss the use of multiple exponential equations of the general form:

$$"a = \sum_{i=1}^{n} c_i e^{-\lambda_i t} \qquad i = 1, 2, \cdots n$$

where a = isotope concentrations of the compound at time t, and c_i and λ_i are constants." In footnote 6 they make the following statement: "When some of the λ values are equal, $e.g.$, $\lambda_1 = \lambda_2 = \lambda_3$, the exponential terms involved assume the form $C_1 e^{-\lambda t} + C_2 t e^{-\lambda t} + C_3 t^2 e^{-\lambda t}$, instead of $C_1 e^{-\lambda_1 t} + C_2 e^{-\lambda_2 t} + C_3 e^{-\lambda_3 t}$, and the conventional method of semilogarithmic plot cannot be used to evaluate this."

Comment upon the dimensional validity of this statement.

Exercise 6

In discussing the so-called "standard" clearance of urea, Smith (98) in a footnote on pages 66 and 67 says, "But, paradoxically, Moller, McIntosh, and Van Slyke confused rather than clarified the problem by calling $U\sqrt{V}/B$ a 'clearance'; it does not signify a virtual volume of cleared blood but is instead the product of a mathematical operation in which a (true) clearance, UV/B, is multiplied empirically by $\sqrt{V}$; hence it is only a presumed clearance, predicated on the assumption that below 2 cc./min. UV/B will decrease in proportion to $\sqrt{V}$, and calculated for convenience to the value of $V = 1.0$." In this quotation, U is the concentration of urea in the urine, B is the concentration of urea in the blood, and V is the volume of urine per minute.

What are the dimensions of $U\sqrt{V}/B$? What mistake is made in the sentence quoted?

Exercise 7

Wolf (121) ascribes the following equation to Adolph (2). Is the equation dimensionally correct?

$$t = 1.3 + L_{H_2O}$$

where t is the time in hours needed to excrete a water-load of L_{H_2O} expressed as per cent of body weight.

Exercise 8

In a paper on changes in renal function with age, Lewis and Alving (69) give the following empirical equation for normal men over 40 years of age:

$$\overline{BN} = 7.56 + 0.1119 \, A$$

where $\overline{BN}$ = blood urea nitrogen in milligrams per 100 cc. of blood
A = age in years

What are the dimensions of 7.56? of 0.1119?

Exercise 9

Criticize the following, taken from West's *Textbook of Biophysical Chemistry* (114):

> *Page 364:* "The term 'velocity' involves a quantity factor and a time factor. In general, it refers to quantity per unit time." "The velocity of a chemical reaction indicates the mols of reactant, or reactants, undergoing change per unit time."

> *Page 365:* "*Unimolecular reactions.* When a substance, such as N_2O_5, decomposes according to a first-order reaction, the following mathematical relations may be worked out. Suppose a = the initial concentration of N_2O_5 in mols per liter and x = the number of mols of N_2O_5 which is changed into products in t seconds. After the reaction has proceeded for t seconds, the concentration of $N_2O_5 = a - x$. The velocity of the reaction at time t is, according to the mass law:

$$V = k(a - x)"$$

> *Page 365:* "k is called the velocity constant or the specific reaction rate."

Exercise 10

A. Write a dimensionally correct *general* equation for surface area as a function of body length and body volume.

B. Suppose that you have made a series of measurements of the surface area, the body length, and the body volume of a number of dogs. How would you go about calculating the most appropriate values for the constants in your general equation?

C. Apply the method you have devised to the following data taken from Table 1 of the paper by Cowgill and Drabkin (24):

Dog no.	4	6	1	7	2	5	3
Length from nose to anus (cm.)	51	62	74	76	98	100	103
Body weight (gm.)	3,390	5,350	5,450	10,150	17,250	25,930	32,640
Surface area (cm.^{2})	2,320	3,284	3,815	5,070	8,104	9,106	10,763

(Assume that body weight in grams is equal to body volume in cubic centimeters.)

Exercise 11

A. Convert 60 mi./hr. to ft./sec.

B. Convert 15 lb./in.^{2} to dynes/cm.^{2}

C. Convert 186,000 mi./sec. to cm./sec.

D. Convert 32 ft./sec.^{2} to (mi./hr.) per min.

Exercise 12

I was recently asked to review a paper concerned with the efficiency of doing work upon a treadmill. One of the equations presented in this paper was

$$\text{``}P_o = 0.0313 \ mvg$$

where P_o = power output of the subject in K-calories per minute

m = mass of subject in pounds

v = treadmill velocity in miles per hour

g = slope of the treadmill which is dimensionless"

How is this equation incorrect? How should it be modified?

Exercise 13

Berliner *et al.* (11) present the following equation for the temperature, t_{x_o}, of a stream of fluid *emerging* from a countercurrent system, at a point X cm. away from the point of entry of the fluid:

$$\text{``}t_{x_o} = t_o + \frac{H}{V} + \frac{\alpha H}{V^2} X$$

where t_{x_o} (is the temperature) at a distance x from the start of the countercurrent system...; t_0 is the temperature of the water as it enters the system; α is a factor related to the efficiency of the heat exchange between inflowing and outflowing fluid; H, the amount of heat added to the system and V, the volume flow." (From discussion of this system elsewhere in the paper, it is evident that H is not "the amount of heat added to the system," but the amount of heat added to the system per minute.) Check this equation dimensionally.

3

AIDS TO MATHEMATICAL WORK

Strictly speaking, the only *indispensable* aids to mathematical work are pencil and paper. To be sure, it is possible—and in moments of boredom it is often entertaining—to ponder some aspect of a mathematical problem with one's hands in one's pockets, though more often than not even this casual sort of mental exercise ends up with a jumble of symbols and figures scribbled on the back of an inevitably too-small envelope. But for serious and sustained mathematical work, it is imperative that every step be written down so that the particular line of thought being followed is clearly recorded and can be reviewed whenever necessary. Pencil and paper are also needed for the continual checking of computations which alone can ensure trustworthy results. While there is no need to hinder one's free flow of thought by being overly fussy about neatness, a modest degree of order in the arrangement of the work helps to avoid errors, and may thus save a good deal of time even in the preliminary stages of a mathematical analysis. It goes without saying that the *final* copy of a *perfected* argument or derivation should be unmistakably clear, and should include precise definitions of every term (see Section 1-4).

3-1. Computers

Whenever possible, a scientist should spare himself the drudgery of numerical work by using a computer. Computers not only avoid the wasting of precious time with tedious arithmetic, but, by and large, they are far less apt to commit blunders then we are. Generally speaking, computers fall into two categories: *analog* computers and *digital* computers, although some combine features of both types.

In an analog computer, certain physical variables (lengths, voltages, resistances, etc.) are made proportional to ("analogous to") the mathematical or biological variables with which we want to deal. Furthermore, these analogous elements in the computer can be so arranged as to mimic

41

the behavior of the corresponding variables in a biological system. Since the variables we wish to study are for the most part continuous, the analogous variables in the computer are also continuously variable and can assume any magnitude within their range of operation. The solution to a problem is often supplied by an analog computer in the form of a continuous curve or set of curves which show how the dependent variables change when the independent variables are manipulated.

In contrast, digital computers, as their name implies, deal with ordinary numbers and give numerical answers to problems. This means that a digital computer designed to handle no more than eight digits would be unable to distinguish between, say, 120,359,250 and 120,359,259. A biologist is not likely to lose much sleep over this limitation. In fact, as far as accuracy is concerned, digital computers are commonly far superior to analog computers. But the two types cannot be compared fairly in this fashion, since, as will be evident from the following brief discussion, analog computers and digital computers are designed to serve different purposes.

3-2. Analog Computers: The Slide Rule

The simplest computer in general use, and one which every scientist should own, is the *slide rule*. A slide rule is a device for adding a length, measured along a fixed scale, to another length, measured along an adjacent moveable scale. If these lengths are proportional to the logarithms of the numbers marked on the scales, addition of lengths will be analogous to addition of logarithms. Addition of logarithms is, in turn, equivalent to multiplication of the numbers marked at the ends of the two lengths (see Appendix B).

The absolute limit of accuracy of a well-constructed slide rule is determined by the observer's visual acuity, *i.e.*, by the smallest difference of scale length which his eye can reliably distinguish. Since this limit for a given observer is the same regardless of the total length of the scale, the percentage error is inversely proportional to scale length. According to Stacy *et al.* ((106) p. 289) the normal eye can just distinguish between two points separated by a distance equivalent to one minute of arc. At a working distance of, say, 10 in., 1 min. of arc corresponds to a linear separation of about 0.0029 in. If 1 log cycle on a slide rule occupies 10 in., 0.0029 in. would represent 0.00029 $\log_{10}$ unit, or an error of about 0.07 per cent. For the same observer, a 20-in. scale would have an error of about 0.035 per cent. With poor luck, the corresponding error for a single multiplication or division (requiring 2 settings and 1 reading) might be 3 times as great, *i.e.*, about 0.2 per cent for a 10-in., or 0.1 per cent for a 20-in. scale. However, with care in setting and reading, slide-rule errors are usually much smaller than the above example, based upon *point-to-point acuity*, would indicate. In a test series of 15 multiplications of 4-digit random numbers (none of which terminated in a zero) the largest error which I obtained with a 20-in. scale was 0.05 per cent, and the average error was about 0.02 per cent. This rather surprising accuracy is due in large part to the close spacing of the vertical lines which mark the scale divisions, and to the equally helpful vertical line on the glass slide. These vertical markings allow settings on a slide rule to share part of the optical advantage of vernier settings and readings which are known to be fantastically accurate, displacements of the order of *one second* of arc being distinguishable under optimal conditions (13).

When serial operations are performed on a slide rule, the percentage error to be expected will naturally be larger, theoretically increasing in proportion to the square root of the number of individual multiplications and divisions. But since measurements of biological variables are seldom accurate to better than 1.0 per cent, the additional error introduced by using a slide rule for calculations is usually negligible. Even when greater accuracy is required, a slide rule check of the computations is a wise precaution against gross errors. Slide rule accuracy should also suffice for checking an equation by numerical substitution (see, for example, Table 14-2). Finally it is worth mentioning that with many slide rules, squares, cubes, square roots, cube roots, reciprocals, logarithms to the base 10, and certain trigonometric functions can be read directly from the various scales.

The principal disadvantage of the slide rule is not its lack of accuracy but its inability to deal efficiently with large quantities of data whose analysis often demands tediously repetitive additions, subtractions, multiplications, and divisions. For such work an electric calculator is all but indispensable (see Section 3-4).

3-3. Electronic Analog Computers

How an analogous electrical circuit in a computer mimics the behavior of a biological system can best be explained by describing a simple example. Let x be some substance, foreign to the body, which is instantaneously distributed throughout extracellular fluid and which is cleared from the plasma at a constant rate and only by the kidneys. Let the volume of extracellular fluid remain constant. Now under these circumstances (as will be fully discussed in Chapter 6), at any instant of time, t, the rate at which the concentration of x is decreasing will be directly proportional to $C_{x.t}$, the concentration of x in extracellular fluid at that time:

$$-dC_x/dt = kC_{x.t} \qquad (3\text{-}1)$$

Moreover, in this example the proportionality constant, k, is equal to the

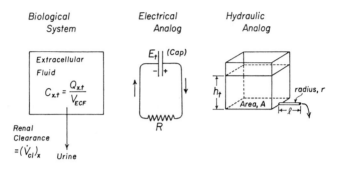

	Rate of Decrease	Is	Proportional to		Magnitude of Thing Decreasing
			(rate constant, k)		
			$\begin{bmatrix} \text{Rate-of-} \\ \text{movement} \\ \text{factor} \end{bmatrix}$	$\begin{bmatrix} 1/\text{Capacity} \\ \text{factor} \end{bmatrix}$	
Clearance of x from ECF	$-dC_x/dt$	$=$	$[(\dot{V}\text{cl})_x]$	$[1/V_{\text{ECF}}]$	$C_{x \cdot t}$
Discharge of capacitor......	$-dE/dt$	$=$	$[1/R]$	$[1/(\text{Cap})]$	E_t
Loss of fluid from reservoir ..	$-dh/dt$	$=$	$[\pi Dgr^4/8\eta l]$	$[1/A]$	h_t

FIG. 3-1. Analogous Biological, Electrical, and Hydraulic Systems

Notice that in each of the three systems, the rate of decrease is proportional to the magnitude of the quantity which is decreasing so that the *proportional* change per unit time is constant. Notice also that the constant of proportionality (the rate constant, k) is determined in all three systems by a *rate-of-movement* factor which shows *how rapidly* the solute x, or the electric current, or the fluid can move, and by a *capacity* factor which, when multiplied by the magnitude of the thing decreasing, shows *how much* x, or electricity, or fluid remains to be moved.

$Q_{x \cdot t}$ = quantity of x in extracellular fluid at time t, h_t = height of the fluid in the reservoir at time t, D = density of the fluid, g = acceleration due to gravity, η = viscosity of the fluid, l = length of the small-bore outflow tube. The remaining symbols are defined either in the diagram or in the text.

ratio of a factor which controls the *rate of movement* of x out of the body (the renal clearance, $(\dot{V}\text{cl})_x$), and a *capacity* factor (the volume of extracellular fluid, V_{ECF}) which indicates how much fluid there is to be cleared of x. k has the dimensions of reciprocal time, and is commonly called a rate constant. *In a precisely analogous way* (Fig. 3-1) at any time, t, the rate of change of the voltage, E_t, across a capacitor of capacitance, (Cap), discharging through a resistance, R, will be directly proportional to the voltage at time t. Furthermore, the rate constant for the capacitor discharge will also be the ratio of a factor controlling *rate-of-movement* $(1/R)$, and a *capacity* factor, (Cap). Accordingly, let us make the capacitance proportional to the volume of the extracellular fluid, the resistance inversely proportional to the clearance of x, and the initial voltage proportional to the initial concentration of x. We can then allow the capacitor to discharge, displaying the curve for the decrease in voltage with time on the face of an oscilloscope or recording it with an oscillograph. This voltage-time curve will faithfully represent the changes in concentration of x in extracellular fluid with the passage of time, because it differs from the concentration-time curve only by the known proportionality factors which relate the biological quantities to the corresponding electrical quantities. If we now wish to see how doubling the clearance of x would affect the rate at which the concentration of x in the extracellular fluid decreases, we need only reduce the resistance to one-half its original value, and let our electronic analog computer trace a new curve.

As is suggested by Figure 3-1, in theory it should be possible to construct a hydraulic analog computer which would serve the same purpose. But for a number of practical reasons, a hydraulic model would be far less accurate and convenient. It is shown merely to illustrate another quite different, yet analogous, set of variables.

Because of its simplicity, the example presented above conveys almost no sense of the real power and versatility of electronic analog computers. As will become evident in later chapters, the loss of a substance from a single compartment by clearance at a constant rate is easy enough to deal with by straightforward mathematical analysis. But in many an important problem we are interested in the distribution of a substance among several different compartments, or with its transformation into other chemical forms, or with both of these at once. It is often difficult or impossible to achieve an explicit mathematical description of such complex systems. But if we have some notion about how the various biological factors controlling distribution and metabolism are related to each other, we may be able to mimic this biological arrangement with correspondingly related electrical elements in an analog computer (19, 102, 104). By varying the capacitances, resistances, etc., of the electrical system, we can then see to what extent the behavior of the electrical analog can be made to match quantitatively the observed behavior of the biological system. In this way, the analog computer may help us decide whether our model of the biological system is reasonable, and, if so, what its parameters are. Furthermore, it allows us to predict how the behavior of the biological system will be altered when one or another of its parameters is changed.

3-4. Digital Computers

For lightening the load of run-of-the-mill arithmetical chores, there is available a wide variety of table-top calculators ranging from old-fashioned hand-operated adding machines to fully automatic electric calculators which almost make pleasurable the computation of such common statistical quantities as standard deviations, regression coefficients, and the like. The time saved by using one of the more versatile modern machines almost always justifies its considerable initial cost, if not for a single laboratory, at least for a department of any size. But the ability of such a machine to perform arithmetical operations swiftly and accurately should not lull the operator into a false sense of security. Mechanical failures are by no means unknown; rough mental checking of answers to see whether they make sense is therefore as essential as when working with pencil and paper. Human errors, in entering numbers on the keyboard and in copying answers from the dials, must always be expected, particularly with inexperienced operators. It is therefore prudent to check your work by performing all calculations twice. When dealing with a large body of data, you would be wise to subdivide the work into smaller parts, checking each of the subdivisions separately and combining them only at the end. Then if an error is identified, it is not necessary to start all over again from the very beginning, because the mistake is known to lie in a particular subdivision.

On an altogether higher level of complexity are the enormous (and enormously expensive!) *electronic digital computers*. These extraordinarily sophisticated machines combine a variety of electronic circuits with a "memory" from which stored information can be recovered whenever necessary. The complexity of the mathematical work which can be performed is astounding and accounts for the popular designation of such a machine as an "electronic brain." But, although the computer, once properly instructed, can complete an intricate mathematical task with remarkable speed and precision and without any human intervention, the appropriate instructions for the machine in the form of a "program" of operation must, of course, be supplied by a human operator. Special knowledge is required to devise a proper program, and the services of both operator and machine are expensive. Nevertheless, electronic digital computers are already being applied to biological problems, and their sphere of usefulness in the analysis of complicated physiological data is sure to expand.

For further information about computers (and for selected references) the reader may wish to consult the articles by Stacy in Volume 3 of *Medical Physics* (104, 105).

3-5. Reference Books and Mathematical Tables

The great majority of the facts, figures, and formulas needed by the biologist for the mathematical description of biological systems can be found in one or two standard reference books. Although those recommended here are the author's personal favorites, they also have the more objective merits of good reputation, ready availability, and low cost.

The Handbook of Chemistry and Physics (58) is a most remarkable compendium of scientific information. Its scope is considerably broader than the title indicates, for in addition to a vast amount of chemical and physical data it contains several hundred pages of mathematical formulas and tables. The mathematical section, like all of the other sections, is revised and extended yearly for each new edition, and it now includes many of the tables used for interpreting statistical tests of significance. No summary description can do justice to the wealth of detailed information in this book. Its extent can be appreciated only by looking carefully at the table of contents. Many references to *The Handbook of Chemistry and Physics* appear throughout this text.

If much use is to be made of logarithms, the graphic table of Lacroix and Ragot (67) is certainly worth buying. Because it is arranged like a slide rule, the graphic table is far easier to read than an ordinary table which has rows and columns of figures. Five-place logarithms can be read directly, and a sixth place can be obtained easily by interpolation between adjacent scale markings. Antilogarithms can be read with equal ease and precision.

Dwight's *Mathematical Tables* (36) is available in an inexpensive paper-bound edition. While many of the tables will not be of much interest to biologists, the extensive table of the exponential functions e^x, and e^{-x}, is often quite useful.

Discussions of the application of physical and mathematical methods to biological problems are so numerous and differ so considerably in breadth and depth that no review of them will be attempted here. However, the reader will certainly find the three volumes of *Medical Physics*, edited by O. Glasser (46), an invaluable source of information and a guide to further reading in many fields.

Though it has no very direct connection with the subject matter of the present text, I should like to commend E. B. Wilson's *An Introduction to Scientific Research* (118) as a book so full of scholarly wisdom and practical good sense that it ought to be in every scientist's library.

3-6. Graphs of the Relationship between Two Variables

Physiological scientists are forever plotting the results of their experiments on graph paper for the very good reason that a graph enables the viewer to see at once the general trend of the results without being distracted by the actual numerical values of the individual observations. Moreover, the relationship between two variables as displayed in a graph often suggests what sort of mathematical analysis will best suit the data.

3-7. The Proper Arrangement of Graphs

Two well-established customs govern the general arrangement of graphs. 1) The values of the *independent variable* are the *abscissas* and are accordingly measured along the lines which are parallel with the horizontal or x-axis. The values of the dependent variable are the *ordinates* and are accordingly measured along the lines which are parallel with the vertical or y-axis. If the variables cannot be identified as independent and dependent, they may be assigned to the two coordinates arbitrarily. 2) The numerical scales must be arranged so that the values along the horizontal scale increase from left to right and the values along the vertical scale increase from below upwards. This makes it possible to tell at a glance whether the slope of a line on the graph is positive (slanting upwards from left to right) or negative (slanting downwards from left to right). These two customs are of no more fundamental importance than is the custom of printing maps with north at the top and east at the right. But to violate them without good reason is as thoughtless and confusing as it would be to print a map of Europe with the Mediterranean at the top and Scandinavia at the bottom.

With arithmetic coordinates, but not with logarithmic or certain other special coordinates, we are free to place the origin of coordinates wherever it suits our convenience, and to make the scale for x and the scale for y as large or as small as is necessary to display the observed points to best advantage. A good rule of thumb is to choose the scales so that the total range of the observed values of y occupies about the same space on the graph paper as the total range of the observed values of x. This arrangement usually brings out the trend of the points clearly without obscuring any deviations of the points in either direction from the line of trend. However, there is no need to follow this rule so slavishly that the points become inconvenient to plot. For example, suppose the effect of a change in body temperature upon heart rate is being studied. If body temperature varies from 36.5 to 40.0°C., and heart rate varies from 60 to 110 beats per minute, it is not necessary to have 50 beats per minute occupy exactly the same number of spaces as 3.5°C. If we allot 20 spaces per degree centigrade, and either 1 or 2 spaces per beat, the general rule will be served quite well enough. Considerable modification of this rule may be required if several variables are to be plotted on the same graph. But it is well to remember that a graph loses its usefulness if the viewer is bewildered by a multiplicity of lines, points, and scales. When the origin of coordinates lies reasonably near the experimental points, it should be shown on the graph, but it would be silly to do so in the above example.

The two types of *logarithmic* graph paper in general use are *semilog* paper with a vertical logarithmic and a horizontal arithmetic scale, and *double-log* or log-log paper with both scales logarithmic. Papers of both types are available with 1, 2, 3, or even more cycles on a sheet, each cycle representing a 10-fold range of numbers (*i.e.*, a single $\log_{10}$ unit). But with any particular paper, the space occupied by a single cycle is fixed, so that our freedom of choice with a logarithmic scale is limited to deciding where the decimal point will be placed. Once we have decided, for example, that the "1" at the beginning of the lowest log cycle is to be 0.01, every other point on the entire scale is fixed. The numeral "1" at the beginning of the next cycle *must* be 0.1 and so on. Remember that these numbers are *not* logarithms; they are the actual values of whatever quantity you wish to plot logarithmically, written for convenience on what is, in fact, a scale of logarithms (see Appendix B). If you were to go to the trouble of looking up the numerical values of the logarithms themselves, you could plot them directly on arithmetic coordinate paper with a perfectly free choice of scale. But the logarithmic papers are entirely adequate for most purposes, and save a great deal of time.

3-8. The Implications of Fitting a Line to the Points on a Graph

When a physiologist has plotted his experimental points on a sheet of graph paper, he is often not content to let them speak for themselves. He draws a smooth line through the points. Whether he realizes it or not, this commonplace act is a first-rate example of inductive reasoning, *i.e.*, reasoning from the particular to the general. Let us see, in a purely descriptive way, how this is true.

I have reason to suppose that a variable, *y*, depends in some way upon another variable, *x*. This supposition is my *working hypothesis*. I devise an experiment in which I can make *x* assume different values, and I observe and record the corresponding values of *y*. I then plot the values of *y* against the values of *x* on graph paper, and draw a line through the resulting points. In drawing the line I am really being guided by three important ideas:

    1. The points are a particular example of a general relationship. The points are not just isolated points but represent a continuous function of *x* and *y*.* It is the general relationship, *i.e.*, the function, and not the points themselves, which interests me.

    2. I have a practical method for getting at this general relationship. If I draw a line to fit the points, and then write an equation for the line, the equation represents my idea of what the underlying functional relationship between *x* and *y* is, at least within the range of the observed values. If I am lucky, the equation will either clearly agree with or clearly contradict my working hypothesis. The equation may even indicate a new hypothesis which can be tested by further experiment.

    3. But I must take due account of the random scatter among the points

---

* Drawing an unbroken line through the points implies that the underlying function can be treated as a continuous function. If this is not true, a continuous line should not be drawn, and some other way of expressing the relation between *x* and *y* must be chosen.

("experimental error") *which is caused by uncontrolled variables.* I realize that many factors besides x must have influenced the observed values of y. Because of the resulting uncontrolled variation among the points, I do not expect the actual points to fall exactly on the smooth line I draw to represent my idea of the underlying functional relationship. In fact, I can place no reliance at all on any line which *exactly* fits the points, because the only way I can estimate the reliability of the line is by studying the extent to which the points scatter around it.

These three ideas underlie any method of fitting a line to a group of experimental points, whether it be "by eye" or mathematically.

3-9. Fitting Experimental Points by an Empirical Equation

The following discussion will be confined to problems in which x and y are the only variables, and where either $y = f(x)$ or $x = f(y)$. Feedback functions (Chap. 5) are not excluded, because it is always desirable, and often possible to study the two functions separately.

The first question to be answered is what general type of equation should be fitted to the points. For any particular group of points there is an infinite number of equations which would agree with the observed values of x and y. While most of these are far too complex to merit consideration, there are often several rather simple equations, each of which would fit the observations satisfactorily (Fig. 3-2). Which one should we choose? First, if our *working hypothesis* suggests a particular form of equation, that is certainly the form we must first use in trying to fit the points. This is not an example of circular reasoning as you might suspect. The general question we ask of any experiment is whether the data agree with the hypothesis being tested. If our hypothesis states that y is an exponential function of x, the specific question we are asking is: "Can the experimental points be fitted by an exponential function of x?" To answer this question we must obviously try to fit an exponential equation to the points; the question would neither be answered nor rendered invalid by finding some other equation which gave a "better" fit. Suppose, for example, that the points happen to lie upon an impressively straight line. This fact may be totally irrelevant to the question at hand, for the points might *also* be fitted satisfactorily by an exponential equation and thus be in perfectly good agreement with our working hypothesis.

If the working hypothesis merely states that there is some unspecified kind of relationship between x and y, any equation which we can find to fit the points will be, in essence, a refinement of the original hypothesis. Under these circumstances we must rely upon our common sense in deciding what kind of equation to use, *i.e.*, what new and more specific hy-

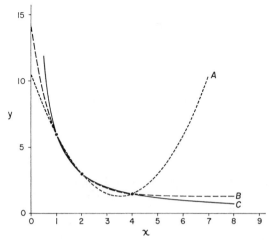

Fig. 3-2. Three Points Equally Well Fitted by Different Curves

The parabola, $y = 10.5 - 5.25x + 0.75x^2$ (*Curve A*); the exponential, $y = 12.93\,e^{-1.005x}$ $+ 1.27$ (*Curve B*); and the rectangular hyperbola, $y = 6/x$ (*Curve C*) all fit the three points (1,6), (2,3), and (4,1.5) perfectly. So also would an infinite number of other curves.

pothesis to propose. Often the general trend of the points will suggest what sort of equation should be tried. Some familiarity with the shape of the curves for a number of simple equations is therefore helpful (62). We may get help from our knowledge of analogous situations. We may get help from thinking about what hypothesis can most easily be tested further, or what equation, extrapolated beyond the present data, would best agree with other known facts. Finally, we must bear in mind the *law of parsimony*, that eminently sensible principle which is also known as William of Occam's razor: *"Entia praeter necessitatem non sunt multiplicanda"* (120). "Except when necessary, things are not to be multiplied." In other words, never invoke additional factors without good reason. If offered a choice between two hypotheses or equations, choose the simpler.

3-10. What Is a Good Fit? The Method of Least Squares

The second question to be answered is, what is meant by a "good" or "satisfactory" or "adequate" fit. This is actually a statistical question to which a detailed answer must be sought elsewhere (99). But it is very important for every physiological scientist to understand the general line of reasoning used. For simplicity, let us assume that x, the independent variable, is under the control of the experimenter who accordingly collects his data by making x assume certain values and observing the corresponding values of the dependent variable, y. (If x cannot be manipulated by the investigator, but, like y, can only be observed, and if x is subject to random errors of sampling, matters become much more complex, and the following discussion is no longer strictly applicable (10).) In the first place, when fitting a curve to such data, we shall be concerned *only* with the variability of y. The reason for this is simply that we are trying to find an equation which will tell how y varies in response to changes in x, not how x varies in response to changes in y. We must therefore accept the values of x as primary or "given," and even when we know that x is itself subject to error, we must ascribe all of the observed variation to y. In the second place, for whatever *general* equation has been chosen, the *specific* equation for the line which best fits the observed points is ordinarily taken to be the equation which minimizes *the sum of the squares of the deviations of the observed values of y from the line* (Fig. 3-3). The best fitting line can be calculated from the observed values of x and y by using the standard statistical method known as the *method of least squares*. Incidentally, by a regrettable accident of etymology the dependence of y upon x is known in statistical jargon as *the regression of y on x*, and the line fitted to the points is known as the *regression line*.

Now suppose we have decided to try to fit the data by a straight line equation of the form

$$y = a + bx \tag{3-2}$$

Using the standard statistical formulas for a least-squares fit, we can easily calculate the best values for the parameters a and b. But why stop there? After all, the equation for a straight line is only a special case of the general polynomial,

$$y = a + bx + cx^2 + dx^3 + \cdots \tag{3-3}$$

and we could improve the fit (*i.e.*, decrease the sum of squares of the deviations of y from the line) by adding another term and fitting the second-order polynomial,

$$y = a + bx + cx^2 \tag{3-4}$$

We could, in fact, obtain better and better fits (smaller and smaller sums of squares of deviations) by using more and more terms and calculating more and more parameters. Indeed, with the use of this general polynomial, if we were to calculate as many parameters as we have points, the resulting equation would fit *exactly* and the sum of squares of deviations would be zero! But the reader will remember that we can put no faith in any such line. How then do we know when to stop adding more terms and

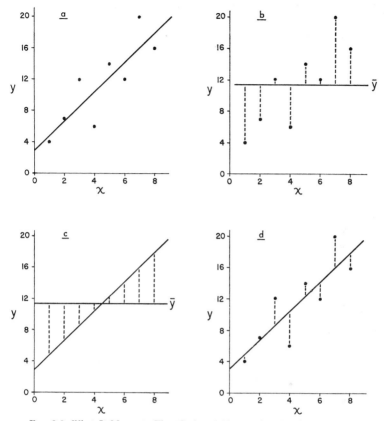

Fig. 3-3. What It Means to Fit a Series of Observations by the Method of Least Squares

a. Eight observations of *y*, the dependent variable, have been plotted against the corresponding values of *x*, the independent variable. The straight line (the "regression line"), $y = 2.96 + 1.87x$, is the "best fitting" straight line calculated by the method of least squares.

b. The *broken lines* show the *total* deviation of each *y* from the mean value of all of the *y*'s, $\bar{y}$, at 11.38 (*solid horizontal line*). If each of these eight deviations is squared, and the squares are added, we obtain 206, the *total sum of squares* of deviations of *y*.

c. Part of the total squared deviations of the *y*'s from their mean is due to the deviations (*broken lines*) of the least-squares line from the mean, $\bar{y}$, at 11.38. If each of these eight deviations is squared, and the squares are added, we obtain 147. This is the sum of squares of deviations which can be ascribed to the dependence of *y* upon *x*, *i.e.*, to the "regression of *y* upon *x*."

d. The remaining part of the total squared deviations of the *y*'s from their mean is due to the random scatter of the observed points about the line. If each of the eight deviations indicated by the *broken lines* is squared, and the squares are added, we obtain 59, the sum of squares for residual or uncontrolled error variation. *The "least-squares line" is the line which makes this sum of squares a minimum.*

Notice that the total sum of squares, 206, is exactly equal to the sum of squares for the regression of *y* on *x*, 147, plus the sum of squares for residual error, 59.

calculating more parameters? To answer this question let us be very simple-minded and inquire why the values of y vary at all. In other words, what are the *sources of variation* which, taken together, account for the observed deviations of the individual values of y from the mean value of y? As *a first source of variation*, we are certain to have "error" variation due to uncontrolled factors which tend to scatter the points. Even if y were not dependent upon x so that we had no line at all, these uncontrolled factors would still cause deviations of the individual y's from their mean value. But if y does depend on x, the tendency of y to change in response to changes in x constitutes *a second source of variation*. If we add a third term, cx^2, so that we fit a curve to the points instead of a straight line, we will have still *a third source of variation* of y, namely, a consistent trend of the observed points away from a simple straight line. There will still be error variations, but now the points will be scattered around a curved line, and the deviations of y due to uncontrolled factors will be deviations from the curve. If we wanted to try to fit a fourth term, we would have a fourth source of variation, and so on. Each of these sources of variation accounts for a portion of the *total sum of squares of deviations of the observed y's from their mean value* (usually abbreviated to "total sum of squares").

Now the portion of the total sum of squares which should be assigned to each source of variation can be calculated precisely. What is more, there are standard statistical methods for comparing the variation associated with any identifiable factor, for example; variations due to departures from linearity, with the variation ascribed to uncontrolled experimental error. It is this comparison which decides how many terms should be used, and how many parameters should be calculated to fit the points. For example, if the departures from linearity are no greater than might well be accounted for by the random scattering of the points due to experimental error, we have no justification for using any terms beyond the two in Equation 3-2 for the straight line. In fact, if there is any doubt about whether the values of y really do change in response to changes in the values of x, we must compare even the variation ascribed to the linear dependence of y on x with the error variation. If it turns out that the variation of y attributed to changes in x is, after all, no larger than can reasonably be accounted for by error variation (*i.e.*, by the deviations of the y values of the points from the line), then we have no right to fit even a straight line to the points because we have no acceptable evidence that significant changes in y were produced by changes in x. For all we can tell from the data, y may not be a function of x at all.

3-11. Degrees of Freedom

To complete this brief discussion of the rationale of curve fitting, it is necessary to introduce the concept of degrees of freedom. Suppose we have made N observations of y. Each observation represents, so to speak, a separate opportunity for the value of y to vary. With N observations, there are therefore N opportunities for y to vary, or N *degrees of freedom for variation*. But in practice, our study of the variability of y is limited to the variation we actually find among the observed values. We can see only how the observed values differ from each other, and there are not N but rather $N - 1$ opportunities for the observed values to differ from each other. To illustrate this, suppose we have made only a single observation of y so that $N = 1$. A single observation tells us nothing at all about how y varies. It tells us nothing about how much the individual values of y differ from the mean of the observed values of y, because the single observation *is* "the mean of the observed values." The single degree of freedom for variation inherent in the single observation has been "used up" or "lost" in fixing the mean. Hence there are $N - 1 = 0$ degrees of freedom left for variation about the mean. Now suppose we have two observations of y. In addition to calculating the mean of the two observations, we have one opportunity for estimating the variability of y by seeing how large is the difference between the two values, or (which is the same thing) by seeing how the two values differ from their mean. With only one difference possible between two values, there are $N - 1 = 1$ degrees of freedom for variation about the mean. And in general, with N individual observations, there are $N - 1$ degrees of freedom for variation about the mean.

But suppose we believe that y is a linear function of x. In accordance with the previous discussion, this means that we think part of the variation in y is due to its dependence upon x. If we have but two observations, our least-squares straight line is bound to fit them precisely, and there will be no opportunity for deviation of the observed y's from the line. Clearly, just as we had to use one degree of freedom for calculating the mean, we have to use an additional degree of freedom for calculating the slope, b, of a straight line. (In the least-squares equation, $y = a + bx$, the intercept, a, in a very real sense represents the mean of y. Let $\bar{x}$ be the mean of x, $\bar{y}$ the mean of y. Since the least-squares line must pass through the point $(\bar{x}, \bar{y})$, we can rewrite the equation as a function of $(x - \bar{x})$ instead of x:

$$y = \bar{y} + b(x - \bar{x}) \qquad (3\text{-}5)$$

Then the intercept *is* the mean of y.) Similarly we must use a total of three degrees of freedom in fitting the curve $y = a + bx + cx^2$. Thus, in general, *one degree of freedom for variation is lost for each parameter calculated from the observations.* If n_{par} parameters are calculated from N observations, the number of degrees of freedom left for estimating the uncontrolled or error variation of y is $N - n_{par}$. But the n_{par} degrees of freedom are not really lost; they are simply taken away from the degrees of freedom for residual error variation and assigned to an *identifiable* source of variation such as the linear dependence of y on x. Thus every source of variation, including the residual error, is characterized by both a sum of squares and a number of degrees of freedom. And the variation of y ascribed to any particular source of variation is properly expressed not by the sum of squares itself, but by the mean square or *variance* which is the sum of squares for that source divided by the degrees of freedom for that source. In deciding whether the variation from a particular source is really significantly greater than the variation due to uncontrolled experimental error, the criterion used is the ratio of the two variances. This procedure is one example of the statistical technique known as *the analysis of variance*, a technique, by the way, of much wider usefulness than has been indicated here.

There is one more general point to be made. Our judgment that the variation due to a given factor is really greater than can reasonably be accounted for by error variation depends not only upon the relative magnitudes of the variances, as expressed by the variance ratio, but also upon how reliable our estimates of the variances are. In particular, if the estimate of the error variance is based on only a very few degrees of freedom, we cannot have much confidence in its accuracy, and we must require the variance ratio to have a relatively large value before we can conclude that the variation ascribed to the given factor is greater than can readily be explained by experimental error. In all properly rigorous statistical tests of significance due allowance is made for the dependence of reliability upon number of degrees of freedom.

3-12. Graphs of Variables Which Are Correlated with Each Other, but Are Not Functionally Related to Each Other

If x and y are correlated with each other simply because both are functions of some third variable (Category 4, Section 4-7) it does not matter which variable we assign to the horizontal and which to the vertical coordinate since neither x nor y is an independent variable. Under these circumstances there is no *mathematical* barrier to our calculating a least-squares line for the regression of y on x (minimizing the sum of squares of deviations of y from the line) just as we did before. But there is a *logical* barrier. For if we calculate *only* this line, we imply that y is a function of x, which it is not. If we calculate the line for the regression of y on x, we are logically compelled to calculate also the line for the regression of x on y which minimizes the sum of squares of deviations of x from the line. We have no reason to minimize the sum of squares for one variable and not for the other. These two regression lines will be different (unless the correlation is perfect) and will intersect at the point whose coordinates are the mean of x and the mean of y. The poorer the correlation, the wider will be the angle formed by the two lines. If the observed relation between x and y is to be compared with the relation predicted by some theory, it may be desirable to calculate *both* regression lines (for an example, see Fig. 10 in reference 107). Although the usual statistical tests for agreement between theory and observation are not applicable here (10), the theoretical line should presumably tend to lie somewhere between the two regression lines. However, if no such comparison between theory and experiment is to be made, it would be better *not* to calculate *either* regression line, but to express the extent of correlation by means of the *coefficient of correlation*. The *square* of the coefficient of correlation is simply the proportion of the total sum of squares (for either x or y) which can be ascribed to correlation, *i.e.*, the proportion which is *not* attributable to experimental error. This proportion is identical whether we calculate it from the sum of squares of deviations of x, or from the sum of squares of deviations of y. The square of the correlation coefficient, varying from unity (perfect correlation) to zero (no correlation), provides a nondimensional measure of correlation which in no way implies that the two variables are functionally related at all. Its square root, the coefficient of correlation itself, is taken as *positive* if the correlation is *direct* (x and y tending to change in the same direction), *negative* if the correlation is *inverse* (x and y tending to change in opposite directions). Regardless of how "high" the coefficient of correlation is numerically, if there is the slightest doubt about whether the observed correlation is significantly greater than might reasonably be ascribed to chance, the question must be settled by subjecting the data to the analysis of variance mentioned above, or to some equivalent statistical test of significance.

3-13. The Paramount Convenience of Straight Lines

Among the many types of relations between physiological variables, straight-line relationships occupy a position of special importance not because they are encountered more frequently than other types (they are not!) but because straight lines are so easily drawn and so readily analyzed mathematically. By the very simple expedient of stretching a bit of strong black thread across the graph paper it is frequently possible to fit a straight line by eye almost as accurately as could be done by the method of least squares. (It is a good idea to mark the point corresponding to the mean of x and the mean of y, because the fitted line must pass through that point.) More often than not, a line which has thus been carefully fitted by eye will be quite good enough for practical purposes. A second major advantage of the straight line is that the arithmetical manipulations needed for applying the method of least squares and for testing the significance of the regression by an analysis of variance are reasonably simple, particularly if it is not necessary to weight the observations (see below).

3-14. Linear Transformations*

Even when the equation expressing the relationship between y and x is curvilinear, it is often possible to rearrange or *transform* the equation so that a straight-line plot can still be used. For the transformed equation, we define quantities x', y', a', and b' in such a manner that

$$y' = a' + b'x' \tag{3-6}$$

By plotting y' against x' and fitting a straight line to the resulting points we can estimate a' and b' which are the parameters of the straight line. Finally we can use these estimates to calculate the parameters of the original curvilinear equation. Curiously enough, if the original equation contains three parameters, it may still be possible to estimate all three from the linear transformation, even though the straight line itself has only two parameters. This point, as well as the general procedure, will be clarified by the following example.

Suppose we know that the functional relationship between x and y should have the general form of a rectangular hyperbola whose asymptotes, a and b, are parallel to the x and y axes:

$$(x - b)(y - a) = c \tag{3-7}$$

We can rearrange this equation to the form

$$y = a + c \left(\frac{1}{x - b} \right) \tag{3-7a}$$

which becomes Equation 3-6 if we define $x' = (1/[x - b])$, $y' = y$, $a' = a$, and $b' = c$. Hence by plotting y against $(1/[x - b])$ we should get a straight line of slope c and intercept a. But we cannot plot y against $(1/[x - b])$ because we do not know the value of b, one of the three original parameters to be estimated from the observations. However, we *do* know that if x and y are related in the way shown by Equation 3-7, a plot of y against $(1/[x - b])$ will give a straight line *only* when b has its proper value. We can therefore try different values of b until we find the one which makes the points fall on a good straight line. From the intercept and slope of this line we can then obtain values for the other two original parameters, a and c. Notice that requiring the points to fit a straight line has enabled us to find, by trial and error, a value for the parameter b

---

* Strictly speaking, these should be called linearizing transformations rather than *linear* transformations.

which at first seemed hopelessly linked with x in the new independent variable, $x' = (1/[x - b])$. The steps outlined above are illustrated in Table 3-1 and in Figure 3-4.

Several linear transformations will be discussed in considerable detail in subsequent chapters.

3-15. The Need for Weighting When Individual Observations Have Different Errors

Suppose we have available three separate estimates, z_1, z_2, and z_3, of some quantity, z, whose true value we wish to determine. If all three estimates are equally reliable, our best approximation of the true value of z is simply the mean of the three individual estimates. But if they are *not* all equally reliable, we should give greatest weight to the estimate which is subject to the smallest error. However, we must not neglect the other two estimates altogether, for, though less reliable, they can still contribute some information to our overall approximation of the true value of z. In fact it may be shown mathematically that the weight which should be given to any one observation is inversely proportional to its variance, *i.e.*, to the *square* of its standard deviation. To obtain the weighted mean, therefore, each individual value is multiplied by a weighting factor equal to the reciprocal of its variance, these weighted values are summed, and the sum is divided by the sum of the weighting factors. For example, if $z_1 = 71 \pm 0.50$ (variance $= 0.25$), $z_2 = 73 \pm 1.00$ (variance $= 1.00$), and $z_3 =$

TABLE 3-1

Values for the points of Figure 3-4

Graph a		Graph b	Graph c	Graph d
"Observed" values		$\dfrac{1}{x - 3.5}$	$\dfrac{1}{x - 2.5}$	$\dfrac{1}{x - 3}$
x	y			
4	14.5	2.000	0.667	1.000
5	7.6	0.667	0.400	0.500
6	5.8	0.400	0.286	0.333
7	5.4	0.286	0.222	0.250
8	4.2	0.222	0.182	0.200
9	4.1	0.182	0.154	0.167
11	3.3	0.133	0.118	0.125
13	3.2	0.105	0.095	0.100
15	2.8	0.087	0.080	0.083

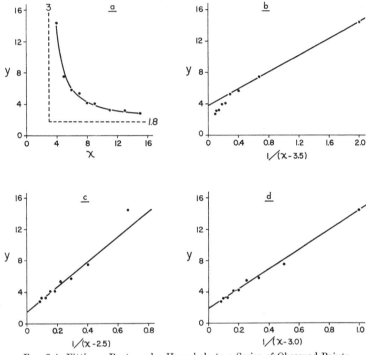

Fig. 3-4. Fitting a Rectangular Hyperbola to a Series of Observed Points

The "observed" values of the dependent variable, y, and the corresponding values of the independent variable, x, given in the first two columns of Table 3-1 are plotted against each other in *Graph a*. In an attempt to fit these points with a rectangular hyperbola of the general form, $y = a + c(1/[x - b])$, of Equation 3-7a, the parameter b was allowed to assume various values until, by trial and error, a value was found which gave a good straight line when y was plotted against $1/(x - b)$.

Graph b. When b is assumed to be 3.5, the points at the lower left fall distinctly below a straight line fitted by eye to the other points.

Graph c. When b is assumed to be 2.5, all of the points except the one at the upper right are reasonably well fitted.

Graph d. When b is assumed to be 3.0, *all* of the points can be fitted reasonably well by a straight line. The rectangular hyperbola

$$y = 1.8 + 12.7(1/[x - 3])$$

corresponding to this line is shown as the *solid curve* in *Graph a*. The *broken lines* in *Graph a* represent the asymptotes, $x = 3$ and $y = 1.8$.

69 $\pm$ 0.60 (variance = 0.36), the weighting factors are $1/0.25$ = 4.0 for z_1, $1/1.00$ = 1.0 for z_2, and $1/0.36$ = 2.8 for z_3. Our best approximation to the true value of z will be $(4.0z_1 + 1.0z_2 + 2.8z_3)/7.8$ = 70.5. The procedure is numerically equivalent to pretending that we have 4.0 values equal to z_1, 2.8 values equal to z_3, and only one value equal to z_2. But of course we have not really increased the number of quantities being averaged, so that the number of degrees of freedom remains the same.

Now when the ordinary method of least squares is used to determine the slope and the intercept of the straight line which best fits a series of observations, each observation is accorded equal weight, just as in calculating an ordinary unweighted mean, each of the individual values has equal weight. The unmodified method of least squares is therefore strictly applicable only when the tendency of y to deviate from the line, *i.e.*, the error of y, is independent of the magnitude of y. This may or may not be true. Some variables may indeed be subject to a reasonably constant error whether the value of the variable be large or small. But with other variables the percentage error, not the absolute error, is constant. If the percentage error is constant, large values of y will tend to deviate more from a fitted line than will small values of y. The ordinary method of least squares will therefore tend to do less than justice to the smaller values of y with their correspondingly smaller deviations, because the sum of squares of deviations from the line (the quantity to be minimized) will be determined preponderantly by the greater absolute deviations of the larger values of y. Under these circumstances, we must again use some system of weighting so that each observation, whether large or small, can exert its proper influence upon the calculated parameters of the fitted line. And again the weight to be given each point is inversely proportional to its variance, *i.e.*, to the square of its tendency to deviate from the line as measured by its standard deviation.

We must return to linear transformations. For we are now prepared to look behind their attractive façade of stretched-thread convenience and to view with appropriate horror a monstrous mathematical pitfall into which the unwary stumble, lured by the prospect of easy least-square solutions to all manner of curvilinear problems. Let us go at once to an example. Suppose we are studying the relation between a dependent variable, y, which is subject to experimental error, and an independent variable, x, which we will assume can be measured or controlled without error. We have strong theoretical grounds for believing that the product of x and y is a constant and that the data should therefore conform to an equation of the general type

$$xy = c \qquad (3\text{-}8)$$

TABLE 3-2

Values for the points of Figure 3–5

"Observed" Data		Transformed Data	
x	y	$1/x$	$1/y$
2	61	0.500	0.016
3	39	0.333	0.026
4	29	0.250	0.034
6	21	0.167	0.048
8	14	0.125	0.071
10	13	0.100	0.077
12	11	0.083	0.091
15	7	0.067	0.143
20	6	0.050	0.167
30	5	0.033	0.200
40	2	0.025	0.500
60	2	0.017	0.500

Equation 3-8 is a simplified form of Equation 3-7 for a rectangular hyperbola in which $a = b = 0$, so that the asymptotes are the x and the y axes. Hence, c is the only parameter to be found from the experimental data. Our 12 "experimental observations" are listed in Table 3-2 and are plotted in Figure 3-5a, where, as expected, they do seem to describe a hyperbolic curve. A linear transformation should help us to evaluate c and thus find the best hyperbola to fit the points. There are two obvious linear transformations of Equation 3-8:

$$y = c(1/x) \tag{3-8a}$$

and

$$x = c(1/y) \tag{3-8b}$$

or

$$(1/y) = (1/c)x \tag{3-8c}$$

so we can either plot y against $1/x$, or $1/y$ against x. Offhand, there seems to be no reason to prefer one kind of plot to the other. The two equations are very much alike. In fact, Equation 3-8a can be changed into Equation 3-8b and vice versa merely by switching the letter symbols. In the previous series of graphs (Fig. 3-4) we plotted y against a transformation of x, so, for the sake of a little variety, let us now plot a transformation of y, $1/y$,

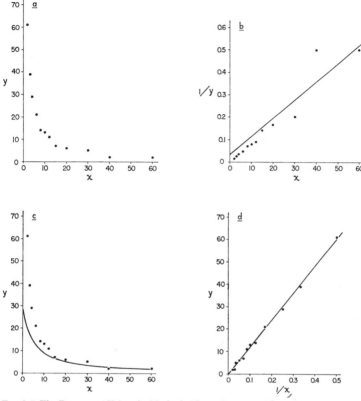

Fig. 3-5. The Dangers of Using the Method of Least Squares without Proper Attention to Weighting

Graph a. The "observed" values of the dependent variable, y, given in Table 3-2 are plotted against the corresponding values of the independent variable, x. It looks as if the points might be fitted well by a rectangular hyperbola of the general form, $xy = c$ (Equation 3-8).

Graph b. The points of *Graph a* have been replotted according to the linear transformation, $(1/y) = (1/c)x$. The straight line represents the result of 1) blindly applying the wrong method of least squares to the transformed data, and 2) giving each point equal weight.

Graph c. When the rectangular hyperbola corresponding to the "least-squares" line of *Graph b* is plotted on the original coordinates of *Graph a*, it fits the points very poorly indeed.

Graph d. Points plotted according to the alternative linear transformation, $y = c(1/x)$, can be fitted closely by a straight line. (This analysis is continued in Figure 3-6.)

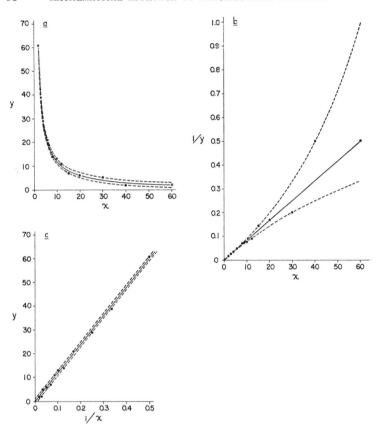

FIG. 3-6. What May Happen to a Constant Error during a Linear Transformation

Graph a. Some of the original "observations" of y (Table 3-2, and Fig. 3-5a) were so chosen as to illustrate error deviations of one unit of y above and below the "true" rectangular hyperbola (*solid curve*). These deviations are shown by the *broken lines*.

Graph b. In the linear transformation of Figure 3-5b, the seemingly small and constant error of y is enormously magnified by taking the reciprocal of y when y is small as is shown by the *broken curves* above and below the "true" relation (the *solid line*).

Graph c. No such distortion of the error occurs when y is plotted against the reciprocal of x, because, in this example, x was measured without error. In this graph, the *solid line* of best fit is the same as in Figure 3-5d. The *broken lines*, representing the same error for y as in *Graph b*, are no further from the "true" line than in *Graph a* because y has not been subjected to any transformation.

against x (Fig. 3-5b). We would be happier if the points scattered a little less, but they do seem to lie roughly on a straight line. We might as well calculate the best straight line by the method of least squares (Fig. 3-5b). Now if we transfer this line back to the original plot—Good Heavens! What a miserable fit! (Fig. 3-5c). Is *that* the best hyperbola? Surely we could have done better by drawing a curve freehand and deducing from it an approximate value of c! What went wrong?

Perhaps, after all, the choice between Equation 3-8a and Equation 3-8b was more important than we thought. Let us try plotting y against $1/x$ as suggested by Equation 3-8a (Fig. 3-5d). Not only is the scatter of points far less, but the least-squares line of Figure 3-5d, translated back to the original points (*solid curve*, Fig. 3-6a), gives us a hyperbola, $xy = 120$, which fits the observations very closely. Indeed it is the very hyperbola around which the author originally distributed the points when making up the "observed values" of Table 3-2. For purposes of illustration, the values of y were either placed directly on this curve, or were allowed to deviate from it by one unit of y in either direction. By indicating these deviations as broken lines above and below the straight lines of Figures 3-6b and 3-6c, the reason for the wretched results of plotting $1/y$ against x is at once made evident. Though the original absolute deviations of y above and below the hyperbolic curve were independent of the magnitude of y, the deviations of $1/y$ from the line in Figure 3-6b are strongly dependent upon the magnitude of $1/y$. The reason is simple. When the *absolute* error of y is constant, the *percentage* error is largest for small values of y. But it is precisely for these small values of y that the values of the transformed variable, $1/y$, are largest! Hence as y becomes smaller and smaller the absolute deviations of $1/y$ rapidly become very large indeed, because large absolute values of $1/y$ are subject to a large percentage error. As a result, the unweighted least-squares line of Figure 3-5b is so greatly influenced by the most unreliable points that it completely misses the points at the lower left whose uncertainty of position, on the scale of this graph, is essentially nil. The distortion of error is so enormous that the point at $x = 2$ should have roughly 1,500,000 times the weight of the point at $x = 60$. Indeed, over 80 per cent of the entire weight of all of the observations together must be given to the single observation at $x = 2$! So even with proper weighting, this method of plotting the data is thoroughly unsatisfactory because most of the observations contribute practically no information about the slope and the intercept of the straight line. In contrast, when y was plotted against $1/x$ (Fig. 3-5d), the deviations of y about the line were not distorted, and no weighting of the points before calculating the least-squares line was necessary. But it must be emphasized that this method of dealing with the data owes its great superiority in this particular example to the fact that x was measured without error. Had x been as subject to error as y was, we would have encountered the same kind of difficulty with both of the linear transformations.

By concentrating our attention in the above example upon the importance of weighting, we have deliberately glossed over a second serious error in the method used to find the preposterous straight line of Figure 3-5b. The least-squares method which was used was indeed the one· usually employed for estimating the parameters of a straight line. But this most common method is appropriate only when both the slope and the intercept of the straight line are to be estimated from the data. In the present example, we are supposed to be trying to fit the data by Equation 3-8 or, more specifically, for Figure 3-5b, by Equation 3-8c. But according to Equation 3-8c, *the straight line must have an intercept of zero*, and it is therefore wrong to estimate the intercept, as if it were unknown, from the data. The correct method is the one which you are asked to consider in Exercise 4-A3. If you use this correct method to calculate the "best" line for the points, you will find that it fits them very much better than the line actually shown in Figure 3-5b. But the preceding discussion of the need for proper attention to weighting remains as valid as before.

It would be quite wrong to conclude from the above example that linear transformations are so tricky and deceptive that they should be entirely avoided. That would be throwing the baby out with the bathwater. But the moral is clear. *Before fitting a line, either by eye or mathematically, to a series of points, consider carefully whether the error variation is reasonably uniform for all regions of the proposed line*. If it is not, due attention *must* be given to the weight of each point when fitting the line. Be particularly cautious when dealing with linear transformations, because then you will be fitting the straight line not to the original variables, x and y, but rather to certain conveniently chosen functions of x and y. The error variation of these functions is quite likely to be different for different values of the functions even when the error variation of the original variables was independent of the magnitude of the variables themselves. If at all possible, avoid the mathematical difficulties of weighting (which may be considerable) by choosing a transformation for which the error variation is reasonably uniform.

3-16. Graphs of Three or More Variables

A relationship between three variables, say x, y, and z, can be displayed graphically in several ways. The most obvious but often the least practical, is to construct some kind of "solid" or three-dimensional graph. A good example is a relief map where x is latitudinal distance, y is longitudinal distance, and z is altitudinal distance, and where the scales are chosen to exaggerate differences in altitude. It is possible to construct a similar three-dimensional graph for any three related variables; but the labor involved is considerable, and the resulting *objet d'art* is likely to be more beloved by its creator than by his audience. Moreover, a three-dimensional graph can be printed on a two-dimensional page only as a photograph or as a drawing in perspective, neither of which is very satisfactory. As a practical alternative to three-dimensional relief, cartographers have long been accustomed to representing altitude by *contour lines*, *i.e.*, lines drawn through points of equal altitude. The interval between successive contour lines is chosen for convenience so that on the average the lines are neither so close together that they clutter up the map, nor so far apart that they fail to show important differences in altitude. In exactly the same way "contour maps" can be used to depict the relationship between any three variables. To make such a graph, one of the variables, let us say z, is assigned a constant value (*i.e.*, is treated as a parameter) and x and y for that particular value of z are plotted against each other in the ordinary way. Then z is assigned another constant value for which x and y are again plotted, thus supplying a second line. This procedure is repeated, each chosen value of z adding another line, until enough lines have been calculated to illustrate clearly how the three variables are related to each other. Since each line connects points of equal value of one of the variables (here z), the line is often named by prefixing *iso*, meaning "the same" or "equal," to a Greek stem which indicates the nature of the variable-held-constant. Thus isobars are lines of equal barometric pressure, isotherms, lines of equal temperature, isochors, lines of equal volume, etc. The *general* term for any such "line of equal something" is "isopleth." Incidentally, one method of constructing a three-dimensional graph is to draw each isopleth on a separate piece of heavy cardboard, cut the pieces out along the isopleths and glue them together so as to build up the proper three-dimensional figure. For this purpose, the difference between any two successive values of the "iso" variable for adjacent isopleths must be constant so that the spacing of units on the vertical scale will be uniform.

Neither the construction of a solid graph nor the plotting of isopleths is suitable if there are *four or more variables*. However, if the variables actually plotted are themselves defined as functions of two or more of the primary variables, a single graph can sometimes be made to convey a surprising amount of information. When possible, it is wise to have the functions which are to be plotted take the form of dimensionless numbers, for this makes it easy to express a wide variety of experimental data in a uniform fashion (68).

Example

Suppose we want to show graphically how the various factors which influence the pulmonary excretion of any inert gas x are related to each other. Assume 1) that no x is present in the inspired air, 2) that x is not being accumulated by nor released from the lung tissue itself, 3) that the exchange of x between blood and air occurs only in the alveoli and not in the dead space, 4) that ventilatory air flow and pulmonary blood flow are parallel and continuous, and 5) that diffusion equilibrium for x is achieved between alveolar air and the arterial blood leaving the lungs. Then it is easy to show that

$$C_{x.\text{ven}}/C_{x.A} = (\dot{V}_A/\dot{V}_B) + S_x \tag{3-9}$$

where

$C_{x.\text{ven}}$ = concentration of gas x in mixed venous blood
$C_{x.A}$ = concentration of gas x in alveolar air
$\dot{V}_A$ = alveolar ventilation
$\dot{V}_B$ = pulmonary blood flow
S_x = solubility of gas x in blood at body temperature

According to Equation 3-9, for any gas which conforms to the stated assumptions, a plot of the (venous blood/alveolar air) concentration ratio against the (ventilation/perfusion) flow ratio should give a straight line whose slope is unity and whose intercept is the solubility of x in blood. This line is independent of the actual concentrations and the actual flows because all of the terms in Equation 3-9 are dimensionless ratios. Indeed by plotting $C_{x.\text{ven}}/C_{x.A}$ against $(\dot{V}_A/\dot{V}_B) + S_x$ rather than against $\dot{V}_A/\dot{V}_B$ it would be possible to plot data for *different* gases on the *same* straight line. The validity of all of the underlying assumptions, taken together, could then be tested by seeing how well the observed values of $C_{x.\text{ven}}/C_{x.A}$ for various gases of different solubilities fit the theoretical straight line when alveolar ventilation and cardiac output are varied. This form of Equation 3-9 would be much better for testing the agreement of theory with observation than would the equivalent forms:

$$\dot{V}_A C_{x.A} = \dot{V}_A \dot{V}_B C_{x.\text{ven}}/(\dot{V}_A + S_x \dot{V}_B) \tag{3-9a}$$

which gives the quantity of x excreted per minute, or

$$\dot{V}_A C_{x.A}/C_{x.\text{ven}} = \dot{V}_A \dot{V}_B/(\dot{V}_A + S_x \dot{V}_B) \tag{3-9b}$$

which gives the pulmonary blood clearance of x. Note that in both of these variants, $\dot{V}_A$ appears as a multiplier on both sides of the equation, so that some degree of correlation is inevitable (Section 4-11).

3-17. Neglecting Minor Factors

An enormous amount of time and trouble can be saved by deliberately neglecting any complicating factors which, *in the particular problem at hand*, are only a nuisance. The italicized restriction is important, for what is negligible in one problem may be of the essence in another. The decision to neglect a given factor must be guided by good judgment coupled with at least order-of-magnitude knowledge of what error will thereby be introduced. If the decision is made consciously, and stated explicitly, no disastrous distortion of facts is likely to result. Really serious errors of omission are usually due to forgetting a factor entirely, or to being forced by mathematical expediency to neglect a factor which is not, in fact, negligible.

Usually the neglected factors have neither very much practical nor very much theoretical importance. But sometimes the theoretical description of a biological system can be simplified substantially by neglecting matters, which, though very important practically, are devoid of theoretical interest. For example, the student beginning his study of respiratory physiology and gas exchange is understandably bewildered by finding that although the inspired air is at room temperature, pressure, and humidity, gas quantities are customarily expressed as volumes of dry gas at standard temperature and pressure (STPD), whereas physiological spaces, such as dead space, contain volumes of gas which must be measured at body temperature and pressure and saturated with water vapor (BTPS). Discussions of the theory of gas exchange would be clarified by pretending that any gas mixture which is to be inspired is already at body temperature and pressure and saturated with water, so that no changes in the volume or composition of the inspired air would occur before it reaches the alveoli where gas exchange takes place. The actual changes in volume and concentration of the inspired gases as room air enters the lungs could then be dealt with as a problem in correcting laboratory data without confusing the underlying theory. As a further simplification, the solubility of gases in blood should be expressed as a dimensionless ratio of concentrations and not as the dimensional Bunsen solubility coefficient (Section 2-8).

3-18. Approximations

Mathematical work can often be simplified by the judicious use of approximations. Some of these are listed in Appendix C which also indicates the ranges within which each approximation can be used without exceeding certain stipulated percentage errors. These approximations may be used both in computations and in simplifying complex equations.

A very useful approximation, too general to be listed in Appendix C, is that a sufficiently short segment of any continuous curve can be treated as a straight line whose slope is approximately equal to the first derivative of the curve at the *mid-point* of the short segment. If the point with co-ordinates x_1, y_1 and the point with the coordinates x_2, y_2 represent the end-points of the segment,

$$(y_1 - y_2)/(x_1 - x_2) = \Delta y/\Delta x \simeq dy/dx \qquad (3\text{-}10)$$

Even though statistical matters are, for the most part, being sedulously avoided in the present text, the author cannot resist remarking upon a statistical approximation which has proved invaluable as an order-of-magnitude check upon the calculation of the standard error of the mean of a series of observations. If N is the number of observations, (range) is the difference between the largest and the smallest observation, and $s_{\bar{x}}$ is the standard error of the mean of the observations:

$$s_{\bar{x}} \simeq \frac{(\text{range})}{N} \qquad (3\text{-}11)$$

Equation 3-11 gives exactly the same results as the rigorous method of calculation when $N = 2$, and yields surprisingly accurate estimates when N is no greater than 10. For larger values of N, the range may still be used to approximate the standard deviation (and hence the standard error of the mean), but the relationship is not so simple as Equation 3-11. Values of the ratio (range/standard deviation), which can be used in such estimations are tabulated in Snedecor's text (99).

EXERCISES. CHAPTER 3

Exercise 1. Estimating the parameters of curves "fitted by eye"

It is important to be able to go easily from a curve on a graph to the corresponding equation, as well as from an equation to the corresponding curve on a graph. Suppose that $y = f(x)$ where y and x are measured experimentally. Let $f(x)$ be of a known form, containing n arbitrary constants (parameters). Further, suppose that you have drawn through the experimental points a rough curve which *approximates* $y = f(x)$. To estimate the n parameters, choose from the curve n points which will represent the upper and lower portions of the curve as well as any curvature or inflections which it has. Each point so chosen supplies a pair of values of x and y. Substitute these numerical values into the general equation, thus obtaining n simultaneous equations in the n unknown parameters. Solve these n equations simultaneously for the n parameters.

In each of the following problems, a generalized form of an equation is given. Calculate the specific values of the n parameters which will make the general equation fit the n points whose coordinates are given. Check your results by plotting the specific equation thus found, together with the points for the data, on graph paper.

Independent variable, x	Dependent variable, y	General equation

A. Time after leaving cocktail party

Independent variable, x	Dependent variable, y	General equation
	Total quantity of alcohol in body	
hr.	gm.	$y = a + bx$
1.0	60.0	
4.0	30.0	

B. Partial pressure of CO_2 in alveoli — Alveolar ventilation

mm. of Hg	L./min.	$y = a + bx + cx^2$
37	5.6	
41	20.4	
49	80.0	

C. Excitation of a nerve fiber by flow of a direct current ("strength-duration curve")

Time during which current flows	Minimum voltage needed to excite single nerve fiber	
msec.	mv.	$y = \dfrac{c}{x - a} + b$
0.25	70	
0.90	40	
2.40	30	

D. A "dose-response" curve

Single dose of norepinephrine given intravenously to an anesthetized cat	Maximum increase in systolic blood pressure (corrected for increase due to saline alone)	
(Q)	(Effect = E)	$E = \dfrac{E_{max}Q^n}{Q_{0.5}^n + Q^n}$
µg. per cat	mm. of Hg	
0.1	10	
10.0	55	
1000.0	120	

where E_{max} is the greatest effect which can be obtained with any dose, however large. $Q_{0.5}$ is the dose needed to produce an effect half as great as E_{max}.

Exercise 2

Look up the following references and comment upon the designated graphs.

A. Rall, D. P., *et al.* (85): Figure 4, page 188.

B. Ross, J. M., *et al.* (92): Figures 2, 3, 4, and 5.

C. Gerber, G. B., *et al.* (44): All the figures, but especially Figure 3.

Exercise 3

In a paper dealing with coronary blood flow in the dog, Alella *et al.* (5) give the following regression equation summarizing the observed relation between coronary flow and myocardial oxygen consumption:

$$\log Y' = 0.928 + 0.986 \log X$$

where

Y' = coronary flow, cc./min./100 gm. heart weight

X = oxygen consumption, cc./min./100 gm. heart weight

Can you suggest an alternative, and more straightforward method of expressing the observed relationship?

Exercise 4. Fitting equations to experimental data by the method of least squares

In applying the method of least squares, note that the *deviation* which is to be squared is the difference between the value of y actually *observed* at a particular value of x, and the value of y *calculated from the fitted equation* for $y = f(x)$ *at the same value of* x. If (x_i, y_i) represents the coordinates of any individual observation,

$$\text{Deviation} = y_i - f(x_i)$$

For example, suppose that we want to fit the equation

$$y = a + bx$$

to a series of experimental points. If the values of x and y for the third observation in the series are x_3 and y_3, the deviation for that particular observation will be $y_3 - [a + bx_3]$.

In each of the following problems, use your knowledge of algebra and calculus to obtain a general "least squares" solution for the values of the parameters which will make the sum of squares of deviations a minimum: Begin by writing the algebraic expression for the deviation of a single value of the dependent variable, y. Next, square the deviation, and sum the squared deviations for all values of y. Finally, letting this sum of squares be symbolized by SS, find values for the parameters, a, b, c, etc., such that SS will be a minimum. To do this, find the partial derivatives, $\partial SS/\partial a$, $\partial SS/\partial b$, $\partial SS/\partial c$, etc., which express how SS changes as, one by one, the parameters are allowed to vary. For SS to be a minimum, all of these partial derivatives must be zero. Therefore, set them equal to zero and solve the resulting equations (which are known as the *normal equations*) simultaneously for the parameters.

A. The coordinates of a series of n observed points are (x_1, y_1), (x_2, y_2), (x_3, y_3), $\cdots$ (x_n, y_n), y being the dependent variable.

    1. What value of the parameter, k, will minimize the sum of squared deviations from the hyperbola, $xy = k$?

    2. What will be the values of a and b for minimizing the sum of squared deviations from the straight line, $y = a + bx$?

    3. Suppose the points are to be fitted by a straight line which is known to pass through the origin of coordinates. What will be the equation for such a line, and will the least-squares solution differ from the one for A-2?

B. How many degrees of freedom are there for variation about the curve in each of the three problems in Section A above?

C. If we make a series of observations of a *single* variable, y (for example, the partial pressure of CO_2 in alveolar air, or the heart rate, or the body weight), in a series of normal human males aged 20 to 30 years, we may want to use the observed values of y to obtain an estimate, $\bar{y}_{est}$, of the true mean value of the variables in all such individuals. This is really equivalent to fitting the observed points by the equation

$$\bar{y}_{est} = a$$

where a is a single parameter to be found from the observations. Given a series of n observations of y: y_1, y_2, y_3, $\cdots$ y_n, find the least-squares value for a.

Exercise 5

An investigator is studying the concentration in blood of a volatile substance. The very crude method he is using has a standard deviation of about ± 40 per cent. After analyzing 247 blood samples with this method, the investigator starts to use a vastly improved method of analysis with a standard deviation of only ± 5 per cent. When he has completed 43 analyses with the new method (covering about the same range of values as the 237 analyses with the old method) he decides to publish his results. In so doing, what use should he make of the 247 analyses by the old method.

Exercise 6

Consider the unhappy linear transformation illustrated in Figure 3-5b. Prove that if the original values of y are subject to a uniform error of $\pm \Delta y$, the weight which should be given to any value of $1/y$ for fitting a line to the transformed data is roughly directly proportional to the fourth power of y, at least when y is substantially larger than Δy.

4

CONSTANTS, VARIABLES, AND FUNCTIONAL
RELATIONSHIPS

Before undertaking the study of any *particular* biological system, it
will be instructive to consider the *general* ways in which one variable can
be related to another. To begin with, we must agree upon the meaning of
certain broad terms.

4-1. Constants and Variables

Every quantity with which a scientist deals is either an absolute constant,
an arbitrary constant, or a variable. An *absolute or universal constant* is a
quantity whose magnitude cannot be changed under any circumstances.
It may be a pure number, for example π, the ratio of the circumference to
the diameter of a circle, or it may have dimensions, for example the speed
of light in a vacuum, or Planck's constant. A *variable* is a quantity whose
magnitude not only can change but *is allowed to change* in the particular
problem under consideration. Note the restriction in this definition. We
cannot define a variable as "anything whose magnitude can change," for
it would then include arbitrary constants as well as variables. An *arbitrary
constant* is a quantity whose magnitude can change, but *is not allowed to
change* in the particular problem under consideration. "Variable-held-
constant" might be an acceptable descriptive synonym for "arbitrary
constant." The term *parameter* is commonly used for the arbitrary constants in an equation which expresses a particular form of relationship
between two or more variables. For example, suppose that a variable, y,
is related to another variable, x, by an equation of the form

$$y = a + bx \tag{4-1}$$

In equation 4-1 the arbitrary constants a and b are the parameters of the
equation. If an equation of a particular form (*e.g.*, Equation 4-1) is to be
fitted to a series of experimental observations of x and y, the parameters
of the equation (*e.g.*, a and b) must be calculated from the data (Chap. 3).

4-2. Functions

Suppose we have a physical system in which there are only two variables, x and y. Suppose these two variables are so related that whenever we change x, y necessarily changes. Then y is said to be *a function of x*. This very simple idea that y is a function of x when the value of y depends *in any way at all* upon the value of x (provided that to each value of the independent variable, x, there corresponds no more than a *single* value of the dependent variable, y) is expressed by the equation

$$y = f(x) \tag{4-2}$$

Equation 4-2 implies nothing about *how* y depends upon x; it merely states that there is a dependence. For all we know from Equation 4-2, y may equal x. On the other hand, y may equal $(2/x) - 4.0$, or $1 - e^{-bx}$, or $3x^4 - 28x^3 + 9x$, or any other expression in which x is the only variable. All too often we do not even know what the exact relationship between y and x is, but Equation 4-2 will still be valid as long as we know that the value of y does depend upon the value of x. When y is a function of x, y is called the *dependent variable*, and x is called the *independent variable*. Note that when we say "y is a function of x," we do not necessarily mean that a change in x directly *causes* the associated change in y. For example, suppose that $C_{\text{in} \cdot t}$ is the concentration of inulin in the plasma of a dog at any time, t, after a single dose of inulin has been administered intravenously. $C_{\text{in} \cdot t}$ will decrease as t increases; if we were somehow able to suspend the passage of time, $C_{\text{in} \cdot t}$ would no longer decrease. $C_{\text{in} \cdot t}$ is thus clearly a function of t. But it would be a senseless perversion of meaning to say that the decrease in the concentration of inulin is caused by the passage of time! The decrease in the concentration of inulin is, in fact, caused by the distribution of inulin from the blood to other tissues in the body and by the renal excretion of inulin. Both of these processes in turn are dependent upon but not directly caused by the passage of time.

Note also that the expression $y = f(x)$ *does not imply* that $x = g(y)$. In the example just given, we could manipulate the concentration of inulin in the dog's plasma by injecting additional inulin, or by injecting a volume of isotonic sodium chloride solution, or by lowering the blood pressure enough to stop glomerular filtration. But no such manipulation of $C_{\text{in} \cdot t}$ could possibly retard or hasten the passage of time. $C_{\text{in} \cdot t}$ is a function of t, but t is *not* a function of $C_{\text{in} \cdot t}$.

Now in all real systems, certainly in all biological systems, the value of a dependent variable, y, will never actually depend upon just one independent variable, x, but upon a whole array of independent variables, u, v, w, x, z, etc. If we wish to express this fact in a perfectly general way, we can write

$$y = f(u,v,w,x,z, \cdots) \tag{4-3}$$

Often the number of important variables can be effectively reduced by designing carefully controlled experiments where, insofar as possible, only one independent factor, let us say x, is allowed to vary. Then, as an approximation to the truth, $y = f(x)$.* Moreover, in a theoretical analysis it is always possible to stipulate that all variables are to be held constant except the two (or occasionally more than two) whose relation to each other is of immediate interest.

4-3. Types of Relationships between Two Variables. A Practical Working Classification

Let us consider *any* two variables whatsoever. To avoid any *a priori* implication that one is an independent, the other a dependent variable, we shall call these two variables A and B rather than x and y. Let us now investigate the relationship between A and B in whatever way seems appropriate and practical. The results of our investigation *must* fall in one of the seven categories which are designated by number in Table 4-1.

4-4. Category 1. There Is No Relation between A and B, and We Observe None

If A is plotted on graph paper against B, the points are either scattered randomly, or conform to the equation $A = k_1$ or $B = k_2$. Nothing further need be said about such a negative result, except that whenever there is any doubt about whether a significant relationship exists, the question must be settled by an objective statistical analysis and not by the hopes and conjectures of the experimenter!

4-5. Category 2. Ordinary Functional Relationships

An ordinary functional relationship between A and B is one which can be completely expressed by a *single* equation. There are two slightly different subcategories:

In the *first subcategory* (2a) one of the variables is unequivocally independent, the other dependent. We can illustrate this kind of relationship with a simple example taken from the physiology of the circulation. Suppose that both the peripheral resistance and the cardiac output remain constant. (These are, of course, arbitrary constants.) Under these circumstances, if we increase heart rate, the pulse pressure (*i.e.*, the difference between systolic and diastolic pressure) will decrease. But if we were to increase pulse pressure, for example, by decreasing the distensibility of the arterial

---

* An approximation because in any real experiment one can never control all of the variables, and at best y will still be subject to "random" variations due to factors which in general cannot even be identified. These variations may be large enough to obscure the fundamental relationship between y and x. Statistical methods of analysis are designed to cope with just such difficulties (see Chap. 3).

TABLE 4-1 (*A practical classification of the relationships between two variables*)

Any two variables, A and B

A significant relation between A and B is *observed*

A significant relation between A and B is *not* observed

A relation between A and B actually does *not* exist
(1)

BAD LUCK!
A and B actually related but relationship was obscured by chance variations in the experimental data
ERROR OF SECOND KIND
(6)

A relation between A and B actually exists

BAD LUCK!
A and B actually not related
Observed relation was due to chance variations in the experimental data
ERROR OF FIRST KIND
(5)

A and B are related only because of the way the data were manipulated mathematically
ARTIFICIAL CORRELATIONS
("Spurious correlations")
(7)

A and B are correlated, but are not functionally related
CORRELATIONS
(4)

Either
$A = f(B)$
or
$B = g(A)$

$A = f(B)$
and
$B = g(A)$

The two equations are *different*
FEEDBACK RELATIONSHIPS
(3)

The two equations are the same

2a

2b

Ordinary
FUNCTIONAL RELATIONSHIPS
(2)

77

walls, the heart rate would not be directly affected. Pulse pressure is a function of heart rate, but heart rate is not a function of pulse pressure.

In the *second subcategory* (2b) the two variables are so related that a primary change in either variable will produce an effect upon the other variable, but the relationship as a whole can still be described by a *single* equation. As an example, consider the concentration of hydrogen ions C_{H^+}, and of hydroxyl ions, C_{OH^-}, in an aqueous solution:

$$(C_{H^+})(C_{OH^-}) = K_w \tag{4-4}$$

where K_w = the ionization constant for water. If we add hydrochloric acid to produce a primary increase in C_{H^+}, the value of C_{OH^-} will decrease according to the relation

$$C_{OH^-} = K_w/C_{H^+} \tag{4-4a}$$

which is the specific equation for $C_{OH^-} = f(C_{H^+})$. If instead we add sodium hydroxide to cause a primary increase in the value of C_{OH^-}, the specific equation for $C_{H^+} = f(C_{OH^-})$ will be

$$C_{H^+} = K_w/C_{OH^-} \tag{4-4b}$$

But Equations 4-4, 4-4a, and 4-4b, are merely variants of one and the same equation. In general, therefore, we may say that if the equations for $A = f(B)$ and for $B = g(A)$ are rearrangements of the *same* equation (Section 12-7), only a single functional relationship, most commonly an *equilibrium*, and not a feedback relationship (Category 3), is involved.

4-6. Category 3. Feedback Relationships

If $A = f(B)$ and $B = g(A)$, and if these two equations are *different* equations, then A and B bear a feedback relationship to one another. Feedback relationships are so interesting and so important in physiology that the next chapter will be devoted entirely to a consideration of their properties.

4-7. Category 4. Correlations

Even when A and B are not *functionally* related to each other, if they are both dependent upon a third variable, C, *i.e.*, if $A = f_1(C)$ and $B = f_2(C)$, they will tend to change simultaneously whenever C changes. Such a relationship between A and B is known as a *correlation*.* When A and B are correlated with each other, if a primary change is made in one of

---

* The term *correlation* is often used loosely to mean any kind of relationship, including a functional relationship, between two variables. It would be possible to avoid confusion by coining some more cumbersome expression, such as *bidependent correlation*, to designate specifically a relationship between two variables which is due to their mutual dependence upon some third variable. But in the present text, the unmodified term *correlation* will be used only for this kind of relationship.

them, there is no resulting change in the other unless the primary change was made by altering C. Hence neither A nor B is an independent variable, and the usual "least-squares" method of fitting observed values of A and B by an empirical equation is not applicable (see Section 3-12). Indeed it may be misleading to write any equation relating A and B because no functional relationship corresponding to such an equation exists. The relation between A and B should, if possible, be expressed by giving explicit equations for the two functions $A = f_1(C)$ and $B = f_2(C)$. But often the two functions are not known, and sometimes C cannot even be identified. Then the relation between A and B can best be summarized by calculating an index of correlation which, for linear relationships, is known as the *coefficient of correlation*. The coefficient of correlation simply shows how well A is correlated with B without implying that there is any functional relationship between A and B (see Section 3-12). Examples of correlation without functional relationship are common: In many species of mammals, including the dog but not the human, there is very close agreement between the renal plasma clearance of inulin and the renal plasma clearance of creatinine. In such species, both inulin and creatinine are filtered through the glomeruli but are not reabsorbed by, nor secreted by the tubules. Accordingly the renal plasma clearance of each is a function of—in fact, a measure of—the glomerular filtration rate. But the two clearances do not affect *each other* in any way. The incidence of one manifestation of a given disease is bound to be correlated with the incidence of any other manifestation of the same disease, but the manifestations themselves may or may not be functionally related. In diabetes mellitus, glycosuria is certainly a function of hyperglycemia. But in vitamin B_{12} deficiency macrocytic anemia and combined degeneration of the spinal cord are not functionally related. Incidentally, classical addisonian pernicious anemia also provides a good illustration of how two seemingly unrelated but associated factors, in this example anemia and atrophic gastritis, may ultimately prove to be functionally related through a long series of intermediate steps:

Macrocytic anemia	is due to	arrested maturation of red blood cells
Arrested maturation	is due to	inadequate supply of vitamin B_{12}
Inadequate vitamin B_{12}	is due to	poor absorption of vitamin B_{12} from the intestine
Poor absorption	is due to	insufficient intrinsic factor from the gastric mucosa
Insufficient intrinsic factor	is due to	atrophic gastritis
Therefore:		
Macrocytic anemia	is due to	atrophic gastritis

If any good purpose were served thereby, we could replace each "is due to" with "is a function of," and each named factor by a symbol, and write: $u = u(v)$, $v = v(w)$, $w = w(x)$, $x = x(y)$, $y = y(z)$, therefore $u = g(z)$. Much of the science of medicine is concerned with disentangling such long chains of functional relationships.

This completes our list of straightforward relationships between two variables. But in real life, where the amount of data available from experiment is always limited, we must consider the unhappy possibility that the true relationship between A and B has been obscured by chance variations. We must even consider the possibility that the true relationship has been obscured by overzealous mathematical manipulation of the data. We must therefore include three additional categories in our practical working classification of the relationships between two variables.

4-8. Category 5. The Relationship Observed Was Due to Chance

A and B are not, in fact, related to each other at all. The relationship which we observed was due to bad luck with a random distribution of results which, in our particular experiment, happened to simulate a real relationship between A and B. Concluding that there is a relationship, when in fact there is none, is sometimes called by statisticians an "error of the first kind" (118). The risk of making an error of the first kind is determined by what "level of significance" the investigator decides his results must achieve before he is willing to accept them as indicating a relation between A and B. To avoid the danger of having this decision influenced by the outcome of the experiment, this level of significance ought to be chosen *before* the experiment is performed.

4-9. Category 6. A Real Relationship between A and B Exists but Was Obscured by Chance

Although A and B are, in fact, related to each other, in our particular experiment the relationship was not evident because it was hidden by the random scatter of the observed data. Concluding from such an experiment that there is no relationship when, in fact, a relationship exists is sometimes called by statisticians an "error of the second kind." The risk of making an error of the second kind is determined by the level of significance chosen, by the number of experimental observations which are made, and by the design of the experiment.

It is essential for the reader to recognize that the preceding discussion of Categories 5 and 6 is wholly inadequate. Only a sound knowledge of statistics enables one to deal comfortably with the vagaries of random variations, and the science of statistics lies beyond the scope of the present book. As a minimum, the reader is urged to consult E. Bright Wilson's brief discussion of the two types of errors mentioned above (118).

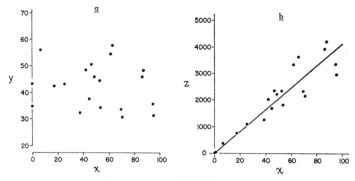

FIG. 4-1. An Artificial Correlation

The values of x and y were drawn independently from a table of random numbers. As would be expected, therefore, x and y show no relationship to each other (*Graph a*). But if a third variable, z, is defined as being the product of x and y, and z is then plotted as a function of x (*Graph b*), an impressive linear correlation between z and x results. This correlation, though real, is wholly artificial and meaningless.

4-10. Category 7. Artificial Correlation

This kind of correlation has sometimes been called a "spurious" correlation. But it is not spurious. As a correlation it is all too genuine. We shall do better to call it an artificial correlation because it is due to artificial manipulation of the data and will still be present even when the two variables which are actually being studied bear no relation at all to each other. Let us see by an example how very easy it is to produce an artificial correlation.

I have under observation two variables, x and y; x can vary from 0 to 100, but y's range, from 30 to 60, is more restricted. I have a strong suspicion that y may be a function of x. I make 20 careful observations of x and y, and plot y against x on graph paper (Fig. 4-1a). To my intense disappointment, there is no evidence of a trend. But I still think y *ought* to be a function of x. Perhaps I should try a slightly different method of plotting the results. For this purpose, I shall define a quantity z such that:

$$y = z/x \tag{4-5}$$

I can now plot z as a function of x (Fig. 4-1b). What a beautiful rectilinear relationship! I can even write an equation for the straight line which I have fitted by eye to the points

$$z = 41.2x \tag{4-6}$$

So x and y were related after all! By Equation 4-5, y is a function of z,[*] and z is obviously a function of x. Hence y must be a function of x. But before you congratulate me on having found just the right function to reveal this hidden relationship, I should perhaps explain briefly how the original observations of x and y were made. I simply copied the technique for collecting data which was used by L. C. Cole (23) in his elegant study of the metabolic rate of the unicorn. The values of x were 20 consecutive numbers from a table of random numbers. The values of y were "observed" in like manner, except that, to make sure y had the stipulated range,[†] only numbers beginning with 3, 4, or 5 were drawn from the table. The random scatter of points in Figure 4-1a is the expected result of this method of obtaining "data." The source of the so-pleasing correlation in Figure 4-1b becomes obvious at once if we *get back to the primary observations* by substituting an honest xy for the wholly gratuitous z. Then Equation 4-6 becomes

$$xy = 41.2x \qquad (4\text{-}7)$$

from which we can immediately remove the identity, $x = x$, by dividing both sides by x, thus obtaining

$$y = 41.2 \qquad (4\text{-}8)$$

Equation 4-8 gives the mean of y as estimated from the straight line of Figure 4-1b. The actual mean was 42.5. Considering the preposterously circuitous route used for obtaining the estimate, the agreement is not bad.

Does the preceding illustration seem far-fetched? Consider, then, the following example which actually occurred in the author's laboratory. The influence of alveolar ventilation upon the pulmonary clearance of ethyl ether was being studied in anesthetized dogs. To explain what happened, the following symbols will be needed:

C_A = concentration of ether in "alveolar" (*i.e.*, end-tidal) air, milligrams per liter

C_{exp} = concentration of ether in the expired air, milligrams per liter

C_{art} = concentration of ether in arterial blood, milligrams per liter

T_{body} = body temperature of the dog, degrees centigrade

$\dot{N}$ = respiratory rate, number of breaths per minute

V_{tid} = tidal volume, liters

$V_{D\cdot eth}$ = volume of the dead space for ether, liters

$\dot{V}_A$ = alveolar ventilation, liters per minute

[*] I hope no reader will let this glib assertion go unchallenged.

[†] Both artificial correlations and inevitable relationships (Section 4-11) are particularly impressive when the quantity common to both plotted variables (here x) has a wide range compared with the other quantities.

$(\dot{V}\mathrm{cl})$ = pulmonary clearance of ether, liters of whole blood per minute

$\dot{Q}_{\mathrm{exp}}$ = quantity of ether expired per minute, milligrams per minute

S = solubility of ether in blood, *i.e.*, (concentration of ether in blood)/(concentration of ether in air) at equilibrium

Of these 11 variables, the first 6 were measured directly. The remaining 5 were calculated as follows:

$$V_{D\cdot\mathrm{eth}} = V_{\mathrm{tid}}(C_A - C_{\mathrm{exp}})/C_A \quad \text{(the Bohr equation)} \tag{4-9}$$

$$\dot{V}_A = \dot{N}(V_{\mathrm{tid}} - V_{D\cdot\mathrm{eth}}) \tag{4-10}$$

$$\dot{Q}_{\mathrm{exp}} = \dot{N}V_{\mathrm{tid}}C_{\mathrm{exp}} \tag{4-11}$$

$$(\dot{V}\mathrm{cl}) = \dot{Q}_{\mathrm{exp}}/C_{\mathrm{art}} \tag{4-12}$$

$$S = f(T_{\mathrm{body}}) \tag{4-13}$$

Equation 4-13 merely indicates that for a given body temperature, the value of S was read from a published graph of the solubility of ether in blood as a function of temperature.

Now when the pulmonary clearance of ether, $(\dot{V}\mathrm{cl})$, was plotted against alveolar ventilation, $\dot{V}_A$, in most experiments there seemed to be some relation between the two variables, but sometimes the relation was by no means as close as had been anticipated from theoretical considerations (Fig. 4-2a). One possible source of error was in determining the concen-

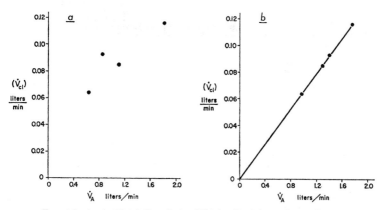

FIG. 4-2. An Artificial Correlation Which "Straightened out" Some Disappointing Data

The mathematical manipulation which placed the scattered points at the *left* on the straight line at the *right* is fully described in the text. Though more subtle, it is no more legitimate than the hocus-pocus of Figure 4-1.

tration of ether in alveolar air. End-tidal samples are not entirely satisfactory for this purpose. In fact, when the Bohr equation is used to calculate
the dead space for carbon dioxide, the alveolar concentration of CO_2 is
often estimated from the partial pressure of CO_2 in the arterial blood on
the assumption that alveolar air and arterial blood are in diffusion equilibrium with each other. When the same assumption was made for ether,
and C_A in Equation 4-9 was replaced by (C_{art}/S), the relation between
the pulmonary clearance of ether and the alveolar ventilation was very
close in every single experiment (Fig. 4-2b). Indeed the correlations were
so very good they roused suspicion that something was wrong with the way
the data were being handled. And so it proved. When the two plotted
variables were expressed *in terms of the variables actually measured*, it was
quite evident that the only genuine variable in the plot of $(\dot{V}cl)$ against
$\dot{V}_A$ was, of all things, body temperature, which was introduced not as an
experimental variable at all but rather as a minor correction factor! As in
the previous example, values for each variable of real interest could have
been assigned at random without spoiling the artificial correlation. For, as
the reader can easily verify, what was being plotted as $(\dot{V}cl)$ was actually
$(\dot{N}V_{tid}C_{exp}/C_{art})$, and what was being plotted as $\dot{V}_A$ was actually
$(\dot{N}V_{tid}C_{exp}/C_{art})[f(T_{body})]$. Further review of the calculations also showed
that Equation 4-12 for pulmonary clearance was in error. C_{art} should be
replaced by the concentration of ether in mixed venous blood, because the
ether is being removed not from blood leaving the lungs but from the mixed
venous blood entering the lungs. Because of this error, even the original
plot of Figure 4-2a really had nothing to do with pulmonary clearance as
a function of alveolar ventilation, but could be reduced to a plot of $(1/C_{art})$
against $(1/C_A)$.

4-11. Inevitable Relationships. Proving the Obvious

An inevitable relationship is one which can be predicted in advance of
experiment from facts already well established. It may be an ordinary functional relationship, belonging to Category 2, or a correlation, belonging to
Category 4. Inevitable relationships are *like* artificial correlations in that the
two variables plotted against each other are *known* to contain one or more
common elements which account for part or all of the observed relationship.
They are *unlike* artificial correlations in that they express some natural
relationship which is often of great theoretical and practical importance.
But however important such a relationship may be, it requires no experimental proof. We waste our time if we collect and publish data which do
no more than illustrate a foregone conclusion. There is no need for us to
be latter-day Magellans embarking upon a costly circumnavigation
merely to prove that the world is round.

Inevitable relationships are often based upon equations which are true *by definition*. As an example, it would be to the last degree ridiculous to measure the circumference and the diameter of several circles of different sizes and then to plot the observed circumference against the observed diameter. It is known, on unimpeachable theoretical grounds, that the ratio of circumference to diameter is a constant, and *by definition*, this constant is π. Hence,

$$(\text{Circumference}) = \pi \, (\text{diameter}) \qquad (4\text{-}14)$$

By plotting (circumference) against (diameter), we would therefore actually be plotting π (diameter) against (diameter), (diameter) being the element which is known to be common to the two variables under study. Although the relationship expressed by Equation 4-14 is of the utmost practical importance, *it requires no experimental proof*.

As a physiological example, let us consider the relation between the rate at which sodium is reabsorbed from the glomerular filtrate by the renal tubules, $\dot{Q}_{\text{Na·reabs}}$, and the glomerular filtration rate, $\dot{V}_{\text{filt}}$. Now although $\dot{Q}_{\text{Na·reabs}}$ cannot be measured directly, conservation *requires* that it be equal to the difference between the rate of filtration of sodium, $\dot{Q}_{\text{Na·filt}}$, and the rate of excretion of sodium in the urine, $\dot{Q}_{\text{Na·}U}$. Accordingly we *define*

$$\dot{Q}_{\text{Na·reabs}} = \dot{Q}_{\text{Na·filt}} - \dot{Q}_{\text{Na·}U} \qquad (4\text{-}15)$$

We can also *define*

$$\dot{Q}_{\text{Na·filt}} = k C_{\text{Na·}P} \dot{V}_{\text{filt}} \qquad (4\text{-}16)$$

where k = the Donnan equilibrium ratio of the concentration of sodium in the filtrate to the concentration of sodium in plasma

$C_{\text{Na·}P}$ = the concentration of sodium in plasma

and we can further define

$$\dot{Q}_{\text{Na·}U} = C_{\text{Na·}U} \dot{V}_{U} \qquad (4\text{-}17)$$

where $C_{\text{Na·}U}$ = the concentration of sodium in the urine

$\dot{V}_{U}$ = the urine flow

Combining Equations 4-15, 4-16, and 4-17 we obtain

$$\dot{Q}_{\text{Na·reabs}} = k C_{\text{Na·}P} \dot{V}_{\text{filt}} - C_{\text{Na·}U} \dot{V}_{U} \qquad (4\text{-}18)$$

Now suppose that we want to find out about the relationship between rate of sodium reabsorption and glomerular filtration rate in normal subjects and in patients with congestive heart failure. One possible approach would be to determine $\dot{V}_{\text{filt}}$ and $\dot{Q}_{\text{Na·reabs}}$ in several such persons, and to plot $\dot{Q}_{\mathrm{Na \cdot reabs}}$ against $\dot{V}_{\mathrm{filt}}$. This was actually done by Mokotoff *et al.* (76). The points for both normals and patients with congestive heart failure fell very close to a single straight line which passed through the origin of co-ordinates, the correlation between the two variables being exceedingly high. *But this experimental approach is quite unnecessary.* Equation 4-18 shows that $\dot{Q}_{\mathrm{Na \cdot reabs}}$ is by definition equal to $kC_{\mathrm{Na \cdot P}}\dot{V}_{\mathrm{filt}} - C_{\mathrm{Na \cdot U}}\dot{V}_{U}$, and, in fact, $\dot{Q}_{\mathrm{Na \cdot reabs}}$ was calculated from the latter expression. The graph is therefore really a plot of $kC_{\mathrm{Na \cdot P}}\dot{V}_{\mathrm{filt}} - C_{\mathrm{Na \cdot U}}\dot{V}_{U}$ against $\dot{V}_{\mathrm{filt}}$ so that $\dot{V}_{\mathrm{filt}}$ is common to both of the variables under study. Consequently, some degree of correlation is inevitable. But we may go further and prove that a very high degree of correlation is inevitable. For even in patients with con-gestive heart failure, the sodium in plasma is maintained at a reasonably normal *concentration* although the total *quantity* of sodium in extracellular fluid is abnormally high. Hence we may write the following approximation:

$$kC_{\mathrm{Na \cdot P}} \simeq (\text{constant}) \tag{4-19}$$

Now the quantity of sodium excreted, $C_{\mathrm{Na \cdot U}}\dot{V}_{U}$, is not at all constant. Depending upon a variety of circumstances, notably sodium intake, it may vary severalfold both in normal subjects and in patients with con-gestive heart failure. However, in both types of individuals, *the quantity of sodium excreted per minute is known to be very small compared with the quantity of sodium filtered per minute.* By neglecting this comparatively small quantity, we can rewrite Equation 4-18 in the following approximate form:

$$\dot{Q}_{\mathrm{Na \cdot reabs}} \simeq kC_{\mathrm{Na \cdot P}}\dot{V}_{\mathrm{filt}} \tag{4-20}$$

Combining Equation 4-19 with Equation 4-20,

$$\dot{Q}_{\mathrm{Na \cdot reabs}} \simeq (\text{constant}) \; \dot{V}_{\mathrm{filt}} \tag{4-21}$$

Thus, by making a few simple deductions from facts which are already well-established, we arrive easily at exactly the same conclusion as was labo-riously reached by a superfluous experimental approach. For Equation 4-21 tells us that if we plot $\dot{Q}_{\mathrm{Na \cdot reabs}}$ against $\dot{V}_{\mathrm{filt}}$, the points are bound to ap-proximate a straight line through the origin. No experimental demonstra-tion of this inevitable relationship is needed. Finally we should note that, besides being inevitable, the observed relation illustrates only *part* of the total relationship which is *completely* specified by Equation 4-18.

The preceding discussion may be generalized as follows: Suppose a vari-able, y, is a *known* function of n other *mutually independent* variables. There will then *inevitably* be a correlation between y and any one of the n variables, any variable necessarily being an element common to itself and to y. Since the entire functional relationship (of which the correlation represents but a part) is already known, it is superfluous to demonstrate any such correlation experimentally. Note, however, that if the n variables are not mutually independent but are related to each other in some unknown way, the known function to which y is equal no longer says all there is to say about the system. Consequently, one may not predict from that known function alone how y will respond to a primary change in one of the n variables upon which it is dependent (for an example, see Exercise 6).

Two rules will help us to identify and to avoid both artificial correlations and inevitable relationships. *First*, whenever one biological variable shows an unexpectedly high correlation with another, we should suspect that at least part of the close relationship may be due to an element which is common to both of the variables. *Second*, when derived variables are being plotted against each other, they should always be expressed in terms of the directly measured quantities from which they were calculated. This will help to reveal any unsuspected common factors. It may then be possible to find some alternative way of plotting the data which will avoid having these common elements influence both of the variables.

4-12. Practical *versus* Theoretical Relationships between Variables

The classification of the relationships between two variables into the seven categories discussed above, and summarized in Table 4-1, is a complete working classification because it comprises *all* of the possibilities which an investigator must, *in practice*, bear in mind when studying the relationship between any two experimental variables. But from a *theoretical* standpoint, the last three categories do not have the same significance as the first four. Category 5 (observed relationship due to chance) is really a subdivision of the first category, no relationship, and Category 6 (true relationship obscured by chance) is really a subdivision of Category 2, 3, or 4. Furthermore, Category 7 is an artifact of illegitimate mathematical prestidigitation—sheer hanky-panky utterly devoid of fundamental meaning. Finally, Category 4, correlation due to mutual dependence upon a third variable, does not really belong in any classification which is restricted to only two variables, but is one type of relationship among three variables. As a prelude to a general consideration of the relationships among several variables, let us undertake a more rigorous examination of the *theoretical* relationships between any two variables.

4-13. Symbol-and-Arrow Diagrams

The following simple representation of functional relationships will be useful. Each variable in the system under consideration will be represented by its usual symbol. If nothing connects a particular pair of symbols directly, the corresponding variables have no immediate functional relationship,

Symbol – and – arrow Diagram	Category
A B	1
A ⟶ B A ⟵ B A ---→ B A ⟵--- B	2 - a
A ⟷ B A ⟵--→ B	2 - b
	3

F_{IG}. 4-3. The 11 Possible Monotonic Relationships between 2 Isolated Variables
Arrows point from the independent toward the dependent variable. A *solid arrow*
means that when the independent variable changes, the dependent variable changes
in the same direction. A *broken arrow* means that when the independent variable
changes, the dependent variable changes in the opposite direction.

though they may be correlated or functionally related via other variables.
If the two symbols are connected by a single arrow, there is a single
functional relationship between them, and the symbol toward which
the arrow points is the dependent variable. (The arrow shows the direction of influence, so to speak.) If the symbols are connected by a
two-headed arrow, either variable may be the independent variable in the
manner explained previously for Category 2-b. Finally, by restricting our
attention to *monotonic functions*,* we can allow a *solid* arrow to indicate
that when the independent variable changes, the dependent variable responds by changing in the *same* direction ("monotonic increasing"), and
a *broken* arrow to indicate that when the independent variable changes,
the dependent variable responds by changing in the *opposite* direction
("monotonic decreasing"). With this notation, the 11 possible relationships
between two variables are summarized in Figure 4-3. With only two variables, the distinction between (for example) a solid arrow pointing from
A to B and a solid arrow pointing from B to A is meaningless, because the
same general *kind* of relationship is indicated regardless of the direction
of the arrow. But with three or more variables, the direction of the arrow may
indeed help to determine the general type of relationship. Furthermore, as
the number of variables increases, so also does the number of *pairs* of variables which may possibly be interrelated. With two variables, there is only

* A function is said to be monotonic when its first derivative does not change sign.
Suppose that $y = f(x)$. If $y_2 = f(x_2)$ is equal to, or greater than $y_1 = f(x_1)$ whenever
x_2 is greater than x_1, the function is "monotonic increasing." If $y_2 = f(x_2)$ is equal
to, or less than $y_1 = f(x_1)$ whenever x_2 is greater than x_1, the function is "monotonic
decreasing."

one pair, but with three variables there are three pairs, with four variables, six pairs, with five variables, ten pairs, or, in general, with N variables, there are $1 + 2 + 3 + \cdots + (N - 1)$ pairs of variables. Since any pair might be related in any one of the ways shown in Figure 4-3, the number of different patterns of interrelationship soon becomes astronomical as the number of variables increases. This rapid increase in complexity merely serves to emphasize the importance of making certain that in the particular system which we want to study we understand exactly how the several variables are related to each other.

4-14. How to Determine What Functional Relationships Link the Variables in a Multivariable System

Mathematical analysis, particularly statistical analysis, can help us to decide whether certain variables are or are not related and what kind of equation can be used to summarize whatever relationship is observed. But the question of whether a given variable in a given system is independent, dependent, or both is not a quantitative mathematical question, and can be answered only by seeing how the system behaves when its component variables are altered one by one. A particular variable is *fully independent* if it is not influenced by any changes which are made in any of the other variables of the system. In a symbol-and-arrow diagram, one or more arrows originate from a fully independent variable, but no arrows point toward it. A particular variable is *fully dependent* if it cannot influence any of the other variables of the system. In a symbol-and-arrow diagram, one or more arrows point toward a fully dependent variable, but none originate from it. A variable which is *partly independent and partly dependent* influences at least one other variable in the system and is also influenced by at least one other variable in the system. In a symbol-and-arrow diagram, such variables have arrows pointing both toward and away from them. In complex biological systems, most variables are partly independent and partly dependent.

Drawing a symbol-and-arrow diagram of a system often helps to summarize complicated interrelationships and to make clear which variables are independent, which are partly independent and partly dependent, and which are entirely dependent. For example, Ohm's law governing the flow of current, I, through a resistance, R, across which the voltage drop is E, is

$$I = E/R \qquad (4\text{-}22)$$

This relationship may be properly diagramed as

$$E \rightarrow I \leftarrow\!-\!-\!- R$$

but not as

$$I \rightarrow E \leftarrow R$$

even though Equation 4-22 can be solved for E:

$$E = IR \tag{4-22a}$$

and used to calculate the voltage drop across a given resistance for a given current flow. The truth is that, as the first diagram correctly indicates, we can change I by manipulating the independent variable E (for example, by adding another voltage source in series), and we can change I by manipulating the independent variable R (for example, by adding another resistance in series). But we *cannot* change E by manipulating I, as the second diagram would have us believe, because the current flow is *due to* the difference in voltage, not vice versa. Indeed, there is no way by which we can possibly change current flow *except* by altering E or R. In this simple system, E and R are both fully independent variables, but I is completely dependent.

4-15. Symbol-and-Arrow Diagrams of Complex Systems

We are now prepared to consider how the general concepts developed above can be applied to real biological systems, where the variables of interest rarely, if ever, occur in isolated pairs but rather as part of a great complex of variables bound together by an intricate network of functional relationships. Figure 4-4 illustrates such a complex. The system depicted

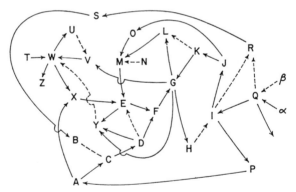

Fig. 4-4. The Complex Relationships among the Variables of a Real Biological System

Each letter in this diagram symbolizes a variable important in the cardiovascular system. It is rarely possible to subject so complicated a system *in toto* to mathematical analysis. The biological scientist must usually content himself with tackling subsystems, isolated in the manner described in Section 4-16.

in Figure 4-4 was not, as the reader might be tempted to suppose, cunningly devised by the author as an impressively bewildering but wholly imaginary example. In fact, Figure 4-4 is a diagram (already considerably simplified) which originally showed some of the actual interrelationships among various factors which influence cardiovascular function. But since we are at present interested only in how to deal with complex systems in general, the real variables of the system (heart rate, coronary blood flow, peripheral vascular resistance, and the like) have been arbitrarily represented by letters of the alphabet.

4-16. Equations for Symbol-and-Arrow Diagrams

At first glance, it might seem foolish to attempt any kind of mathematical analysis of a system as complicated as the one shown in Figure 4-4. And so it would be if we were obliged to deal with the system as a whole. Fortunately, however, we can usually tackle it piecemeal, first trying to find the simple equations which express how each single variable is related to the comparatively few variables upon which it is *immediately* dependent. It may then be possible to combine these elementary equations so as to achieve a useful description of some coherent portion of the system, if not of the system as a whole.

Since every variable (for example, V) toward which one or more arrows point is *dependent* upon the variables from which those arrows originate (for example, G and U), it may be possible to write an equation to describe this relationship. (Occasionally, more than one equation may be needed.) The equation does *not* include any factors (for example, W) which are dependent upon the variable for which the equation is being written. Now suppose that we want to investigate the relationships among a particular subgroup of variables, let us say U, V, and W. To isolate the subsystem containing U, V, and W from the remaining portion of the larger system, we *disregard* any arrows which point *away from* the subsystem, but we *retain* all arrows (and the symbols from which they originate) which point *toward* the subsystem. The subsystem may then be written as

$$Y \dashrightarrow W \leftarrow T$$
$$\swarrow \quad \nwarrow$$
$$U \dashrightarrow V \leftarrow G$$

and we hope that this simplified system can be completely described by three equations corresponding to

$$U = f_1(W) \tag{4-23}$$

$$V = f_2(G, U) \tag{4-24}$$

$$W = f_3(T, V, Y) \tag{4-25}$$

Notice that as far as this subsystem is concerned, G, T, and Y are fully independent variables (*i.e.*, no arrows point toward them). Yet in the whole system of Figure 4-4, it is possible for the variables of the subsystem to influence G via X, E, and F, or Y via X and E. In working with the subsystem, therefore, we must be careful to arrange the experimental conditions so as to preclude any such influences, for example, by not allowing W to affect X. Furthermore, if at all possible, we should arrange the experiment so that we can control each of the independent variables, G, T, and Y, by holding two of them constant while we study the effect of changes in the third upon the U, V, W subsystem.

It should be obvious that symbol-and-arrow diagrams of this sort do not, by themselves, solve any problems. Furthermore, they are not always easy to construct, particularly when a single variable, perhaps time, serves as an independent variable for many other variables in the system. However, symbol-and-arrow diagrams often do help to clarify the interrelationships among a group of biological variables, especially when the variables form a feedback loop as U, V, and W do in the example just discussed. We shall therefore make considerable use of such diagrams in the next chapter which is devoted to biological feedback systems.

EXERCISES. CHAPTER 4

Exercise 1

Comment on the following statement made by Berliner *et al.* (11) in a paper discussing the mechanism by which the kidney forms a hypertonic urine: "The direction and magnitude of the effect of these various factors on the concentration achieved may be indicated as

$$U_{osm} = F\left(P_{osm}, \Delta T_{NaCl}^1, \alpha, \frac{1}{V^n}\right)"$$

(For present purposes, we do not need to know the meanings of the symbols which are used in this equation.)

Exercise 2

For each pair listed below, draw a symbol-and-arrow diagram so as to specify what kind of relationship exists between the two variables.
   A. 1. Mean barometric pressure in a given geographical region
      2. Mean hematocrit of healthy residents of that region
   B. 1. Length of the side of a square
      2. Length of the diagonal of a square
   C. 1. Number of cigarettes habitually smoked per day
      2. Incidence of cancer of the lung among persons smoking that number

    D. 1. Threshold voltage needed to excite a nerve
       2. Duration of current flow needed to excite the nerve
    E. 1. Steady-state concentration of inulin in plasma during a continuous intravenous infusion of inulin
       2. Renal plasma clearance of inulin
    F. 1. Annual death rate from tuberculosis in the U. S. A. between 1935 and 1960
       2. Population of Canada between 1935 and 1960
    G. 1. Partial pressure of carbon dioxide in alveolar air
       2. Partial pressure of carbon dioxide in arterial blood
    H. 1. Concentration of monobasic phosphate in blood plasma
       2. Concentration of dibasic phosphate in blood plasma

Exercise 3

Suppose you drew 100 pairs of numbers from a table of 10,001 random numbers which lists all the numbers from 0 to 10,000, inclusive. Let n_1 and n_1' be the first pair of numbers drawn, n_2 and n_2', the second pair, etc. If you pretended that n_1, n_2, n_3, etc., represent observed values of an independent variable, x, and that the products, n_1n_1', n_2n_2', n_3n_3', etc., represent the corresponding values of a dependent variable, y, what kind of scatter of points would you get by plotting these "variables" against each other. What would be the approximate equation for the "least-squares" straight line fitted to the points?

Exercise 4

A. Berglund *et al.* (9) (among others) have measured coronary blood flow, $\dot{V}_{cor}$, and the difference in oxygen concentration between arterial blood and venous blood from the coronary sinus, $C_{O_2 \cdot art} - C_{O_2 \cdot ven}$, under a variety of circumstances. The oxygen consumption of the heart was calculated from these measurements as: $\dot{Q}_{O_2} = \dot{V}_{cor}(C_{O_2 \cdot art} - C_{O_2 \cdot ven})$. In their experiments there was a very high correlation between oxygen consumption and coronary blood flow, both of which varied over more than a 3-fold range, while the arteriovenous oxygen difference remained relatively constant. They conclude that "during normal circumstances, myocardial oxygen consumption is a major determinant of coronary blood flow."

    1. Is the high correlation between $\dot{Q}_{O_2}$ and $\dot{V}_{cor}$ an example of an inevitable relationship?

    2. Do you think that the conclusion quoted is valid?

B. In criticizing reports such as the one cited above, Scott and Balourdas (95) point out that if random sampling numbers with the same mean and standard deviation as the observations of coronary blood flow and arteriovenous oxygen difference are substituted for the real observations, equally high correlations between coronary blood flow and oxygen consumption are obtained. Scott and Balourdas conclude that: "Correlations of variables in the field of circulation in general, and the heart in particular, frequently involve two variables having a common element. The common element appears to be responsible for the relatively high r values [*i.e.*, coefficients of correlation] obtained, as indicated by the very similar results obtained with the use of random numbers. This invalidates r values of this type as evidence of a functional relationship between the two variables in question." Is this a well-founded criticism?

Exercise 5

A pharmacologist stimulated strips of frog ventricle electrically under various circumstances in an attempt to find out what factors determined the maximum tension, $(\text{Ten})_{\max}$, developed by the strip. As a measure of the *rate of development* of tension, he took the maximum slope of the tension-time curve, $(d(\text{Ten})/dt)_{\max}$. He was able to show that this maximum slope, and the total time of rise of tension, t_{rise}, were not related to each other but varied independently. He also found that the average rate of rise of tension, $(\text{Ten})_{\max}/t_{\text{rise}}$, was a constant fraction of the maximum rate of rise of tension.

In writing up this work for publication, should the investigator include a graph showing that there was a good "correlation" between $(\text{Ten})_{\max}$ and t_{rise}?

Exercise 6

Draw a symbol-and-arrow diagram to show the relationship between heart rate, N_{beat}, stroke volume, V_{beat}, and systemic output, $\dot{V}_{\text{syst}}$, in the isolated, denervated, mammalian heart-lung preparation receiving blood by gravity flow into the right atrium from a reservoir of blood at constant height. Is there an inevitable relationship between heart rate and systemic output in this system?

Exercise 7

Poiseuille's Law expresses the relationship between the following factors:
r = the radius of a cylindrical tube of small but uniform bore
l = the length of the tube
η = the viscosity of the fluid flowing through the tube
$\dot{V}$ = the rate of flow through the tube
P_α = the pressure at the beginning of the tube
P_ω = the pressure at the end of the tube

Write a symbol-and-arrow diagram to show the relations of these several factors. Which are independent, which dependent?

5

FEEDBACK RELATIONSHIPS. HOMEOSTASIS

In Chapter 4 we undertook a general survey of the various ways in which two variables, A and B, might be related to each other. It was there pointed out that when $A = f(B)$ and $B = g(A)$, and the two equations are *different* equations, the relationship between A and B is a *feedback relationship* (Category 3 of Table 4-1 and Fig. 4-3). The present chapter will be devoted to a more detailed consideration of feedback relationships. Although the discussion will be elementary from the mathematical standpoint, it will nevertheless reveal some of the unique properties of feedback functions, and will suggest a method whereby even the beginner can calculate the effectiveness of feedback mechanisms, including many homeostatic mechanisms, quantitatively.

5-1 Feedback Loops, Positive Feedback, Negative Feedback

Consider the symbol-and-arrow diagrams for Category 3 in the lower part of Figure 4-3. In each of these diagrams, A and B are linked to each other in a *feedback loop* by the two functional relationships represented by the arrows. The upper arrow in each diagram shows that B is a function of A, so that for this functional relationship A is the independent and B the dependent variable. But for the functional relationship represented by the lower arrow in each diagram, these roles are reversed, A being the dependent and B the independent variable. Hence, for the feedback loop as a whole, we cannot identify either variable as independent or dependent; we can only call them *interdependent*.

To appreciate the significance of this interdependence, let us focus our attention upon a single loop, taking first the one at the upper left which has two solid arrows. Let us suppose that besides being a function of B, A is also dependent upon some external factor which is not itself a member of the loop. What will happen if we suddenly change the value of this external variable so as to cause a *primary increase* in the value of A? The upper solid arrow shows that in response to this primary increase in A, B must also increase. But this secondary increase in B will, in turn, cause a *further increase* in A according to the relationship depicted by the lower arrow. The operation of the feedback loop thus *tends to enhance or magnify the primary change* in A. This is called *"positive feedback."* Similarly, if A and B are connected by two *broken* arrows (*e.g.*, upper right of the four loops in Fig. 4-3), any *primary increase* in A will cause a secondary *decrease* in B. But this *decrease* in B will, in turn, tend to *increase* A still further. Again, the primary change in A has been enhanced by the operation of the feedback loop, so that this too is positive feedback. In contrast, if one arrow is solid and one broken, as in the diagram at the lower left of Figure 4-3, a primary *increase* in A will cause a secondary *increase* in B according to the functional relationship shown by the upper solid arrow. But now this *increase* in B will *decrease* A, so that the operation of the feedback loop *tends to oppose or minimize the primary change* in A. This is called *negative feedback*. The reader should have no difficulty in showing that the diagram at the lower right also represents a negative feedback loop.

In maintaining the constancy of the *milieu intérieur*, the mammalian body again and again takes advantage of the ability of negative feedback loops to minimize the effects of external changes. We should not be surprised, therefore, to find that most (though not quite all) *homeostatic mechanisms* are negative feedback mechanisms. Because of their commanding importance in physiological regulations, most of the subsequent discussion will be devoted to negative feedback mechanisms. But we must not forget that positive feedback also plays an important role in normal physiology, especially in certain psychological phenomena such as precoital sex play. Moreover, the "vicious circles" so prominent in pathological processes are all examples of positive feedback.

5-2. The Advantage of Confining Attention to Steady States

For a complete mathematical description of a feedback mechanism we would need to derive a series of equations which would specify the precise magnitude of every variable in the system at every instant of time, both for steady states and during the transient adjustments which the system must make in response to evanescent disturbances or while passing from one steady state to another. Unfortunately, the time-dependent *transient* responses of biological feedback systems are frequently too complex for straightforward mathematical analysis. To describe the system quantitatively, the investigator must then resort to imitating the biological behavior with a more-or-less elaborate electrical analog. Considerable progress has already been made in the application of analog computers to the analysis of homeostatic mechanisms (52, 108). But the ordinary physiological scientist, who is not familiar with the technical terms used in cybernetics and in servo-system engineering, is more likely to be perplexed than to be instructed by these analyses.

Much of the mathematical complexity of feedback systems disappears if one confines attention to *steady states*, *i.e.*, states in which every variable in the system has had time to adjust to a particular set of constant conditions and is therefore no longer changing with the passage of time. In the present chapter, an elementary mathematical analysis of the steady-state characteristics of homeostatic systems will be presented. While this approach lacks the elegance and completeness of the more sophisticated analyses, it will provide a simple and precise way of calculating how effectively a given negative feedback mechanism minimizes the effects of some external change which disturbs the homeostatic equilibrium. The steady-state analysis also shows that for many biological systems the effectiveness of homeostasis is not constant but depends upon the magnitude of the external factors which tend to alter the homeostatic equilibrium.

5-3. Characteristics of a Simple Negative Feedback System

For the sake of simplicity, let us first consider the steady-state behavior of an electrical feedback system which can be described by two straight-line equations. The analysis of this example will suggest how more complicated homeostatic mechanisms can be dealt with. The electrical system is diagrammed in Figure 5-1. A constant signal of S millivolts contributes to the input, I millivolts, of a direct-current amplifier which multiplies the input by a constant factor or "gain," G. The total output, "O" millivolts, of the amplifier is divided into a constant fraction, F, which is "fed back" into the input, and the remaining fraction, $1 - F$, which represents the net output of the amplifier. To provide *negative* feedback, the F"O" millivolts which are fed back must have their sign changed so as to oppose the signal. The input voltage is thus the algebraic sum of the original signal voltage, S, and the voltage, $-F$"O", which is fed back. In the lower part of Figure 5-1 the circuit is depicted by symbols and arrows as a feedback loop in the manner of Figure 4-3. Since we are now dealing with a specific system, we shall also include the factors S, G, and F, which, though external to the loop itself, nevertheless help to determine the values of I and "O" and therefore appear in the feedback equations. We shall now write these equations. For "O" as a function of I, we have

$$"O" = GI \tag{5-1}$$

while for I as a function of "O" we have

$$I = S - F"O" \tag{5-2}$$

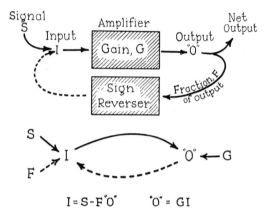

$$I = S - F\overset{\cdot}{O}\overset{\cdot}{} \qquad \overset{\cdot}{O}\overset{\cdot}{} = GI$$

Fig. 5-1. A Simple Electrical Illustration of a Negative Feedback Loop

The essential features of the circuit shown in the upper portion of the figure have been redrawn in the lower portion as a symbol-and-arrow diagram, beneath which are given the two feedback equations. Note that when the loop is regarded as a homeostatic system designed to limit changes in the input, I, only those arrows pointing toward one of the variables in the loop need be considered in the mathematical analysis. For example, the *Net Output* shown in the upper part of the figure does not concern us at all. For a numerical example of how this system operates, see Figure 5-2.

Now it is obvious that the system can be in a steady state only when Equations 5-1 and 5-2 are simultaneously satisfied. Simultaneous solution yields

$$I = \frac{S}{1 + FG} \tag{5-3}$$

for the steady-state value of I, and

$$\text{``}O\text{''} = \frac{SG}{1 + FG} \tag{5-4}$$

for the steady-state value of "O".

Now suppose we disturb this system by adding an increment, ΔS millivolts, to S so as to establish a new constant signal of $(S + \Delta S)$ millivolts. Equation 5-1, being independent of S, remains unchanged, but Equation 5-2 now becomes

$$I = (S + \Delta S) - F\text{``}O\text{''} \tag{5-5}$$

Accordingly, I and "O" must assume *new* steady-state values determined by simultaneous solution of Equations 5-1 and 5-5:

$$I_{new} = \frac{S + \Delta S}{1 + FG} \qquad (5\text{-}6)$$

$$``O"_{new} = \frac{(S + \Delta S)G}{1 + FG} \qquad (5\text{-}7)$$

Now let us imagine that this system is a homeostatic system designed to maintain I as constant as possible. How effective is it? In other words, to what extent is the system able to decrease the change in I which would occur in the absence of feedback (i.e., in an "open-loop" system)? With no feedback, Equation 5-2 would simply be $I = S$, so that the change in I without feedback, ΔI_{open}, would be equal to the primary change in S:

$$\Delta I_{open} = \Delta S \qquad (5\text{-}8)$$

or,

$$\Delta I_{open}/\Delta S = 1 \qquad (5\text{-}8a)$$

But with the feedback loop closed, so that feedback can occur, the actual change, ΔI_{closed}, is obtained by subtracting the original steady-state value of I (Equation 5-3) from the new steady-state value of I (Equation 5-6):

$$\Delta I_{closed} = I_{new} - I = \frac{S + \Delta S}{1 + FG} - \frac{S}{1 + FG} = \frac{\Delta S}{1 + FG} \qquad (5\text{-}9)$$

or,

$$\Delta I_{closed}/\Delta S = \frac{1}{1 + FG} \qquad (5\text{-}9a)$$

By dividing Equation 5-9a by Equation 5-8a we can now calculate how much the primary or "open-loop" change in I for a given change in S has been magnified by the feedback:

$$\text{Magnification} = \frac{\Delta I_{closed}/\Delta S}{\Delta I_{open}/\Delta S} = \frac{1}{1 + FG} \qquad (5\text{-}10)$$

But the very *raison d'être* of a homeostatic negative feedback system is to *minimize* change. Hence it would seem reasonable to define:

$$\text{Minification} = \frac{1}{\text{magnification}} = 1 + FG \qquad (5\text{-}11)$$

as a better measure of how well this homeostatic system is working. But now suppose that either F, or G, or both were zero, so that there was, in fact, no feedback. The homeostatic system would then be completely ineffective, offering no opposition at all to a change in I caused by a change in S. Under these circumstances, the homeostatic effectiveness of the system would clearly be zero, but the minification would be unity. As a more satisfactory expression for homeostatic effectiveness, let us therefore subtract unity from the minification, and define:

$$\text{Homeostatic index} = \text{minification} - 1 = FG \qquad (5\text{-}12)$$

The *homeostatic index*[*] will be zero when the homeostatic system is completely ineffective, but will approach infinity as the system approaches perfect compensation for a change in S. Finally, we should note that G is the slope of one feedback equation (Equation 5-1), while $-F$ is the slope of the other feedback equation (Equation 5-2). The product of these slopes is $-FG$, and the homeostatic index is simply the negative product of the slopes:

$$\text{Homeostatic index} = -(\text{product of slopes}) \qquad (5\text{-}12a)$$

The above argument can easily be extended to any pair of straight-line equations which describe a negative feedback relationship between two variables. We may therefore summarize the preceding discussion by stating that *if a homeostatic system consists of two variables in a negative feedback loop which is described by two straight-line equations, the homeostatic index is the negative product of the slopes of the two equations.*

Figure 5-2 illustrates the system just discussed. In this and several subsequent figures in which both feedback equations are plotted in the same graph, one function must necessarily be plotted in an unorthodox way with the independent variable on the ordinate and the dependent variable on the abscissa. These unorthodox lines are shown as broken lines, and the reader should note that for the broken line, a slope (sign ignored) which looks small is actually large, whereas a slope which looks large is actually small.

The majority of biological feedback mechanisms differ from the simple electrical circuit described above in two important respects: 1) the relationships between the variables in the feedback loop are often curvilinear rather than rectilinear; 2) the loop is likely to contain more than two variables. The problems posed by these complications must be dealt with before the simple product-of-slopes concept can be extended to real homeostatic mechanisms.

5-4. Curvilinear Feedback Relationships

Suppose that the equations which express the interrelationships between the variables A and B in the feedback loop are not straight-line equations, but are equations for some more complicated curvilinear relationship such as is shown in Figure 5-3. It is obvious that the steady-state values of A and B will still be determined by simultaneous solution of the two equations

---

* "Homeostatic index" is synonymous with the engineering term "open-loop gain" for a constant input signal and negative feedback.

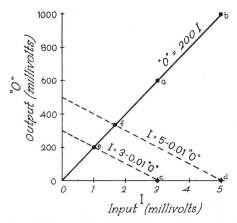

FIG. 5-2. How the Negative Feedback System of Figure 5-1 Operates

Suppose the signal, S, changed from 3 to 5 mv., an increase of 2 mv. If feedback could not occur because F, the fraction of the output "fed back," was zero, I would change from point a to point b, an "open loop" increase of 2 mv., equal to the original change in S. Similarly, if feedback could not occur because G, the gain of the amplifier, was zero, I would change from point c to point d, again an "open loop" increase of 2 mv., equal to the original change in S. But if $F = 0.01$ and $G = 200$, the steady-state value of I for a given value of S will be determined by simultaneous solution of the two feedback equations, $i.e.$, by the point of intersection of the lines. When S increases from 3 to 5 mv., the steady-state value of I will then move from point e to point f, an increase of only $\frac{2}{3}$ mv. The change in I is thus reduced by the operation of the feedback loop to $\frac{1}{3}$ of what it would have been if unopposed by feedback. The magnification of the change is therefore $\frac{1}{3}$, the minification, 3, and the homeostatic index, 2. The homeostatic index may also be calculated as the negative product of the slopes, in this example, $-(-0.01)(200) = 2$.

as illustrated graphically by the point of intersection of the two curves in Figure 5-3.*

Consider the two arbitrary curves of Figure 5-3 at and $immediately$ $adjacent$ to their point of intersection. Within this limited (properly, infinitesimal) range, the curves can be replaced by the straight lines, a, a' and b, b', which are tangent to the curves at their point of intersection. But if the curves can be replaced by these straight lines, the simple argument for rectilinear functions remains valid. Hence, at a given point of intersection, $i.e.$, for a given steady state, the homeostatic index is the

* It is, of course, theoretically possible that the equations might have more than one simultaneous solution within the physiologically permissible range. This would pose curious problems of stability which the author believes would rarely be encountered in physiological systems.

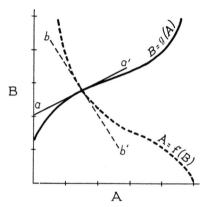

Fig. 5-3. A Curvilinear Negative Feedback System

The two curves, for B as a function of A (*solid*), and for A as a function of B (*broken*), represent two arbitrary feedback equations. In the steady state, A and B must have the values indicated by the point of intersection of the curves. At that point, the homeostatic index is the negative product of the first derivatives of the curves, *i.e.*, the negative product of the slopes of the tangents, a,a' and b,b', to the curves at the point of intersection.

negative product of the slopes of the tangents at the point. Now the slope of the tangent to a curve at a point is merely the first derivative of the equation of the curve at that point. We are thus led to the following generalization: When two variables constituting a homeostatic feedback loop are in a steady state, the homeostatic index is the negative product of the first derivatives of the two feedback equations at that steady state:

$$\text{Homeostatic index} = -\frac{dg(A)}{dA} \cdot \frac{df(B)}{dB} \qquad (5\text{-}13)$$

where $A = f(B)$ and $B = g(A)$

The reader will see at once that Equation 5-12a is merely a special case of Equation 5-13.

5-5. Feedback Loops with More Than Two Variables

In Figure 5-4I a hypothetical negative feedback system is depicted which consists of four variables and the four equations linking them. If this were a physiological mechanism, it would probably not be possible to study more than two of the variables quantitatively. Now it is clear that the entire loop can be reduced to two variables by combining some of the equations so as to eliminate the other variables. For example, by combining Equations 5-15 and 5-16, we can eliminate D, thereby obtaining Equation 5-18 and reducing the loop to the three-variable system shown in Figure

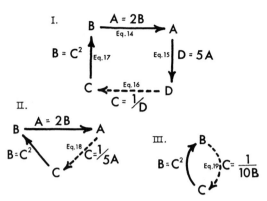

FIG. 5-4. Reduction of a Multivariable Negative Feedback Loop to a Two-Variable Loop

See text for explanation.

5-4II. Similarly, we can eliminate A by combining Equations 5-14 and 5-18 to obtain Equation 5-19 thus reducing the entire loop to the two variables B and C (Fig. 5-4III), which may then be dealt with in the manner outlined above.

Fortunately, it matters not at all which two variables happen to be retained when reducing a multivariable loop to a two-variable loop, provided that the set of equations characterizing the loop have but one simultaneous solution within the physiologically permissible range. The reader can easily satisfy himself that whatever two variables are chosen for study, the homeostatic index of the system depicted in Figure 5-4I is exactly 2. For example, suppose that we can study only variables B and C, and that we have no knowledge at all about Equations 5-14, 5-15, and 5-16. By varying C and observing the resulting changes in B, we would be able to find Equation 5-17:

$$B = C^2 \tag{5-17}$$

Similarly, by varying B and observing the resulting changes in C, we would find Equation 5-19:

$$C = \frac{1}{10B} \tag{5-19}$$

The negative product of the first derivatives of Equations 5-17 and 5-19 (*i.e.*, the homeostatic index for the steady-state values of B and C) is

$$\text{Homeostatic index} = -(2C)\left(-\frac{1}{10B^2}\right) = \frac{C}{5B^2} \tag{5-20}$$

By solving Equations 5-17 and 5-19 simultaneously, we find that the steady-state values of B and C are $B = 10^{-2/3}$; $C = 10^{-1/3}$. Substituting these values in Equation 5-20 we find that

$$\text{Homeostatic index} = (1/5) \, (10^{-1/3}) \, (10^{4/3}) = 2 \qquad (5\text{-}21)$$

If we apply these same steps to any of the other pairs of variables in the four-variable system of Figure 5-4I, we will obtain exactly the same value for the homeostatic index. *The homeostatic index of a multivariable feedback loop at any particular steady state can be calculated from the functional equations relating any two of the variables in the loop.* It is the effectiveness of the loop as a whole, not of some portion of the loop, which is expressed by the negative product of first derivatives.

5-6. How to Obtain the Feedback Equations

The method outlined above for calculating the homeostatic index is useful only when the two feedback equations are known. As will be evident from the actual examples discussed later, it is often possible to derive a theoretical feedback equation from the interrelationships known to exist among the various factors influencing the feedback loop. But it may be necessary to obtain one or both of the feedback equations, let us say the equation for $A = f(B)$, by actually observing how the steady-state value of the dependent variable, here A, changes when a new steady-state value of the independent variable, B, is induced by altering one or more factors which directly influence B but are not themselves members of the loop. Any variables external to the loop which directly influence the value of A must be held constant. The observed steady-state values of A are then plotted against the corresponding steady-state values of B and the points are fitted by a suitable empirical equation. It is not necessary to interrupt the feedback loop. However, one must be able to measure both A and B, and one must also be able to manipulate the independent variable B, so as to produce changes in the dependent variable A, over a sufficiently wide range.

If the homeostatic system includes a *localized* sensing mechanism which detects changes in one of the variables of the feedback loop, it may be advantageous to use a somewhat different approach which *does* require interruption of the feedback loop. As a concrete example, consider the body temperature, T_{body}. Let T_{hypo} be the temperature of the temperature-sensitive region of the hypothalamus. In experimental animals, it is possible to heat the hypothalamus *locally* so as to increase T_{hypo} by a known amount, and to measure the steady-state lowering of T_{body} which results. Let us now regard body temperature and hypothalamic temperature as two separate members of a feedback loop. An empirical feedback equation for

T_{body} as a function of T_{hypo} may be obtained by varying T_{hypo} and measuring T_{body}. The second feedback equation is obviously $T_{hypo} = T_{body}$. By employing this technique, von Euler (113) has found that a small increase in T_{hypo} will result in a large decrease in T_{body}, the indicated homeostatic index being 8 to 10. To generalize, suppose that a variable, A, in a feedback loop can be split into two parts: A_{sens}, the intensity of A at some localized sensing mechanism, and A_{body}, the intensity of A in the rest of the body. A_{sens} and A_{body} may then be regarded as two separate members of the feedback loop, the equation for $A_{sens} = f(A_{body})$ usually being $A_{sens} = A_{body}$, and the equation for $A_{body} = g(A_{sens})$ being obtained by empirical observation of how A_{body} changes in response to localized manipulations of A_{sens}.

5-7. An Alternative Method for Calculating the Homeostatic Index

Even when we cannot find explicit equations for two feedback functions in a homeostatic loop, it may still be possible to calculate the homeostatic index.

Suppose we can manipulate some variable, α_{ext}, external to the loop, so as to cause measurable changes in one of the variables in the loop, for example, A. α_{ext} is not itself a member of the feedback loop but is some other variable in one of the feedback equations, for example, S in Equation 5-2. A must be a member of the loop and must either itself be directly influenced by α_{ext} or be linearly dependent on the member which is directly influenced by α_{ext}. All external variables other than α_{ext} must be held constant. Suppose further that we can measure the change in A in response to a change in α_{ext} both when the feedback loop is "open" (no feedback) and when the loop is "closed" (with feedback). By writing Equation 5-10 in a general form and using differentials so that it will be applicable to curvilinear systems, we obtain the following definition of the magnification due to feedback:

$$\text{Magnification} = (dA/d\alpha_{ext})_{closed}/(dA/d\alpha_{ext})_{open} \qquad (5\text{-}22)$$

The same reasoning as before leads us from this definition to an equation analogous to Equation 5-12:

$$\text{Homeostatic index} = [(dA/d\alpha_{ext})_{open}/(dA/d\alpha_{ext})_{closed}] - 1 \qquad (5\text{-}23)$$

In using Equation 5-23, it is important to realize that $(dA/d\alpha_{ext})_{open}$ and $(dA/d\alpha_{ext})_{closed}$ must both be calculated for the *same* value of A. Otherwise we would not be comparing the open-loop and closed-loop responses under comparable conditions. It is also important to realize that although $(dA/d\alpha_{ext})_{open}$ is often unity, it is not necessarily unity. For example, suppose we had a system governed by the following two feedback equations:

$$A = C^2 - 2B \tag{5-24}$$

and

$$B = (1/D) + A \tag{5-25}$$

where A and B are the variables in the feedback loop, and C and D are variables external to the loop. Suppose that we keep D constant. Even if Equations 5-24 and 5-25 are unknown to us, we may be able to change the external variable, C, and observe the consequent changes in A. As the observed equation for the closed-loop response, we would thus obtain

$$A = (C^2/3) - K \tag{5-26}$$

where K (which actually equals $2/3D$) would be observed merely as an empirical constant. By taking the first derivative of this equation we would find the closed-loop value of dA/dC:

$$(dA/dC)_{\text{closed}} = 2C/3 \tag{5-27}$$

Suppose we then carelessly *assumed* that $(dA/dC)_{\text{open}} = 1$. We would erroneously conclude that the homeostatic index (calculated from Equation 5-23) was $(3/2C) - 1$. In fact, however, the true value of $(dA/dC)_{\text{open}}$ is the first derivative of Equation 5-24 with respect to C, namely, $2C$. The correct homeostatic index calculated from Equation 5-23 is thus $(2C)(3/2C) - 1 = 2$. This is identical with the value which would be obtained by using Equation 5-13.

It is by no means always easy to open a homeostatic loop experimentally without altering the whole system so profoundly as to invalidate a comparison between $(dA/d\alpha_{\text{ext}})_{\text{closed}}$ and $(dA/d\alpha_{\text{ext}})_{\text{open}}$. Fortunately, however, it may be possible to predict, on solid theoretical grounds, what the open-loop response of A to a change in α_{ext} would be.

5-8. The Derivation of Equations for Feedback Relationships Which Cannot Be Studied Directly

Let us combine Equations 5-13 and 5-23:

$$[(dA/d\alpha_{\text{ext}})_{\text{open}}/(dA/d\alpha_{\text{ext}})_{\text{closed}}] - 1 = -\left[\frac{dg(A)}{dA}\right]\left[\frac{df(B)}{dB}\right] \tag{5-28}$$

If three of the four first derivatives in Equation 5-28 are known and are substituted into the equation, one may solve for the remaining first derivative. The original unknown equation of which it is the derivative may then be found (to within a constant) by integration.

The concepts presented above will now be illustrated by four physiological examples drawn from diverse fields. The symbol-and-arrow diagrams for all four examples will be found in Figure 5-5.

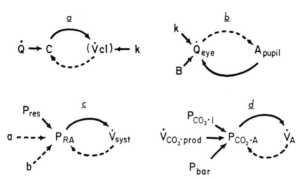

Fɪɢ. 5-5. Symbol-and-Arrow Diagrams for the Four Homeostatic Systems Discussed in the Text

_a_. The metabolism of the thyroid hormone (Section 5-9). _b_. The pupillary response to light (Section 5-10). _c_. Cardiac output in the Starling heart-lung preparation (Section 5-11). _d_. The respiratory response to carbon dioxide (Section 5-12). The proportionality constants, k, in systems _a_ and _b_ are, of course, entirely different.

5-9. The Metabolism of Thyroid Hormone

Let $\dot{Q}$ be the supply of thyroid hormone (micrograms of hormonal iodine per day) in a human subject. Let $(\dot{V}\text{cl})$ be the liters of plasma cleared of thyroid hormone (chiefly by metabolism) per day, and let C be the concentration of hormonal iodine in plasma in micrograms per liter. Now in a steady state, the quantity of hormone cleared from plasma daily, $(\dot{V}\text{cl})C$, must be equal to the daily supply, $\dot{Q}$:

$$(\dot{V}\text{cl})C = \dot{Q} \quad \text{or} \quad C = \dot{Q}/(\dot{V}\text{cl}) \tag{5-29}$$

Moreover, there is evidence that the clearance of thyroid hormone in man is directly proportional to the concentration of hormone in plasma (88). Letting k be the constant of proportionality,

$$(\dot{V}\text{cl}) = kC \tag{5-30}$$

Equations 5-29 and 5-30 show that $(\dot{V}\text{cl})$ and C are members of a negative feedback loop. The homeostatic index of this loop, calculated as the negative product of the first derivatives of the equations, is

$$\text{Homeostatic index} = -(-\dot{Q}/(\dot{V}\text{cl})^2)k = k\dot{Q}/(\dot{V}\text{cl})^2 \tag{5-31}$$

Substituting in Equation 5-31 the value of k from Equation 5-30,

$$\text{Homeostatic index} = \dot{Q}/(\dot{V}\text{cl})C \tag{5-32}$$

Substituting in Equation 5-32 the value of $(\dot{V}\text{cl})C$ from Equation 5-29,

$$\text{Homeostatic index} = \dot{Q}/\dot{Q} = 1 \tag{5-33}$$

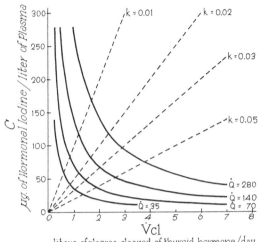

FIG. 5-6. The Metabolism of Thyroid Hormone in Man as a Feedback System

The concentration of hormonal iodine in plasma, C, as a function of the volume of plasma cleared of thyroid hormone per day, ($\dot{V}$cl), is shown by the *solid curves* for four rates of hormone secretion, $\dot{Q}$, (micrograms of hormonal iodine secreted per day). The clearance of hormone, in turn, is proportional to the concentration of hormonal iodine as illustrated by the broken lines which represent four different values of the proportionality constant, k, (liters², micrograms⁻¹, days⁻¹). The negative product of the first derivatives, and hence the homeostatic index, is unity at every point of intersection. When interpreting such a graph it is important to remember that for the *broken lines*, a slope which appears to be large is actually small.

The homeostatic index of this system is therefore always unity, even though one of the two feedback equations (Equation 5-29) is curvilinear. This peculiarity is illustrated graphically in Figure 5-6 where the two equations have been plotted for various values of k and $\dot{Q}$. It is evident that as the slope of the straight line (Equation 5-30) *increases*, the negative slope of the tangent to the curve (Equation 5-29) at a point of intersection *decreases*, so that the negative product of first derivatives at every point of intersection is unity. Since a homeostatic index of unity corresponds to a magnification of $\frac{1}{2}$, the operation of the feedback loop will decrease any small change in the concentration of hormone to half of what it would be without feedback. This reduction by $\frac{1}{2}$ remains constant whether the rate of hormone production is high or low.

5-10. The Pupillary Reaction to Light

The constriction of the pupil of the eye in response to light may be treated as a homeostatic adjustment designed to limit the quantity of light entering the eye per unit of time, $\dot{Q}_{eye}$. In the following analysis, the area of the pupil in square millimeters, A_{pupil}, will be used as the measure of pupillary size.

Crawford (26) has studied the steady-state diameter of the pupil in 10 normal subjects as a function of the brightness of a wide illuminated screen facing the subject. The brightness, B, was varied in 16 steps from 1.1×10^{-8} candles per square foot to 21 candles per square foot. Under these conditions, it is obvious that the quantity of light entering the eye per unit of time must be proportional to the brightness of the screen and to the area of the pupil:

$$\dot{Q}_{eye} = kBA_{pupil} \qquad (5\text{-}34)$$

where k is a proportionality constant with appropriate dimensions. (For convenience, we will express $\dot{Q}_{eye}$ in such units that the numerical value of k is unity when screen brightness is measured in candles per square foot and area of the pupil is measured in square millimeters.) Crawford gives values for screen brightness and for mean pupillary diameter from which A_{pupil} and $\dot{Q}_{eye}$ may readily be calculated. The observed dependence of A_{pupil} upon $\dot{Q}_{eye}$ is closely approximated by the following empirical equation:

$$A_{pupil} = (160\dot{Q}_{eye}^{-0.4} + 94.8)/(4\dot{Q}_{eye}^{-0.4} + 15.8) \qquad (5\text{-}35)$$

The homeostatic index of this system is given by the negative product of the first derivatives of Equations 5-34 and 5-35,

$$\text{Homeostatic index} = -(kB) \ [-860\dot{Q}_{eye}^{-1.4}/(4\dot{Q}_{eye}^{-0.4} + 15.8)^2] \qquad (5\text{-}36)$$

or, since by Equation 5-34, $kB = \dot{Q}_{eye}/A_{pupil}$,

$$\text{Homeostatic index} = [1/A_{pupil}][860\dot{Q}_{eye}^{-0.4}/(4\dot{Q}_{eye}^{-0.4} + 15.8)^2] \qquad (5\text{-}37)$$

The homeostatic index is never large, and varies with pupillary diameter in the manner shown by Figure 5-7. The maximum ability of the pupil to compensate for changes in the quantity of light incident upon the cornea occurs with a moderately small pupil, corresponding to moderately bright illumination. Even at its maximum, however, the homeostatic index is small because a rather large change in $\dot{Q}_{eye}$ is needed to cause a comparatively small change in the size of the pupil.

5-11. The Regulation of Cardiac Output in the Starling Heart-Lung Preparation

In the Starling heart-lung preparation, blood is pumped by the left ventricle through an artificial "systemic" resistance and is then collected in an open reservoir, the venous reservoir, from which it flows by gravity through a tube of moderate resistance into the right atrium via a cannula in the vena cava. The "venous supply" to the heart can be varied by

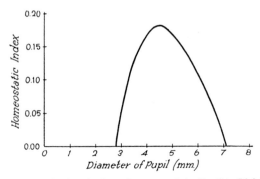

FIG. 5-7. The Steady-State Response of the Pupil to Light

The homeostatic index, calculated from Crawford's mean data for normal human subjects (26), is plotted as a function of the pupillary diameter.

raising or lowering the venous reservoir so as to alter the head of pressure, P_{res}, from the surface of the blood in the reservoir to the right atrium. It might also be altered by changing the resistance to flow between reservoir and right atrium, but usually this resistance is arbitrarily adjusted to give a reasonable cardiac output at the beginning of the experiment and is not deliberately changed thereafter.

In this system, with systemic resistance held constant, the greater the pressure of blood in the right atrium, P_{RA}, the greater will be the diastolic distension of the heart, and hence (according to Starling's "Law of the Heart") the greater will be the systemic cardiac output, $\dot{V}_{syst}$.* On the other hand, the greater the cardiac output, the less will be the pressure in the right atrium. Cardiac output and right atrial pressure are thus members of a negative feedback loop. Now the volume of blood per minute flowing back to the heart from the venous reservoir, $\dot{V}_{ven}$, will be directly proportional to the difference between P_{res} and P_{RA}, and inversely proportional to the resistance, R, between the reservoir and the atrium:

$$\dot{V}_{ven} = (P_{res} - P_{RA})/R \qquad (5\text{-}38)$$

But in the steady state, the venous return, $\dot{V}_{ven}$, must be equal to the systemic output. Hence,

$$\dot{V}_{syst} = (P_{res} - P_{RA})/R \qquad (5\text{-}39)$$

The resistance to flow, R, may be constant when the systemic output is

* The total output from the left ventricle also includes the coronary flow, but the coronary flow does not contribute to the blood returning to the right atrium from the venous reservoir, and hence need not be considered in the present discussion.

small, but with larger outputs the resistance is usually found empirically to be a linear function of the pressure difference:

$$R = a + b(P_{\text{res}} - P_{RA}) \qquad (5\text{-}40)$$

where, for a given experiment, a and b are constants whose values depend upon the physical characteristics of the blood and of the channel through which the blood flows from the reservoir to the heart. Combining Equations 5-39 and 5-40 and solving for P_{RA} as a function of $\dot{V}_{\text{syst}}$,

$$P_{RA} = P_{\text{res}} - [a\dot{V}_{\text{syst}}/(1 - b\dot{V}_{\text{syst}})] \qquad (5\text{-}41)$$

Equation 5-41 is one of the two feedback equations. It has nothing at all to do with the heart itself, but merely describes the physical characteristics of the part of the system which lies between the reservoir and the heart. In contrast, the other feedback equation is determined by how much blood the heart is able to pump at various right atrial pressures. It is therefore greatly influenced by the condition of the myocardium, *i.e.*, by the "competence" of the heart. This second equation must be determined empirically by measuring cardiac output at various right atrial pressures. It has been described as linear (84) but is actually curvilinear. Moreover, the nature of the functional dependence of cardiac output upon right atrial pressure seems to vary with the condition of the myocardium; so that a type of equation which satisfactorily fits data obtained from a competent heart may not be at all suitable for data obtained from the same heart in failure (Fig. 5-8). Finally, the system is never in a truly steady state because the Starling heart-lung preparation gradually deteriorates with the passage of time and cardiac failure becomes more and more severe. Under these circumstances, the labor of finding an empirical equation to describe a particular set of observations is scarcely justified, and the experimenter may well be content to fit a curve to the points by eye, and to estimate graphically the slope of the tangent to the curve at any desired point. The negative product of this estimated slope, $(d\dot{V}_{\text{syst}}/dP_{RA})_{\text{estim}}$, and the first derivative of Equation 5-41 will then be an estimate of the homeostatic index at that point:

$$\text{Homeostatic index} = -[-a/(1 - b\dot{V}_{\text{syst}})^2][d\dot{V}_{\text{syst}}/dP_{RA}]_{\text{estim}} \qquad (5\text{-}42)$$

Wollenberger and Krayer (122) proposed the "competence index" as a measure of the ability of the heart to respond to increases in venous supply with corresponding increases in output. The competence index is defined as:

$$\text{Competence index} = (\Delta P_{\text{res}} - \Delta P_{RA})/\Delta P_{\text{res}} = 1 - (\Delta P_{RA}/\Delta P_{\text{res}}) \qquad (5\text{-}43)$$

where ΔP_{RA} is the "closed loop" increase in P_{RA} caused by raising the pressure head of the reservoir by an increment of ΔP_{res}. (Note that, in

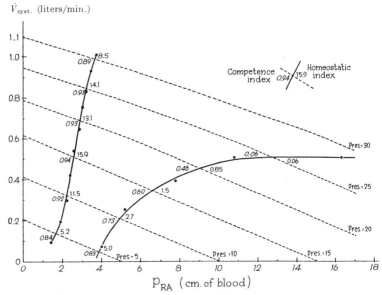

$\dot{V}_{\text{syst.}}$ (liters/min.)

FIG. 5-8. The Starling Heart-Lung Preparation as a Homeostatic System

The animal was a dog weighing 9.0 kg. The *broken lines* show the right atrial pressure, P_{RA}, as a function of the systemic cardiac output, $\dot{V}_{\text{syst}}$, for various reservoir pressures, P_{res}, (centimeters of blood above the right atrium). For flows greater than about 0.6 L. per minute, the lines are curved because the resistance to flow increased as the pressure difference, $P_{\text{res}} - P_{RA}$, increased. The *solid curve* on the *left* represents the systemic output of the heart as a function of right atrial pressure just after the heart had been isolated. Each point represents the mean of two steady-state determinations. The *solid curve* on the *right* represents the systemic output of the heart as a function of right atrial pressure after severe heart failure had been produced by adding pentobarbital sodium to the blood perfusing the heart. Because the failure was progressive, each point represents but one determination. The values of the homeostatic index and of the competence index shown at each point of intersection were calculated as described in the text. (I am greatly indebted to Dr. Joseph M. Benforado for performing this experiment and for permitting me to reproduce the results.)

the absence of feedback, the "open loop" ΔP_{RA} would be equal to ΔP_{res}.) If the two feedback equations were linear, the ratio $\Delta P_{RA}/\Delta P_{\text{res}}$ would correctly express the magnification due to feedback. The competence index would then be $(1 - \text{magnification})$ and would convey exactly the same information as the homeostatic index which has already been defined as $(\text{minification} - 1)$:

$$\text{Competence index} = \text{homeostatic index}/(1 + \text{homeostatic index}) \quad (5\text{-}44)$$

But since the feedback equations are in general curvilinear, the competence index, like the homeostatic index, will vary continuously as P_{RA} and $\dot{V}_{syst}$ change, so that the homeostatic effectiveness of the system cannot be calculated properly from such finite changes as appear in Equation 5-43. If the competence index were redefined in terms of differentials instead of gross changes it would be a perfectly valid measure of the effectiveness of the Starling heart-lung preparation as a homeostatic system. However, as Price and Helrich (84) have pointed out, the competence index is not a suitable measure of *cardiac* competence, for it is as much influenced by the resistance of the tubing between reservoir and heart as it is by the ability of the heart to pump blood. It would seem desirable to find some more specific measure of cardiac competence which would be influenced only by the effectiveness of the heart as a pump.

5-12. The Respiratory Response to Carbon Dioxide

Certain aspects of the chemical control of breathing are still poorly understood, but there can be no doubt that more than one feedback loop is involved. The present discussion will be confined to the homeostatic hyperventilation which occurs when the concentration of carbon dioxide in the alveolar air (and in the arterial blood) increases. (Although the effect of an increase in the metabolic production of CO_2 upon this same feedback system will also be discussed, the reader must *not* infer that this particular feedback system actually governs the respiratory response to the increased metabolic production of CO_2 during exercise, for example. There is ample evidence that the respiratory response to exercise is strongly influenced by at least one other feedback loop which is not discussed here at all.) The two variables in this feedback loop which are most easily measured are the alveolar ventilation per minute, $\dot{V}_A$, and the partial pressure of CO_2 in alveolar air, $P_{CO_2 \cdot A}$. For simplicity let us assume that the respiratory exchange ratio is unity, that the inspired gas mixture is at body temperature and saturated with water vapor, and that the partial pressure of oxygen in the inspired gas is about 150 mm. Hg. Now in the steady state, the volume of CO_2 expired per minute, $\dot{V}_{CO_2 \cdot E}$, must equal the volume of CO_2 inspired per minute, $\dot{V}_{CO_2 \cdot I}$, plus the volume of CO_2 produced by metabolism per minute, $\dot{V}_{CO_2 \cdot prod}$ (all volumes being at ambient temperature and pressure):

$$\dot{V}_{CO_2 \cdot E} = \dot{V}_{CO_2 \cdot I} + \dot{V}_{CO_2 \cdot prod} \qquad (5\text{-}45)$$

With appropriate correction for the ineffective ventilation of the dead space, Equation 5-45 may be written in terms of partial pressures as

$$(P_{CO_2 \cdot A}/P_{bar})\dot{V}_A = (P_{CO_2 \cdot I}/P_{bar})\dot{V}_A + \dot{V}_{CO_2 \cdot prod} \qquad (5\text{-}46)$$

where $P_{CO_2 \cdot I}$ = the partial pressure of CO_2 in the inspired gas

P_{bar} = the barometric pressure

Multiplying Equation 5-46 by P_{bar}, and solving for $P_{CO_2 \cdot A}$,

$$P_{CO_2 \cdot A} = P_{CO_2 \cdot I} + (\dot{V}_{CO_2 \cdot prod} P_{bar} / \dot{V}_A) \qquad (5\text{-}46a)$$

Equation 5-46a is the feedback equation for $P_{CO_2 \cdot A}$ as a function of $\dot{V}_A$. The second feedback equation, for $\dot{V}_A$ as a function of $P_{CO_2 \cdot A}$, must be determined empirically by observing how alveolar ventilation per minute varies when $P_{CO_2 \cdot A}$ is increased by inhalation of gas mixtures containing various concentrations of CO_2. In the normal human, this relationship is said to be linear, at least over a considerable range, and may be expressed by the following equation:

$$\dot{V}_A = -135 + 3.5\, P_{CO_2 \cdot A} \qquad (5\text{-}47)$$

where $\dot{V}_A$ is in liters per minute, and $P_{CO_2 \cdot A}$ is in millimeters of Hg. The slope, (3.5 L. per min.-mm. Hg), was estimated from the average slope for minute volume as a function of $P_{CO_2 \cdot A}$ when alveolar oxygen tension was normal in the subjects studied by Lloyd et al. (70). The intercept, −135 L. per minute, was calculated by assuming that in a normal individual at rest, $\dot{V}_A$ would be 5 L. per minute and $P_{CO_2 \cdot A}$ would be 40 mm. Hg. According to Equation 5-46a the normal carbon dioxide production corresponding to these values would be 0.263 L. per minute at a barometric pressure of 760 mm. Hg.

The homeostatic index for this loop is the negative product of the first derivatives of Equations 5-46a and 5-47:

$$\text{Homeostatic index} = -(-\dot{V}_{CO_2 \cdot prod} P_{bar} / \dot{V}_A{}^2)(3.5) \qquad (5\text{-}48)$$

The two feedback equations, 5-46a and 5-47, have been plotted in Figure 5-9. In the upper portion of the figure, $P_{CO_2 \cdot A}$ has been varied by increasing the concentration of CO_2 in the inspired air (at 760 mm. Hg) from 0 to 6 per cent, the production of CO_2 being kept constant at 0.263 L. per minute. It is evident that as the concentration of CO_2 inspired increases, the negative slope of the curve for $P_{CO_2 \cdot A}$ as a function of $\dot{V}_A$ (Equation 5-46a) at its intersection with the straight line for Equation 5-47 decreases very rapidly. Accordingly, the homeostatic index rapidly becomes smaller and smaller as the concentration of CO_2 in the inspired gas mixture increases. In the lower portion of the figure, $P_{CO_2 \cdot A}$ has been altered by increasing the production of CO_2 from the assumed normal value of 0.263 L. per minute (Curve N) to 12.5 times normal (Curve 12.5N), the concentration of CO_2 in the inspired air being kept constant at 0 per cent. While the absolute value of the slope of the curve for $P_{CO_2 \cdot A}$ as a function of $\dot{V}_A$

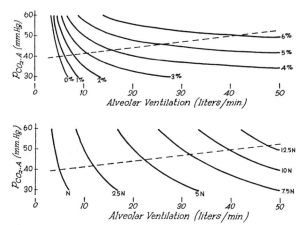

Fig. 5-9. The Ventilatory Response of Normal Humans to an Increase in the Partial
Pressure of Carbon Dioxide in Alveolar Air (or Arterial Blood)

In the upper portion of the figure, the *solid curves* represent $P_{CO_2 \cdot A}$, the partial
pressure of CO_2 in alveolar air, as a function of alveolar ventilation when the $P_{CO_2 \cdot A}$
is increased by inhalation of the indicated percentages of CO_2. In the lower portion
of the figure, the *solid curves* represent $P_{CO_2 \cdot A}$ as a function of alveolar ventilation
when $P_{CO_2 \cdot A}$ has been elevated by an increase in the production of CO_2. N indicates
"normal" production of CO_2, while $2.5N$, $5N$, etc., indicate a rate of CO_2 production
which is 2.5 times "normal," 5 times "normal," etc. Note that the curves for 0%
CO_2 in the *upper* graph and for "normal" CO_2 production in the *lower* graph are
identical. The *broken straight line* for alveolar ventilation as a function of $P_{CO_2 \cdot A}$ is
also the same for both graphs.

again decreases as the "load" of CO_2 increases, the decrease is not nearly
so marked as it is when the extra CO_2 comes from the inspired air.

The ability of this feedback system to maintain its homeostatic effectiveness better when $P_{CO_2 \cdot A}$ is increased by increasing the production of CO_2
than when $P_{CO_2 \cdot A}$ is increased by increasing the concentration of CO_2 in
the inspired air is further illustrated by Figure 5-10. With no CO_2 in the
inspired air, and with the "normal" CO_2 production previously assumed,
the homeostatic index is about 28. As $P_{CO_2 \cdot A}$ increases, the homeostatic
index falls regardless of whether the additional CO_2 was derived from
metabolism alone (*upper curve*) or from the inspired air alone (*lower
curve*). But the rate of decrease is quite different for the two sources of
CO_2. For example, if the steady-state $P_{CO_2 \cdot A}$ were increased from 40 to
46 mm. Hg by an increased metabolic production of CO_2, the homeostatic
index would still be about 6, so that the effect of any further infinitesimal
increase in either CO_2 production or concentration of CO_2 in the inspired

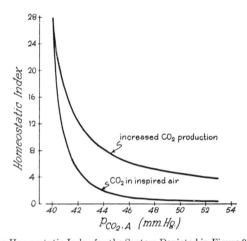

Fig. 5-10. The Homeostatic Index for the System Depicted in Figure 9 as a Function of the Partial Pressure of Carbon Dioxide in Alveolar Air (or Arterial Blood)

The homeostatic index decreases much more rapidly when $P_{CO_2 \cdot A}$ has been elevated by an increase in CO_2 in the inspired air than when $P_{CO_2 \cdot A}$ has been elevated by an increase in CO_2 production.

air would be reduced to $\frac{1}{7}$ of the effect which it would have if unopposed by the feedback loop. But if the *same* steady-state $P_{CO_2 \cdot A}$ were due to the inhalation of CO_2, the homeostatic index would be only 1, so that the effect of a further infinitesimal increase of either CO_2 production or CO_2 in the inspired air would only be reduced by feedback to one-half of its effect in the absence of feedback.

Why is this homeostatic mechanism able to minimize changes in $P_{CO_2 \cdot A}$ so much better when the changes are due to increased CO_2 production? The answer becomes obvious if one remembers that hyperventilation is an essential feature of the homeostatic adjustment. When the additional CO_2 is produced by metabolism, hyperventilation merely increases the quantity of CO_2 which is "blown off" during expiration. But if the additional CO_2 is due to CO_2 in the inspired air, hyperventilation *also* increases the quantity of CO_2 which is drawn into the lungs during inspiration. The greater the hyperventilation, the greater the amount of CO_2 inspired per minute. Thus the compensatory mechanism tends to defeat its own purpose and rapidly loses its effectiveness as $P_{CO_2 \cdot I}$ increases.

5-13. Limitations of the Present Analysis

While the preceding mathematical description of homeostasis has the advantage of being valid for nonlinear systems, it suffers from the serious disadvantage of not being applicable at all to transient states in which time is an independent variable. However, many of the homeostatic mechanisms of the mammalian body are well enough damped so that ordinarily they do not exhibit oscillatory behavior. With such systems, the steady-state analysis will supply the information which is of the greatest physiological interest.

It is also important to realize that the mathematical analysis described above has thus far been applied only to simple single-loop systems in which it has been assumed that all variables external to the loop can either be controlled or remain constant. The refinements needed to apply this method to more complex systems which contain several interlinked or interrelated loops have not yet been worked out. Therefore, at present the preceding analysis can be applied to complex systems with multiple loops only when each loop can be isolated from the others and treated as a subsystem in the manner described in Section 4-16.

5-14. Positive Feedback Systems

In order to apply the foregoing discussion to positive feedback systems, it is only necessary to realize that positive feedback is designed to magnify, rather than to minimize, the effect of any external variable tending to disturb the steady state. For the effectiveness of such systems, the appropriate measure is not the homeostatic index, defined as (minification − 1), but rather the quantity, (magnification − 1):

$$(\text{Magnification} - 1) = \frac{(\text{product of first derivatives})}{(1 - \text{product of first derivatives})} \quad (5\text{-}49)$$

We must consider three cases:

1. *The product of first derivatives lies between zero and unity.* In this case the magnification is positive and finite. Under these circumstances, the feedback enhances the effect of an external change, but the system is nevertheless still able to establish a new steady state.

2. *The product of first derivatives is unity.* In this case, no feedback exists because the curves representing the two functions are either separate but parallel, so that they have no simultaneous solution, or they coincide, so that they in fact represent one and the same equation, the relationship then belonging to Category 2-b.

3. *The product of first derivatives is greater than unity.* In this case, the system is in a state of unstable equilibrium so that any change, however small, will be enhanced without limit and at a faster and faster rate, so long as the original equations remain valid. We may term this *explosive feedback*. A particularly terrifying example of

explosive feedback is the atomic bomb in which the two variables of the feedback loop are the neutron flux and the rate of fission of uranium atoms.

From the above discussion it should be evident that the simple steady-state analysis can be applied to positive feedback only when the product of first derivatives lies between zero and unity.

EXERCISES. CHAPTER 5

Exercise 1

As an example of feedback, Goldman (47) presents an elaborate mathematical analysis of "A Buffer Solution as a Feedback Mechanism." What are the two feedback equations for a buffer solution?

Exercise 2

Ross et al. (92) perfused the hind leg of the dog with blood of varying oxygen saturation and measured the resulting blood flow at constant perfusion pressure. As an index of autoregulation, they calculated the "oxygen available to the tissues" as the product of blood flow and percentage of hemoglobin saturation, expressing this as percentage of control. For example, in one experiment the oxygen saturation was maintained at 60 per cent of normal. Had there been no change in blood flow, the "oxygen available to the tissues" would have been 60 per cent of normal, or an "initial degree of abnormality" of $100 - 60 = 40$ per cent. But because of a compensatory increase in blood flow, the actual "oxygen available to the tissues" was 93 per cent of the control value, giving a "final degree of abnormality" of 7 per cent. The authors state, "The effectiveness of a control system can be depicted by the gain or amplification of the system. By dividing the initial degree of abnormality (40%) minus the final degree of abnormality (7%) by the final degree of abnormality (7%) a gain of about 5 is found $[(40 - 7)/7 = 4\frac{5}{7}]$." In one experiment the calculated "gain" was reported to be infinite (see their Fig. 5), and (though the authors do not mention it) in other experiments (see their Fig. 3) the "gain" must have been *negative*, the "oxygen available to the tissues" being substantially *higher* during anoxia than during the control period! Is this an example of "overcompensation," or is there some other explanation for these curious results?

Exercise 3

In the text it was shown that when the four-variable loop of Figure 5-4 was reduced to the two variables, B and C, the homeostatic index was 2.

Show that this same index is obtained when the loop is reduced to any other pair of variables.

Exercise 4

Consider a substance (such as inorganic phosphate) which has a renal "threshold" for excretion. Assume that there are no extrarenal routes of loss from the body. Assume that none of the substance is bound to plasma protein and that the concentration in the glomerular filtrate is the same as the concentration in plasma, C_P. Assume further that the quantity of the substance which is excreted per unit time in the urine, $\dot{Q}_U$, is zero until the quantity filtered per unit of time, $\dot{Q}_{filt}$, exceeds the "tubular reabsorptive capacity," $\dot{Q}_{abs \cdot max}$, which we will take to be constant. Show that when the substance is given by continuous intravenous infusion at a rate of $\dot{Q}_{in}$, the plasma concentration and the renal clearance, ($\dot{V}cl$), are members of a feedback loop, and derive a general equation for the homeostatic index of this loop. What will the homeostatic index be when $\dot{Q}_{filt}$ *just* exceeds $\dot{Q}_{abs \cdot max}$, and how can you explain this value?

Exercise 5

Making assumptions similar to those used in Exercise 4, and using a similar approach, show that when the substance being infused is not re-absorbed by the tubules but is excreted by them, the concentration in plasma, C_P, and the renal plasma clearance, ($\dot{V}cl$), are members of a *positive* feedback loop when the tubular transfer maximum, $\dot{Q}_{tub \cdot max}$ is exceeded. Derive an equation for the quantity, (magnification -1), as an index of the effectiveness of this positive feedback loop.

Note: Many of the topics touched upon in this and the previous chapter are discussed much more fully in: D. S. Riggs, *Control Theory and Physiological Feedback Mechanisms*, The Williams & Wilkins Company, Baltimore, 1970.

6

EXPONENTIAL GROWTH AND DISAPPEARANCE

6-1. General Characteristics of Exponential Change

In the mathematical analysis of biological systems one often encounters some variable, let us say, y, which is increasing or decreasing at a rate which is directly proportional to the magnitude of the variable itself. This proportionality can be expressed as a simple equation by defining an appropriate proportionality constant:

$$\pm dy/dt = ky_t \tag{6-1}$$

where y_t = the value of y at any instant of time, t

t = time, measured in any convenient units

dy/dt = the infinitesimal increment (plus) or decrement (minus) of y at time t during an infinitesimal interval of time

k = a constant of proportionality

Since k has the dimensions of reciprocal time, *i.e.*, of a rate, it is usually called a rate constant. If dy/dt is positive (as we will assume for the time being), Equation 6-1 states that at any instant of time the rate of increase of y is proportional to y. If we divide both sides of Equation 6-1 by y_t we obtain

$$(dy/dt)/y_t = k \tag{6-1a}$$

which states that the rate of increase of y *per unit of y*, in other words the *proprotional rate of increase* of y, is constant and, in fact, is equal to the rate constant, k. Accordingly, we may henceforth think of k the as *instantaneous rate of change of y expressed as a proportion* of the y present at any particular instant. For example, if k is 0.03 per hour, and $y_{t_1} = 22.0$ is the particular value of y at the particular instant of time t_1, then *at that particular instant y* will be increasing at a rate of $dy/dt = ky_{t_1} = (0.03)$ $(22.0) = 0.660$ units of y per hour. But because y is continuously increasing, a short time later at t_2 y will have increased to, let us say, 22.1. Hence

120

the instantaneous rate of increase of y at t_2, (namely, $dy/dt = ky_{t_2} = (0.03)(22.1) = 0.663$ units of y per hour) will be slightly larger than at t_1. But both at t_1, and at t_2, and at every other instant of time, the *proportional rate* of increase is constant at 0.03 of y_t per hour. (Though we often prefer to say "3 per cent per hour," we must remember that "3 per cent" means "3 divided by 100." In all mathematical expressions this division must either be indicated numerically $(3/100)$, or actually performed (0.03), so that "3 %" will be written as a proportion and be in no danger of appearing as "3.")

Upon integration, Equation 6-1 becomes

$$y_t = y_0 e^{kt} \tag{6-2}$$

where y_t = the value of y at any time, t

y_0 = the value of y at time zero

e = the base of natural logarithms

k = the rate constant as previously defined

Equation 6-2 enables us to calculate a specific value for y at any given instant of time provided we know the values of y_0 and k. A logarithmic transformation of Equation 6-2 is often convenient. Taking natural logarithms of both sides of Equation 6-2,

$$\ln y_t = \ln y_0 + kt \tag{6-3}$$

or,

$$\ln (y_t/y_0) = kt \tag{6-3a}$$

If we want to convert to $\log_{10}$, we must multiply both sides of Equation 6-3 by the factor 0.4343 so as to change the base from e to 10:

$$\log_{10} y_t = \log_{10} y_0 + 0.4343 kt \tag{6-3b}$$

or, if we define a new rate constant $k' = 0.4343k$,

$$\log_{10} y_t = \log_{10} y_0 + k't \tag{6-4}$$

However, unlike k, k' does not have the simple meaning of proportion per unit of time. For this reason it is usually better to work with natural logarithms, using Equation 6-3 and 6-3a.

Because the independent variable t appears in Equation 6-2 in the exponent, y is said to be an exponential function of t. Furthermore, Equations 6-3 and 6-4 are linear equations such that if $\ln y$ (or $\log_{10} y$) be plotted against time, one obtains a straight line whose intercept is $\ln y_0$ (or $\log_{10} y_0$) and whose slope is k (or k'). Hence y is commonly said to be increasing logarithmically.

The material just presented is of fundamental importance. It is essential for the reader to understand clearly the precise meaning of terms such as "instantaneous rate of change," "proportional rate of change," and "exponential growth." Before going any further, therefore, these concepts will be illustrated by a concrete analogy—compound interest—which is familiar to everybody.

6-2. Compound Interest as an Analogy for Exponential Growth

Suppose that at time zero we place y_0 dollars in a bank which agrees to pay us interest at a rate of ($100k$) per cent per year, compounded annually. k is thus the proportional rate of interest, or proportional rate of increase, per year. We will allow y_0 to grow for a period of 10 years ($t = 10$). Table 6-1 shows exactly how the amount of money in the bank at the end of each compounding period is calculated. The interest (*column D*) is obtained by multiplying the money already in the bank (*column C*) by the proportional rate of interest, k. The interest is added to the amount already in the bank so as to obtain the total in *column E*. This total is then carried down to the next row and entered in *column C* so that it will serve as the amount at interest during the next compounding period. From the first three rows of the table, it is evident that in general (*fourth row*) the money in the bank at the end of the nth annual compounding period will be $y_0(1 + k)^n$ dollars. If y_0 is 100.00 dollars and k is 0.04 per year, then the total at the end of 10 years will be 148.02 dollars. The amount of money in the bank at every instant of time is shown in Figure 6-1. It increases in a stepwise way, changing only once a year.

Now suppose we place our y_0 dollars in a bank which pays interest at the same annual rate but compounds the interest z times a year. Then the interest added at the end of each compounding period will no longer be k times the amount in the bank, but (k/z) times the amount in the bank. Table 6-1 again shows how the amount at the end of each compounding period can be calculated. But now our general expression for the amount of money at the end of any compounding period is

$$y_{(t\text{-end period})} = y_0[1 + (k/z)]^n \tag{6-5}$$

However, since

$$n = (t.\text{end period})z \tag{6-6}$$

where (t.end period) is the time in years at the end of any compounding period, we can write Equation 6-5 in the form

$$y_{(t.\text{end period})} = y_0[1 + (k/z)]^{z(t.\text{end period})} \tag{6-7}$$

Equation 6-7 is a perfectly general equation for compound interest. For example, if interest is compounded quarterly ($z = 4$), and we place 100

TABLE 6-1
Growth of money at compound interest

A	B	C	D	E
Compounding period number	Time at end of period n	Money in bank at *beginning* of period n	Interest added at end of period n	Money in bank after interest has been added
n	(t, end period)			$y_{(t,\text{ end period})}$
	years	*dollars*	*dollars*	*dollars*
1	1	y_0	ky_0	$y_0 + ky_0 = y_0(1+k)^1$
2	2	$y_0 + ky_0$	$ky_0 + k^2y_0$	$y_0 + 2ky_0 + k^2y_0 = y_0(1+k)^2$
3	3	$y_0 + 2ky_0 + k^2y_0$	$ky_0 + 2k^2y_0 + k^3y_0$	$y_0 + 3ky_0 + 3k^2y_0 + k^3y_0 = y_0(1+k)^3$
n	n			$y_0(1+k)^n = y_0(1+k)^{(t,\text{ end period})}$
1	$1/z$*	y_0	$(k/z)y_0$	$y_0 + (k/z)y_0 = y_0[1+(k/z)]^1$
2	$2/z$	$y_0 + (k/z)y_0$	$(k/z)y_0 + (k/z)^2y_0$	$y_0 + 2(k/z)y_0 + (k/z)^2y_0 = y_0[1+(k/z)]^2$
3	$3/z$	$y_0 + 2(k/z)y_0 + (k/z)^2y_0$	$(k/z)y_0 + 2(k/z)^2y_0 + (k/z)^3y_0$	$y_0[1+(k/z)]^3$
n	n/z			$y_0[1+(k/z)]^n = y_0[1+(k/z)]^{z(t,\text{ end period})}$

*z = number of compounding periods per year.

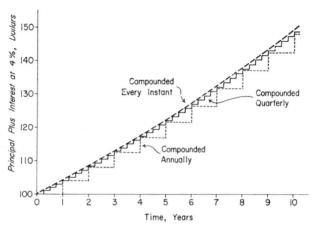

Fig. 6-1. Growth of $100 Left at 4.0 Per Cent Compound Interest for 10 Years

Compounded annually, the money will increase but once a year as shown by the *broken-line steps*. Compounded quarterly (*solid-line steps*), the money will increase in smaller increments but more frequently, and the money in the bank at 10 years will be slightly greater than when compounded once a year. The *broken curve* represents continuous exponential increase at the same rate (4.0 per cent per annum). Exponential growth is precisely equivalent to growth at compound interest when the interest is compounded *at every instant of time*. The yield at the end of 10 years is very slightly larger than when the interest was calculated quarterly. However, it is evident from the figure that the curve for continuous exponential growth lies very close to the amounts of money in the bank just after the addition of interest at the ends of the finite compounding periods.

dollars at 4 per cent interest as before, it will grow in smaller, but more frequent steps to a final amount of $100[1 + (0.04/4)]^{(10)4} = 100(1 + 0.01)^{40} = 148.88$ dollars at the end of 10 years—slightly more than when interest at the same rate was compounded annually. The stepwise growth toward this final amount is also shown in Figure 6-1.

Finally, imagine a bank which compounds interest not annually, nor quarterly, nor even once a month, but *continuously at every instant of time*. For such a bank, z in Equation 6-7 must be imagined to increase without limit toward infinity. It can be shown that as z approaches infinity, $[1 + (k/z)]^{(t.\text{end period})z}$ approaches $e^{k(t.\text{end period})}$. But when interest is thus compounded continuously, the compounding periods are infinitely frequent and infinitely brief so that *every* instant of time represents the "end" of a "compounding period." Hence for the special case of continuous compounding we can replace the clumsy (t.end period) by a simple and unrestricted "t" and Equation 6-7 accordingly becomes

$$y_t = y_0 e^{kt} \qquad (6\text{-}2)$$

which is nothing but the integral form of the equation for exponential growth discussed earlier. *Exponential growth is precisely equivalent to growth at compound interest when the interest is compounded continuously.* The smooth curve of exponential growth which would result if we could place 100 dollars at 4 per cent interest compounded continuously is also shown in Figure 6-1.

Notice that exponential growth is actually a simpler concept than step-wise addition of interest at intervals, for it can be fully characterized by specifying the original amount at time zero, the proportional rate of growth (*i.e.*, the rate of interest) and the elapsed time. We need not specify length of compounding period as a bank must do. Notice also that the final amount of money in the bank after 10 years of exponential growth (149.18 dollars) is only a little larger than the final amount when interest was calculated quarterly. Intermittent addition of interest is thus a reasonably good approximation of exponential growth (and vice versa) provided the stepwise increments are sufficiently small.

6-3. Exponential Disappearance

The preceding discussion has been limited to exponential growth so that it could be illustrated simply and directly by the example of compound interest. But although there are indeed some important biological examples of exponential growth, the great majority of exponential changes encountered in nature are decreases, not increases. A familiar example of exponential disappearance is the decay of a radioactive isotope by spontaneous disintegration into some other element. For the description of simple exponential disappearance no new mathematical concepts are needed because the only difference between exponential increase and exponential decrease is in the *direction* of the change. Hence all of the equations previously written for exponential growth will serve equally well for exponential disappearance if we merely take the sign of the change, dy, as *negative* in Equation 6-1 so as to indicate that y is *decreasing*:

$$-dy/dt = ky_t \qquad (6\text{-}1)$$

Equation 6-2 then becomes

$$y_t = y_0 e^{-kt} \qquad (6\text{-}8)$$

so that when y is decreasing exponentially, the exponent of e is negative. Accordingly, Equation 6-3, the linear logarithmic transformation of Equation 6-2, becomes

$$\ln y_t = \ln y_0 - kt \qquad (6\text{-}9)$$

so that the slope, $-k$, of the straight line obtained by plotting $\ln y$ as a function of t is negative.

There is one very important difference between exponential growth and exponential disappearance. As long as the exponential function remains valid, y can increase more and more, and faster and faster, with no limit whatsoever. (It might amuse you to calculate how rich you would be today had you been able to invest 1.00 dollar at 2 per cent interest compounded continuously, in a very stable bank at the beginning of the Christian era (Exercise 1)). But if y is undergoing simple exponential disappearance, as time increases, y must necessarily approach a limiting value of zero (Fig. 6-2). True, zero is approached asymptotically, so that, in theory at least, it is not actually reached until infinite time has elapsed. But in practice, a factor which is decreasing exponentially toward zero becomes negligibly small within a finite period of time. In contrast, exponential growth can terminate only when conditions change so that the exponential equation is no longer valid.

6-4. Ways of Expressing the Rate of an Exponential Change

There are two general ways of expressing the rate of the exponential change in y which is specified by the equation $y_t = y_0 e^{\pm kt}$. The simplest is to give the rate constant, k, either directly as "proportion per unit time," or as $100k$, "per cent per unit time." The second is to give the time required for y to change by a specified proportion. This time can be calculated by solving Equation 6-9 for t:

$$t = \frac{\ln (y_t/y_0)}{-k} \tag{6-9a}$$

The choice of what proportion to specify is arbitrary, and usage unfortunately differs. For an exponential decrease, there are two common practices. One is to give the *half-time* (also called the *half-life*), $t_{1/2}$, which is the time required for y to decrease to one-half of its original value. Then $y_t/y_0 = 1/2 = 2^{-1}$, and by Equation 6-9a

$$t_{1/2} = -\ln 2/-k = 0.693/k \tag{6-10}$$

The other is to give the *time constant*, $t_{1/e}$, which is the time required for y to decrease to $1/e$ of its original value. Then $y_t/y_0 = 1/e = e^{-1}$, and by Equation 6-9a

$$t_{1/e} = -\ln e/-k = 1/k \tag{6-11}$$

The time constant is thus the reciprocal of the rate constant, k. In biochemistry, the time constant is often called the *turnover time*, because if some metabolic process is removing compound X from a "metabolic pool" of X at a constant proportional rate k, but the amount of X in the pool is kept constant by influx of an equivalent amount of fresh X, $1/k$ is the time needed for the process to metabolize an amount of X just equal to the total quantity of X present in the pool. In this sense the whole pool will have been "turned over" metabolically during the time $1/k$; but of course many of the individual molecules originally in the pool at time zero ($1/e$ of them, in fact) will still be present at time $1/k$, even though "the whole pool has turned over." Most biologists prefer to use the half-time.

Since the half-time is, by definition, the time needed for a 50 per cent decrease of y, it is tempting to say that "y is decreasing at the rate of 50 per cent per $t_{1/2}$," or "$(50/t_{1/2})$ per cent per unit time." But the validity of this statement is severely limited. The "proportion per unit time" $0.5/t_{1/2}$ is not at all equivalent to the "proportion per unit time" k. By Equation 6-10, k is, in fact, equal to ln $0.5/t_{1/2}$, not $0.5/t_{1/2}$; k is the proportion per unit time at *every instant* of time; $0.5/t_{1/2}$ is an *average* proportion per unit time which is valid *only* for a 50 per cent decrease of y.

Example

(See Fig. 6-2.) Suppose y has an initial value of 32 at time zero, and decreases exponentially with a half-time of 5 hours. From this we can say immediately that at 5 hours, y will be 16, at 10 hours, 8, at 15 hours, 4, etc. More generally, at the end of any 5-hour period, y will be half what it was at the beginning of that particular period. So it would be correct to say that during any *5-hour* period the *average* rate of decrease per hour is 10 per cent *of the value at the start of the period*. Note the important italicized restrictions in this statement. Note in particular that we cannot apply the "10 per cent per hour" to any time period other than 5 hours. For example, it would be wrong to conclude that 10 per cent of y_0 will disappear in the first hour. Indeed, it is far better not to translate the half-time into any such "per cent per unit time" at all, but to use the half-time to calculate k, the proportional rate which is valid at *all* times:

$$k = 0.693/t_{1/2} \qquad (6\text{-}10a)$$

The value of k so obtained (in this example 0.1386 per hour) may then be substituted into Equation 6-8 or Equation 6-9 for calculating y at *any* time.

It is often just as satisfactory to use semilog graph paper, plotting y_0 on the log scale at time zero, and $(1/2)y_0$ at time $t_{1/2}$. A straight line drawn through these two points will be the line whose equation is the $\log_{10}$ transformation of Equation 6-8:

$$\log_{10}y_t = \log_{10}y_0 - k't \qquad (6\text{-}12)$$

where, as before, $k' = 0.4343\ k$.

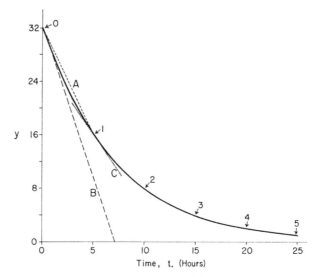

FIG. 6-2. Simple Exponential Decrease

The *solid curve* shows how a quantity, y, would decrease exponentially with time from an initial value of 32 toward a final value of zero when the half-time is 5 hours. The *numbered arrows* mark the ends of successive half-times. Notice that the value of y at the nth arrow is accordingly just one-half the value of y at the $(n - 1)$th arrow. Indeed, the value of y at the end of *any* 5-hour period is one-half the value of y at the beginning of that period.

The distinguishing feature of exponential change is that at every instant of time, the rate of change, here $-dy/dt$, is proportional to the magnitude of the thing changing, here y:

$$-dy/dt = ky$$

where the proportionality constant, k, is simply the rate of change of y per unit of y, *i.e.*, the proportional rate of change. For example, at time zero, the rate of change (slope of the tangent, *line B*) is -4.44 units of y per hour. Since y is 32 at time zero, k is $-4.44/32$ per hour or -0.139 per hour. At 5 hours, the rate of change of y with time (slope of the tangent, *line C*) is half as great, being only -2.22 units of y per hour. But at 5 hours y is *also* only half as great, so that the *proportional* rate of decrease, k, $(-2.22/16 = -0.139)$ is precisely the same as at time zero, or, indeed, at any other point on the curve.

Students are often tempted to argue that since half of the y initially present disappears in 5 hours, the rate of decrease is 50 per cent per 5 hours, or 10 per cent per hour. *Line A* is drawn to illustrate this erroneous argument. *Line A* has a slope of -3.2 units of y per hour, or 10 per cent of the *initial* value of y per hour. But *line A* does not correspond at all to the actual manner in which y decreases. For half the initial amount to disappear exponentially during a 5-hour period, the proportional

Values of y for any time may be read directly from such a graph.

Some of the points discussed thus far are illustrated in Figure 6-2.

For ease in converting rate constants to half-times and vice versa by means of Equations 6-10 and 6-10a, it is well to memorize the factor ln $2 = 0.693$. It is also important to become familiar with the notation $y_t = y_0 \exp(-kt)$ which is often used in place of $y_t = y_0 e^{-kt}$ so as to avoid superscripts which are difficult to print, especially when k is a fraction or some complicated sequence of symbols.

6-5. When Is Exponential Change Likely to Be Encountered?

Before proceeding to a somewhat more complex mathematical analysis of exponential change, it is desirable to discuss briefly several general situations which commonly lead to exponential growth or disappearance. The reader will thus be able to see at once that exponential change is indeed so frequent as to justify calling it "the compound interest law of nature" (see Reference 74, page 56 ff.). Each of the general situations will be illustrated by a concrete example chosen for its simplicity, not necessarily for its biological importance.

6-6. I. The Factor Causing a Change Is Itself Decreased by the Change

Example

Consider a capacitor of fixed capacitance (Cap) carrying an initial charge, q_0. When the switch is closed (Fig. 6-3) the voltage difference ΔE, across the capacitor causes current to flow through the fixed resistance, R. But this flow of current, in turn, decreases the charge on the capacitor. The flow of current also decreases the voltage difference because at any instant of time the voltage difference, ΔE_t, is directly proportional to the charge, q_t :

$$\Delta E_t = q_t/(\text{Cap}) \qquad (6\text{-}13)$$

By Ohm's law, the current flowing at any instant of time, I_t, will be directly proportional to the voltage difference at that time, and inversely proportional to the resistance:

$$I_t = \Delta E_t/R \qquad (6\text{-}14)$$

But current is simply the quantity of electric charge flowing per unit

rate of disappearance at every instant of time must be -0.139, or -13.9 per cent per hour, not -10 per cent per hour.

If *line B*, representing the initial rate of decrease, is produced downward, it intersects the baseline at 7.22 hours. This is the time required for y to decrease to $1/e$ of its initial value. As explained in the text, in biochemical studies with radioisotopes this time is often called the "turnover time."

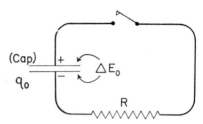

Fig. 6-3. The Discharge of a Capacitor

When the key is closed, the initial charge, q_0, on the capacitor will begin to flow through the resistance, R, under the influence of the potential difference, ΔE. Since the current, dq/dt, is at all times proportional to q, the discharge of the capacitor is an example of simple exponential decrease.

time. Hence in this circuit the flow of current from the capacitor at any instant of time is equal to $-dq/dt$, the instantaneous rate of decrease of the charge on the capacitor:

$$I_t = -dq/dt \qquad (6\text{-}15)$$

Combining Equations 6-13 and 6-14,

$$I_t = q_t/(\text{Cap})R \qquad (6\text{-}16)$$

Combining Equations 6-15 and 6-16,

$$-dq/dt = q_t/(\text{Cap})R \qquad (6\text{-}17)$$

Since both the capacitance and the resistance are constant, Equation 6-17 is of the form of Equation 6-1 and may therefore be integrated to the exponential form:

$$q_t = q_0 \exp[-t/(\text{Cap})R] \qquad (6\text{-}18)$$

It is worth noting that although Equation 6-18 is written in terms of the charge on the capacitor, the current, I_t, and the voltage difference ΔE_t, also decrease exponentially at exactly the same proportional rate.

In this example, as in many others, a *gradient* (here a gradient of electrical potential) causes something to *flow* (here electrons) in such a way as to decrease the gradient. But the reader should be warned that sometimes the relation between flow and gradient is not so simple, and the change which then occurs is not a simple exponential change. (See, for example, the discussion of Fick's law of diffusion in Chapter 7.)

6-7. II. Clearance of a Solute from a Compartment at a Constant Rate

Suppose that a solute, S, is distributed throughout a well-mixed fluid compartment whose volume remains constant. If S is removed from (*i.e.*, "cleared from") a constant fraction of the fluid in the compartment per unit of time, the *total amount of S* in the compartment and, volume being constant, the *concentration of S* in the compartment will decrease exponentially with time.

Example

Let S be a drug which is distributed uniformly throughout the extracellular fluid of the body. Let this "volume of distribution" of S, $(V\text{dist})_s$, remain constant. Let the total plasma clearance of S (*i.e.*, the volume of plasma cleared of S per unit of time by excretion, metabolism, and storage of S), $(\dot{V}\text{cl})_s$, remain constant. Then if $C_{s.t}$ be the concentration of S in the extracellular fluid, and $Q_{s.t}$ be the quantity of S in the extracellular fluid at any time, t,

$$Q_{s.t} = C_{s.t}(V\text{dist})_s \tag{6-19}$$

During an infinitesimal interval of time, dt, the infinitesimal quantity of S removed (cleared) from extracellular fluid, $-dQ_s$, is equal to the volume cleared during the interval, $(\dot{V}\text{cl})_s \, dt$, multiplied by the concentration:

$$-dQ_s = (\dot{V}\text{cl})_s \, dt \, C_{s.t} \tag{6-20}$$

Combining Equations 6-19 and 6-20 by eliminating $C_{s.t}$,

$$-dQ_s/dt = [(\dot{V}\text{cl})_s/(V\text{dist})_s]Q_{s.t} \tag{6-21}$$

Since both $(\dot{V}\text{cl})_s$ and $(V\text{dist})_s$ are constant, Equation 6-21 is of the form of Equation 6-1 and may therefore be integrated to

$$Q_{s.t} = Q_{s.0}\exp[-(\dot{V}\text{cl})_s t/(V\text{dist})_s] \tag{6-22}$$

Note that $(\dot{V}\text{cl})_s/(V\text{dist})_s$ corresponds to the rate constant, k, of Equation 6-1, and should therefore be a proportion per unit time. And so it is; for (clearance/volume of distribution) is simply the proportion of the total volume of the compartment (here, the extracellular fluid compartment) which is cleared of S per unit time.

Most drugs disappear from the body exponentially in this fashion. Other common illustrations of the same general situation are the "washing out" of a solute from a well-mixed compartment by a steady flow of pure solvent through the compartment, and the disappearance of a metabolite labeled with a radioactive isotope from a "pool" of the normal, nonradioactive metabolite when the pool is maintained in a steady state by a constant rate of turnover (Chap. 9).

In the following two situations, the change which occurs is not actually continuous since it is due to the sudden multiplication or destruction of discrete units. However, if the number of single units is large enough, the change may be treated, for all practical purposes, as a continuous exponential change.

6-8. III. Random Disappearance of Individual Units in a Large Group

Consider a large number, N_t, of similar units present at time t. Any particular unit is subject to destruction by some random event whose time of occurrence is impossible to predict. However, suppose the probability, P, that any particular unit will be destroyed within a given interval of time is constant, and is the same for each individual unit. Then as long as N_t remains sufficiently large, it may be treated as an exponential function of time.

There is still controversy about the logical meaning of the term "probability." However, for present purposes, we may define the probability, P, that a single event will occur as the ratio of the number of events which actually do occur (here ΔN) to the total number of opportunities for the event to occur (here N_t) when the total number of opportunities tends toward infinity:

$$P = \Delta N/N_t \quad \text{as} \quad N_t \rightarrow \infty \tag{6-23}$$

If we solve Equation 6-23 for ΔN (taking it as negative to indicate a decrease), and divide both sides by Δt,

$$-\Delta N/\Delta t = (P/\Delta t)N_t \tag{6-24}$$

But if N_t is exceedingly large, it may be treated as a continuous function of time. Δt may then be allowed to become smaller and smaller, and Equation 6-24 may be written*

$$-dN/dt = (P/dt)N_t \tag{6-25}$$

Since (P/dt) is a constant, Equation 6-25 is of the form of Equation 6-1 and may be integrated to

$$N_t = N_0 e^{-(P/dt)t} \tag{6-26}$$

Now although Equations 6-23 to 6-26 are *strictly* accurate only when $N \rightarrow \infty$, they will be sufficiently valid whenever N is so large that the irregularities due to the random destruction of individual units are "lost in the statistical average" when N is plotted as a function of t.

Example

Radioactive isotopes undergo spontaneous destruction ("decay") because of some kind of nuclear rearrangement which occurs at random, but with a constant probability. For example, the isotope of iodine with a mass number of 131 has a half-life of about 8 days. The proba-

* See footnote at bottom of page 167.

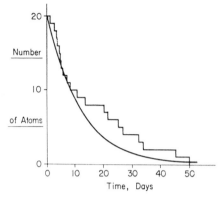

FIG. 6-4. Disappearance of 20 Atoms of I^{131}

The irregular stepwise decrease in the number of atoms remaining is poorly represented by the smooth curve for exponential decay because so few atoms are involved.

bility per unit time, P/dt, that any particular atom of I^{131} will disintegrate is very close to 10^{-6} per second. If we had a sample of I^{131} containing only 20 atoms at time zero, and if we were able to record the exact time of disappearance of each of the atoms, we would observe an irregular stepwise decrease of N of the sort which is shown in Figure 6-4. Equation 6-26 gives a poor approximation to the actual rate of disappearance, because the number of individual atoms is too small to justify the assumption of continuous change upon which Equation 6-26 is based. But suppose we had a sample containing 2×10^9 atoms of I^{131}. If we were to confine our attention only to the disappearance of the first 20 of these atoms, we would again observe an irregular stepwise decrease, this time occurring in a fraction of a second (Fig. 6-5a). But now a single atom represents so small a proportion of the total number that on any sensible scale of time and number (Fig. 6-5b) these irregularities are utterly inconsequential, and Equation 6-26 describes the decrease of N perfectly adequately.

Any first-order reaction occurring under constant conditions (Chap. 10) will lead to an exponential decrease of the reactant in the manner described above.

6-9. IV. Growth of a Population with a Constant Net Rate of Reproduction

Consider a large population consisting of N self-reproducing units. As long as net reproduction (*i.e.*, the excess of births over deaths) per unit of

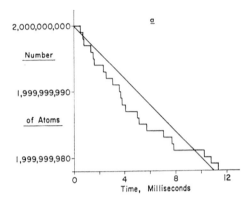

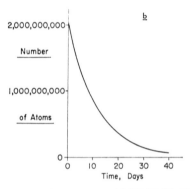

FIG. 6-5. Disappearance of 2,000,000,000 Atoms of I[131]

If we were to confine our attention to the decay of the first 20 atoms (*Graph a*), we would observe the same kind of irregular stepwise decrease as in Figure 6-4. But if, as is usual, we consider the whole group of 2 (10^9) atoms, we can describe their disappearance very well by a smooth exponential curve (*Graph b*) with a half-time (half-life) of 8 days.

population and per unit of time is constant, the population, N, will change exponentially.

By definition, the instantaneous net rate of reproduction per unit of population is $(dN/dt)/N_t$. If this is constant,

$$(dN/dt)/N_t = k \tag{6-27}$$

or,

$$dN/dt = kN_t \qquad (6\text{-}27a)$$

which is of the now familiar form of Equation 6-1.

Example

Consider the increase in the number, N, of bacterial cells growing in a favorable medium. For simplicity, assume that the mortality is nil, and that the interval between successive binary divisions (the *generation time*) is precisely half an hour, all cells present dividing synchronously. Then the number of cells present just after any division will be given by Equation 6-5 if we let

$$y_{(t.\text{end period})} = N_{(t.\text{after division})}$$
$$k/z = 1 \; (i.e., 100 \text{ per cent interest every half hour!})$$
$$n = t/t_{\text{gen}} \text{ where } t_{\text{gen}} \text{ is the generation time}$$

Then

$$N_{(t.\text{after division})} = N_0 2^{(t/t_{\text{gen}})} \qquad (6\text{-}28)$$

Now in fact, successive division of all cells will *not* occur with absolute regularity and the cells will be out of step. As a result, the increase will not take place in discrete jumps which double the population every half hour but (providing again that N is large enough) will take place at a steady exponential rate given by the integral of Equation 6-27a:

$$N_t = N_0 e^{kt} \qquad (6\text{-}29)$$

We can still speak of a *mean generation time* which will be the time needed for the population to double. The mean generation time for exponential growth is the exact equivalent of the half-time for exponential disappearance. For the sake of generality, we would do better to call it the doubling time, t_{double} :

$$t_{\text{double}} = \ln 2/k = 0.693/k \qquad (6\text{-}30)$$

It is scarcely necessary to warn the reader that the rate of increase (or decrease) of an actual population is usually influenced by so many unpredictable factors that exponential growth is a rarity except under carefully controlled laboratory conditions.

6-10. Quantities Which Increase so as to Approach an Asymptote Exponentially

Thus far, we have restricted the discussion to situations in which the quantity under study either increased indefinitely (positive rate constant, k) or decreased toward zero (negative rate constant, $-k$). For example, the charge on a capacitor which is discharging through a fixed resistance approaches zero as time approaches infinity (Equation 6-18). But suppose we start with zero charge and follow the change of q with time as a steady external voltage, ΔE_{ext}, charges the capacitor through the resistance, R (Fig. 6-6). It is clear that Equation 6-13 will still be true. But Equation 6-14 must be changed, for now the voltage gradient which makes current flow through the resistance is not ΔE_t but is the difference between ΔE_{ext} and ΔE_t :

$$I_t = (\Delta E_{ext} - \Delta E_t)/R \qquad (6\text{-}31)$$

Furthermore, the charge on the capacitor now *increases* with time, so that dq is positive, and Equation 6-15 becomes

$$I_t = dq/dt \qquad (6\text{-}15a)$$

Combining Equations 6-13, 6-31, and 6-15a,

$$dq/dt = (\Delta E_{ext} - [q_t/(\text{Cap})])/R \qquad (6\text{-}32)$$

But at time infinity

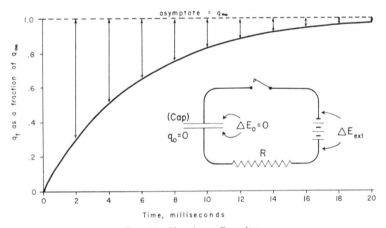

FIG. 6-6. Charging a Capacitor

The circuit is the same as in Figure 6-3, except that a battery has been added, providing a constant external voltage difference of ΔE_{ext} to charge the capacitor (which has no charge initially) when the key is closed. The charge on the capacitor increases at a rate which is proportional to the difference between the external voltage and the voltage already achieved across the capacitor. Notice that the quantity which is decreasing exponentially is the *difference* between the asymptotic charge, q_∞, which will be reached at time infinity, and q_t, the charge already achieved at any time, t. This quantity is indicated by the *vertical arrows* in the graph.

$$\Delta E_\infty = \Delta E_{ext} \tag{6-33}$$

and by Equation 6-13

$$\Delta E_\infty = q_\infty/(\text{Cap}) \tag{6-34}$$

Combining Equations 6-32, 6-33, and 6-34,

$$dq/dt = (q_\infty - q_t)/(\text{Cap})R \tag{6-35}$$

Now since q_∞ is a constant,

$$d(q_\infty - q_t)/dt = -dq/dt \tag{6-36}$$

Combining Equations 6-35 and 6-36,

$$-d(q_\infty - q_t)/dt = (q_\infty - q_t)/(\text{Cap})R \tag{6-37}$$

Since $1/(\text{Cap})R$ is a constant, Equation 6-37 is of the form of Equation 6-1 and can be integrated to

$$(q_\infty - q_t) = (q_\infty - q_0)e^{-[1/(\text{Cap})R]t} \tag{6-38}$$

In order to compare Equation 6-38 for *charging* a capacitor with Equation 6-18 for *discharging* a capacitor, it will be convenient to rewrite Equation 6-18 in the form

$$(q_t - q_\infty) = (q_0 - q_\infty)e^{-[1/(\text{Cap})R]t} \tag{6-18a}$$

which is identical to Equation 6-18 because for *discharge* of the capacitor, q_∞ is zero. Now both Equation 6-38 and Equation 6-18a can be reduced to the *same* equation:

$$(q_\infty - q_t)/(q_\infty - q_0) = e^{-[1/(\text{Cap})R]t} \tag{6-38a}$$

Equation 6-38a deserves careful study. Notice that q_∞, the asymptote being approached as time increases, is a finite quantity when the capacitor is being charged, but is zero when the capacitor is discharging. The opposite is true of q_0 which is finite for discharge, zero for charging. Notice also that the absolute magnitude (sign disregarded) $|q_\infty - q_0|$ represents the total change which q would undergo in infinite time, whereas $|q_\infty - q_t|$ represents the amount of change which q has *not yet* undergone at time t. The ratio $(q_\infty - q_t)/(q_\infty - q_0)$ is positive regardless of the direction in which q is changing. This ratio is *the proportion of the total exponential change which has not yet occurred at time t*, and is equal to $e^{-[1/(\text{Cap})R]t}$ whether the capacitor is being charged or is discharging.

Now for *discharge* of the capacitor we can correctly say that the charge is decreasing exponentially with time. But in charging the capacitor, the charge, q, *increases*. Is it then correct to say that q increases exponentially? Hardly, for the unqualified term "exponential increase" would mean a process characterized by a positive exponent of e. If we say that q is increasing exponentially toward an asymptote, we are not likely to be misunderstood. But strictly speaking, it is not q itself but the quantity $(q_\infty - q_t)$ which is changing exponentially, and it is decreasing, not increasing. If we want to plot such an exponential change in logarithmic form so as to obtain a straight line we must plot the logarithm of $(q_\infty - q_t)$, not the logarithm of q itself, as a function of time.

6-11. A General Equation for Exponential Decrease

The example of charging the capacitor at once suggests a generalized equation for exponential decrease. We have just seen that it is not q itself which changes exponentially, but rather the difference between q and the asymptote toward which it is rising. This difference is indicated by the *vertical arrows* in Figure 6-6. Now let us generalize by letting y be *any* quantity which is rising or falling in this fashion toward an asymptote, y_{asymp}. Then *the quantity which is decreasing exponentially will always be the absolute value of the difference between y_{asymp} and y*. To avoid negative differences which cannot be plotted logarithmically we will define the difference as $+(y_{\text{asymp}} - y)$ when y is less than y_{asymp}, and

$$-(y_{\text{asymp}} - y) = +(y - y_{\text{asymp}})$$

when y is greater than y_{asymp}. But whether y is greater than or less than the asymptote, the following generalized version of Equation 6-38a will be valid:

$$(y_{\text{asymp}} - y)_t / (y_{\text{asymp}} - y)_0 = e^{-kt} \tag{6-39}$$

or, solving for y_t,

$$y_t = y_{\text{asymp}.t} + (y_0 - y_{\text{asymp}.0})e^{-kt} \tag{6-39a}$$

Now for charging the capacitor, $y_{\text{asymp}} = q_\infty = E_{\text{ext}}(\text{Cap})$ which is a constant, so that on the graph of q against time (Fig. 6-6) it is represented simply by a horizontal line. But y_{asymp} is not necessarily a constant. Indeed y_{asymp} can be *zero*, or a *positive or negative constant*, or any *function of time which has the dimensions of y*. As an illustration, consider Figure 6-7 in which the *broken line* is an arbitrary curve whose exact equation is not known but which is a continuous function of time. Let this curve be the asymptote toward which y, represented by the *solid curve*, is rising. For the broken curve we may write

$$y_{\text{asymp}} = f(t) \tag{6-40}$$

Combining Equations 6-39a and 6-40,

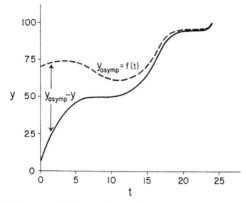

FIG. 6-7. Exponential Approach to a Curvilinear Function of Time

The quantity which is decreasing exponentially is the *difference* between the asymptote, illustrated by the *broken curve*, and the actual value of y (*solid curve*) at the same instant of time. This difference, at time 1.5, is indicated by the *vertical arrows*.

$$y_t = f(t) + [y_0 - f(0)]e^{-kt} \qquad (6\text{-}41)*$$

Equation 6-41 for y as a function of time is the equation for the solid curve in Figure 6-7. But since $f(t)$ can stand for *any* function of t, the equation is also quite general. At risk of being tiresome, it may again be pointed out that the quantity changing exponentially is *not* y, but $(y_\text{asymp} - y)$ which, in this example, is $[f(t) - y]$, indicated in Figure 6-7 by the *vertical arrow*. A plot of $\log_{10} y$ against time (Fig. 6-8, *broken curve*) would merely be confusing. But a plot of $\log_{10}[f(t) - y]$ against time (Fig. 6-8, *solid curve*) is a straight line with a slope of $-0.4343k$.

6-12. Sums of Exponentials

Since the asymptote $f(t)$ may be *any* function of time with the dimensions of y, there is nothing to prevent $f(t)$ from being itself an exponential function of time. For example, suppose that

$$f(t) = f(0)e^{-k_1 t} \qquad (6\text{-}42)$$

where k_1 is another rate constant. Combining Equations 6-41 and 6-42,

$$y_t = f(0)e^{-k_1 t} - [f(0) - y_0]e^{-kt} \qquad (6\text{-}43)$$

Since $f(0)$ and y_0 are both constants, we may simplify Equation 6-43, and

* $f(0)$ simply means the value of $f(t)$ when $t = 0$.

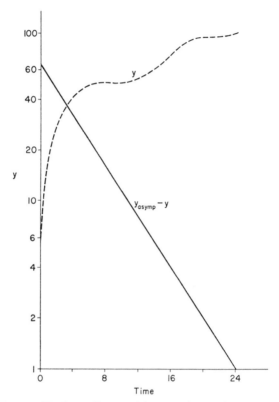

Fig. 6-8. Plotting an Exponential Approach to an Asymptote on
Semilogarithmic Paper

If the values of y from Figure 6-7 are plotted on a logarithmic scale against time on an arithmetic scale (*broken curve*) they make no sense at all. But if the values of $y_{asymp} - y$ are similarly plotted (*solid line*) they describe a straight line whose slope indicates the rate at which y is approaching its asymptote. To obtain a straight line on semilogarithmic paper, one must always plot the *difference* between y and its asymptote.

at the same time prepare for subsequent work, by rewriting it in the form

$$y_t = Ae^{-k_1 t} + Be^{-k_2 t} \tag{6-44}$$

where $A = f(0)$
$ B = -(f(0) - y_0) = (y_0 - f(0))$
$ k_2 = k$

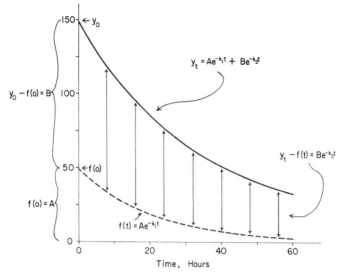

FIG. 6-9. The Meaning of a Double Exponential Curve

In this figure, the *solid curve* represents the sum of two exponential functions of time. Let us regard the first of these, $Ae^{-k_1 t}$ (*broken line*), as the function of time, $f(t)$, toward which y is falling exponentially. The *difference* between the value of y at any time t, y_t, and this asymptote, $f(t)$, is indicated by the *vertical arrows* in the graph. But this difference, $y_t - f(t)$, or, to be specific, $y_t - Ae^{-k_1 t}$, is itself the *single* exponential function of time, $Be^{-k_2 t}$. This difference will therefore give a straight line if plotted on semilogarithmic paper, just as did the difference between y and its asymptote in Figure 6-7 when plotted in Figure 6-8.

Notice that the choice of which exponential term to regard as the asymptote is entirely arbitrary. In fact, if y is equal to the sum of n exponential terms, the sum of any $n - 1$ of these terms may be regarded as the asymptote which y is approaching at the single exponential rate given by the remaining term.

Equation 6-44 is illustrated in Figure 6-9. From Equation 6-44 it is evident that when the asymptote is itself an exponential function, y is made up of the sum of two exponentials *either* of which might be considered the asymptote for the other. Each exponential term is characterized by an initial magnitude, in this example A or B, and each term declines toward zero at the rate specified by the appropriate rate constant.

The above argument can easily be extended to more than two terms, and we can accordingly write a general equation for y as the sum of a series of exponentials:

$$y_t = \pm Ae^{-k_1 t} \pm Be^{-k_2 t} \pm Ce^{-k_3 t} \pm \cdots \pm Z \qquad (6\text{-}45)$$

where Z is included to provide for the possibility that the final constant asymptote may not be equal to zero. Notice that at time zero, Equation 6-45 reduces to $y_0 = \pm A \pm B \pm C \pm \cdots \pm Z$, while at time infinity it reduces to $y_\infty = \pm Z$. Equations of this type are encountered quite frequently in the mathematical analysis of biological systems. For example, two or more quite independent processes with different rates, usually occurring in different regions of the body, may each contribute exponentially to the overall change in the concentration of a drug or a metabolite in the plasma. For a specific example, see Chapter 13.

6-13. Fitting an Exponential Equation to Experimental Observations

Suppose that some underlying theory, or, less happily, the mere trend of the experimental points suggests that an observed variable, y, is changing exponentially with time. We may then want to find an empirical equation to describe the data—a problem of curve fitting which will serve to reinforce the previous general remarks about fitting empirical equations to observed data (Chap. 3). In the following discussion we shall assume that the exponent of e is negative so that the data can be fitted by an equation of the general form of Equation 6-39:

$$(y_t - y_{\text{asymp}.t}) = (y_0 - y_{\text{asymp}.0})e^{-kt} \qquad (6\text{-}39b)$$

where y_{asymp} may be zero, or a constant, or a function of time. For curve fitting, a linear transformation of Equation 6-39b,

$$\log_{10} | (y_t - y_{\text{asymp}.t}) | = \log_{10} | (y_0 - y_{\text{asymp}.0}) | - k't \qquad (6\text{-}46)$$

is most useful. Logarithms to the base 10, and accordingly k', equal to $0.4343k$, are used in Equation 6-46 only because a great deal of time can be saved by using semilog graph paper which always has a $\log_{10}$ scale. Our object is to find an asymptote, y_{asymp}, such that a straight line will be obtained when the logarithm of the difference between the asymptote and the observed values of y is plotted against time. The desired parameters of Equation 6-39 can then be calculated from the slope, k', and the intercept, $\log_{10} | (y_0 - y_{\text{asymp}.0}) |$, of the straight line.

As an example, consider the data in Table 6-2. We will assume that time is measured without error, but that the dependent variable, y, has a known standard deviation of about ± 1.6 units. The *first step* is to plot the observed values of y against time on ordinary arithmetical coordinate paper and to draw a smooth freehand curve through the points (Fig. 6-10). From this curve, it looks as if y might be rising toward a constant asymptote of about 90. (Of course, y might be rising toward some function of time as an asymptote, rather than toward a constant value. But unless this is sug-

TABLE 6-2

Values for the points of Figures 6-10, 6-11, and 6-12

Time, t	Observed, y	$90 - y$	$94 - y$	$97 - y$	$100 - y$	$104 - y$	$108 - y$
0.0	39.0	51.0	55.0	58.0	61.0	65.0	69.0
0.5	47.0	43.0	47.0	50.0	53.0	57.0	61.0
1.0	50.0	40.0	44.0	47.0	50.0	54.0	58.0
2.0	59.0	31.0	35.0	38.0	41.0	45.0	49.0
3.6	66.5	23.5	27.5	30.5	33.5	37.5	41.5
5.0	73.5	16.5	20.5	23.5	26.5	30.5	34.5
6.0	80.5	9.5	13.5	16.5	19.5	23.5	27.5
7.4	84.0	6.0	10.0	13.0	16.0	20.0	24.0
9.6	86.5	3.5	7.5	10.5	13.5	17.5	21.5

gested by the underlying theory, the law of parsimony requires us to choose the simplest asymptote which will fit the points adequately.) The *second step* is to plot $(90 - y_t)$, *i.e.*, the difference between the assumed asymptote, 90, and the observed values of y, on the logarithmic scale of a suitable sheet of semilog graph paper, time being plotted on the arithmetic scale (Fig. 6-11). It is excellent practice to indicate the standard deviation for each point directly on the semilog graph. In the present example, short horizontal bars, connected by a vertical line have been placed at a distance of 1.6 units (one standard deviation) above and below each point. This serves two purposes. First, it reminds us that in fitting a line to the points, comparatively little weight should be given to the smaller values of $\log_{10}$-$(y_{asymp} - y_t)$ which are subject to a relatively large error. Second, it helps us to decide whether a particular line fits the points satisfactorily. A well-fitting line ought to pass within ± 1 standard deviation of the majority (but by no means necessarily all) of the points. With these two considerations in mind, the *third step* is to try to draw a straight line by eye through the points. From Figure 6-11 it is evident that when the asymptote is assumed to be 90, all three points at the lower right tend to lie somewhat below the line. This suggests that the estimate of 90 for the asymptote is too low. Accordingly the *fourth step* is to revise our estimate of the asymptote, and to repeat the second, third, and fourth steps as often as is needed to obtain a straight line which fits the points satisfactorily. If no satisfactory straight line can be found to fit the points, it is likely that an asymptote more complex than a constant, possibly a second exponential, must be used.

Even when the very first estimate of the asymptote yields a straight line, it is wise to vary the asymptote over a range of values so as to obtain some

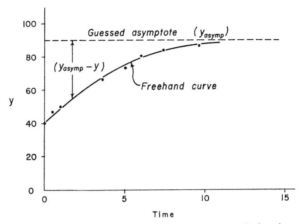

FIG. 6-10. Fitting a Single Exponential Curve to a Series of
Observed Points. First Step

The observations of Table 6-2 are here plotted on ordinary graph paper. As a first step, draw a freehand curve through the points and try to guess what asymptote is being approached. Remember always that it is the *difference* between y and its asymptote which is decreasing exponentially, so that if, for each point, the value of y is subtracted from the proper asymptote, the series of differences should fall on a reasonably good straight line when plotted on semilogarithmic paper (Fig. 6-11).

idea of how precise the estimates of the parameters are. For the present example, the results of using asymptotes of 90, 94, 97, 100, 104, and 108 are depicted in Figure 6-11. As we have already seen, 90 seems to be a little too low. On the other hand, 104 and 108 seem to be somewhat too high, since the lines miss four of the nine points by more than one standard deviation. However, neither 90 nor 104 gives a really bad fit, for when the corresponding curves are plotted, together with the original observations, on ordinary graph paper (Fig. 6-12), each curve describes the trend of the points reasonably well. For an asymptote of 90, the straight-line equation is

$$\log_{10}(90.0 - y_t) = \log_{10}(90.0 - 40.0) - 0.1056t$$

whence the exponential equation for y_t itself would be

$$y_t = 90.0 - 50.0e^{-0.243t}$$

Similarly, for an asymptote of 104,

$$\log_{10}(104.0 - y_t) = \log_{10}(104.0 - 40.5) - 0.0650t$$

whence

$$y_t = 104.0 - 63.5e^{-0.150t}$$

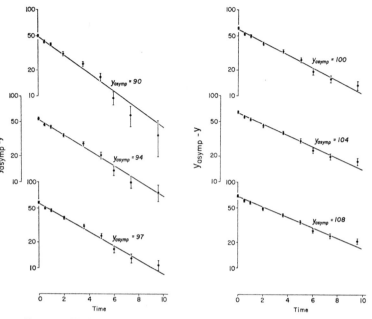

FIG. 6-11. Fitting a Single Exponential to a Series of Observed Points.
Subsequent Steps

In each graph, the difference between an assumed asymptote and each of the observed points in Figure 6-10 has been plotted on a logarithmic scale against time on an arithmetic scale. The *vertical bar* through each point represents plus-and-minus one standard deviation. When the assumed asymptote is 90, the points do not seem to be as well fitted by a straight line as when a somewhat higher asymptote is chosen. On the other hand, an asymptote of 104 or 108 seems a little too high.

From this analysis, we may conclude that y is rising from an initial value of about 40 toward an asymptote which probably lies within or near the range of 90 to 104, with the corresponding half-times ranging from about 2.8 hours for the asymptote of 90 to about 4.6 hours for the asymptote of 104. Notice that the ranges for asymptote and for half-time are not independent. For example, the data would not be fitted at all well by a curve rising toward an asymptote of 90 with a half-time of 4.6 hours.

It would be possible to calculate, with due attention to proper weighting, the "best" straight line for each asymptote by the method of least squares. One could then choose as the "best" asymptote the one which minimized the sum of squares of deviations of the *original* values of y (*not* their logarithms) from the exponential curve plotted on ordinary arithmetical co-

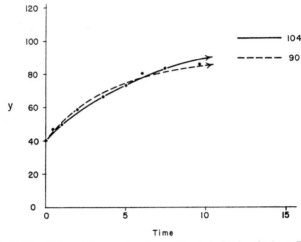

Fig. 6-12. Two Different Exponential Curves Plausibly Fitting the Same Points

The curves corresponding to the straight lines for guessed asymptotes of 90 and 104 in Figure 6-11 are here plotted, as in Figure 6-10, on arithmetical coordinate paper together with the original points. Although they have considerably different asymptotes (shown at the *top right*) and different half-times, both curves fit the points reasonably well.

ordinate paper. But the substantial labor required for such calculations would not be justified by the very slight increase in the precision with which the parameters could be estimated. Figure 6-13 calculated from the lines of Figure 6-11 which were fitted by eye shows how slowly the error variance changes as the asymptote is changed. Even for asymptotes of 90 and 104 the standard deviation of the points from the curve is not much larger than for the intermediate asymptotes. Moreover, as far as the author is aware, statistical methods for estimating the reliability of parameters calculated in this manner have not yet been devised. For these reasons, we may as well content ourselves with some simple statement, such as was given above, about the general range within which the parameters probably lie. Such a statement is properly, if disappointingly vague, and avoids any implication that the trial-and-error analysis has somehow established a unique set of values for the parameters.

6-14. Fitting Experimental Observations by a Multiple Exponential Function

In the previous example the data could be fitted by a single exponential function of time together with a suitably chosen constant asymptote. In-

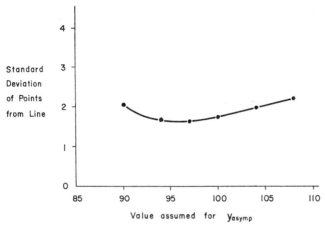

FIG. 6-13. The Difficulty in Deciding Which Is *"The"* Line of Best Fit

In spite of their different parameters, the lines of Figure 6-11 fit the observed points almost equally well, as shown by the closely similar values for the standard deviation of the points from the exponential curves corresponding to the several assumed asymptotes.

deed several such functions were found to fit the data almost equally well. Often, however, a series of experimental observations, which we have reason to believe may represent some kind of exponential function of time, cannot be fitted by an equation with a single exponential term. We may then want to see whether the data can be fitted by a sum of exponentials of the general form of Equation 6-45, particularly if there are theoretical grounds for believing that several processes with widely differing rates are contributing to the overall change.

For simplicity, let us first consider the double exponential function specified by Equation 6-44:

$$y_t = Ae^{-k_1 t} + Be^{-k_2 t} \tag{6-44}$$

If k_1 and k_2 were equal, this would reduce to the single exponential:

$$y_t = (A + B)e^{-kt} \qquad (k_1 = k_2 = k) \tag{6-47}$$

which would, of course, give a straight line when plotted against time on semilog graph paper. But even when k_1 and k_2 are not identical, if they are similar in magnitude, a semilog graph of actual data will probably not differ noticeably from a straight line unless the experimental observations of y are unusually extensive and accurate (Fig. 6-14a). However, if $|k_1|$ is substantially larger than $|k_2|$, the *early* behavior of y will be determined

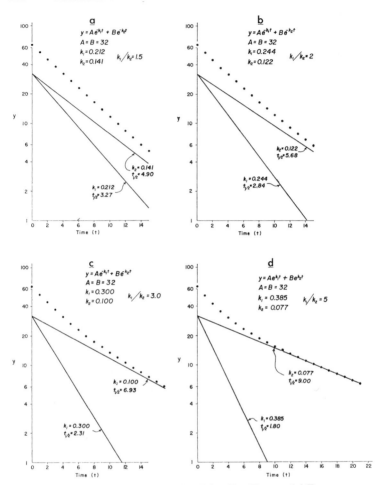

FIG. 6-14. The Difficulty of Identifying Two Exponential Terms
with Similar Rate Constants

In each graph, the *lower solid line* represents a first exponential term with a shorter half-time; the *upper solid line* represents a second exponential term with a longer half-time. The sum of these two exponential terms has been plotted as a series of points at hourly intervals, and these points represent the values of the dependent variable, y. Even when the rate constants differ by as much as 1½-fold or 2-fold (*upper graphs*), the observed values of y are likely to fall so close to a straight line on semilogarithmic paper that the experimenter will be able to fit them quite satisfactorily by a *single* exponential. But if the rate constants differ by a factor of three predominantly by the first exponential term, $Ae^{-k_1 t}$, in accordance with which y will at first decline rather rapidly. Due to its rapid rate of decrease, the first term will soon become negligibly small, and thereafter y will be governed almost entirely by the second exponential term, $Be^{-k_2 t}$. Now if y is plotted on semilog paper against time, we will obtain a biphasic curve whose last straight portion represents for all practical purposes only the second term of Equation 6-44, *i.e.*, the term with the longer half-time (Fig. 6-14$\underline{d}$). The relative magnitudes of the coefficients A and B will also help to determine how soon the first term becomes "negligibly small," but the rate constants are usually more important.

It is easy to extend the above argument to a sum of more than two exponential terms. As is customary, let us arrange the terms in order of decreasing absolute magnitude of the rate constants, *i.e.*, of increasing half-times. Then if we can observe y for a long enough period of time, the slowest rate of exponential decrease, represented by the last exponential term, may still be influential after all of the other terms have become negligibly small. Of course, if the final exponential decrease is "infinitely slow," the final term will be simply a constant.

Now in trying to fit a multiple exponential to experimental points, we do not usually know *a priori* how many exponential terms will be required. Nor do we know whether the last few points are still being influenced by more than one exponential term. Nevertheless, the above discussion suggests a method whereby we can *attempt* to identify successive exponential terms when confronted with actual observations. We plot the observed values of y against time on semilog paper. We then *assume* that the last few points represent the "slowest term," and we find its coefficient and rate constant from the intercept and slope of a straight line drawn through these points. We can then read from this straight line (or calculate) a value of the last term for each of the earlier points lying above the line. By subtracting these values from the *observed* values of y, we obtain a series of differences which we hope can be fitted by an exponential function with one less term than the original points required. We next plot these differences against time on semilog graph paper. If the trend of the points for the differences *curves downward* as time increases we conclude that the straight line which we originally drew to represent the lowest rate should not, after

(*lower left*), the trend of the points will differ noticeably from a straight line if the points are sufficiently accurate and extensive. A 5-fold difference in the rate constants (*lower right*) produces a still more marked curvature in the trend of the points; in fact, the influence of the "fast" exponential term upon the later points is now negligible.

Having studied these graphs, the reader will be able to understand why, when a multiple exponential function is used to fit biological data, the rate constants of successive terms almost always differ from each other by a factor of at least three!

all, pass through the final points but should lie somewhat below them and have a smaller slope. In other words, we conclude that the last observations were still being influenced by more than one term. We must revise our estimates of the parameters of the last term accordingly. If the trend of the points is a *straight line*, we conclude that only two exponential terms are needed to fit the original observations. We have already found the parameters of the final term, and we can now calculate the parameters of the initial term from the straight line which fits the differences. If the trend of the points *curves upward* as time increases, we conclude that at least three exponential terms are needed to fit the experimental observations, *i.e.*, the last term, which has already been found, plus at least two more for the series of differences. To find the next-to-last term, we treat the series of differences exactly as we did the original observations. By repeating this procedure as often as necessary we should (in theory at least) be able to identify term after term, with shorter and shorter half-times, until by successive subtraction we finally arrive at a series of differences which can all be fitted by a single straight line on semilog graph paper. This line should then represent the initial exponential term. The procedure will now be illustrated by two specific examples.

Example 1

As a first example, let us analyze the "data" which were plotted as points in Figure 6-14c. Since the plotted values were calculated precisely from the two component exponential terms, they are subject only to rounding errors. We shall therefore be able to concentrate on the mechanics of the procedure without being bothered by the random deviations which inevitably affect real observations. The values of y to be fitted by an exponential function are given in Table 6-3 and are plotted on semilog graph paper in Figure 6-15. The last few points are fitted almost perfectly by *line A*. We will tentatively assume that *line A* represents the last exponential term. Hourly values read from this straight line have been subtracted from the original values in Table 6-3 to obtain a set of differences. These differences are plotted in Figure 6-15 as *open circles*. The lower points obviously trend *downward* away from the straight *line B* drawn through the upper four points. This suggests that *line A* for the "slowest" component should be redrawn somewhat below the original points and with a smaller slope. However, the original points from 10 to 17 hours show so little departure from linearity that we are not inclined to move the line very far. Perhaps *line A'* will serve our purpose. The hourly values read from *line A'* have also been subtracted from the corresponding original values in Table 6-3 and the resulting differences are plotted as *crosses* in Figure 6-15. The straight *line B'* fits the crosses so well that there seems no

TABLE 6-3
Values for the points of Figure 6-15

Time, t	y_t	y_t Minus Corresponding Value from *Line A* in Fig. 6-15	y_t Minus Corresponding Value, from *Line A'* in Fig. 6-15
0	64.0	24.0	29.9
1	52.8	17.1	22.0
2	44.0	12.0	16.3
3	36.9	8.3	11.9
4	31.2	5.6	8.6
5	26.6	3.7	6.3
6	22.9	2.4	4.5
7	19.8	1.5	3.2
8	17.3	1.0	2.4
9	15.2		1.8
10	13.4		1.3
11	11.9		1.0
12	10.5		
13	9.4		
14	8.4		
15	7.5		
16	6.7		
17	6.0		

need for further revision. We conclude that the original points can be very satisfactorily fitted by the equation

$$y_t = 30e^{-0.313t} + 34e^{-0.103t}$$

whose first term has been calculated from *line B'* and whose second term has been calculated from *line A'*. In this particular example, we know that the "true" equation is

$$y_t = 32e^{-0.300t} + 32e^{-0.100t}$$

so that our analysis has yielded nearly "correct" values for the parameters.

Note well, however, that this happy result is due very largely to the unrealistic precision of the data. Had y been subject to even a small standard error, we would probably have been delighted to find that the *open circles* in Figure 6-15 fell so nearly on a straight line. We would probably have accepted *line A*, and some line such as B'' for the *open*

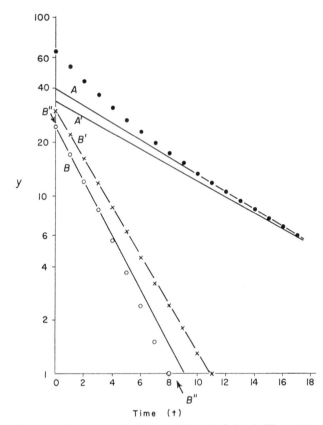

Fig. 6-15. Fitting a Double Exponential to the Points in Figure 6-14c

The steps used, and the meaning of the various points and lines, are fully explained in the text.

circles, without question. *Lines A* and *B″* would have given us the equation

$$y_t = 24.2e^{-0.376t} + 40.0e^{-0.112t}$$

which actually does fit the observed points remarkably well, even though its parameters are quite different from the "true" parameters. *Example 2*

The data in Table 6-4 have also been derived from a known function. However, random error in the measurement of y has been simu-

TABLE 6-4
Values for the points of Figure 6-16

Time, t	y_t
hours	
0.2	60.9
0.4	58.5
0.7	51.7
1.0	50.5
1.2	47.5
1.4	44.2
1.8	42.5
2.5	36.1
3.0	31.3
4.0	27.1
5.0	24.8
6.0	21.9
8.0	19.0
10.0	17.7
12.0	14.8
14.0	13.0
16.0	12.3
18.0	9.9
21.0	11.7
24.0	9.3

lated by adding to each "exact" value of y a number chosen at random from a normally distributed group of numbers with a mean of about zero and a standard deviation of about ± 1.0 unit of y. This makes the present example far more realistic than the previous one.

Let us begin by plotting y as a function of time on ordinary arithmetical graph paper (Fig. 6-16). The general trend of the points, indicated by the freehand curve drawn through them, suggests that y may be decreasing exponentially toward some asymptote greater than zero. The law of parsimony requires us to consider first the possibility that y is decreasing at a single exponential rate. Since the last four points do not show a significant downward trend, we might try their mean (about 11) as a tentative constant asymptote. But the quantity $(y_t - 11)$ plotted against time on semilog graph paper cannot be fitted by a single straight line, because the later points tend to stray above a line fitted to the earlier ones (Fig. 6-17). This suggests that the asymp-

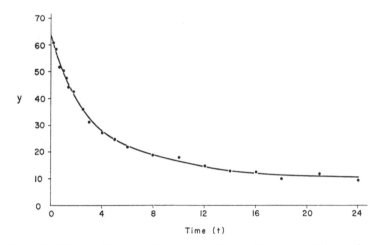

Fɪɢ. 6-16. Fitting an Exponential Function to a Series of Experimental Observations

The "experimental values" of y given in Table 6-4 are plotted against time on arithmetical coordinate paper, and a freehand curve has been drawn through the points. In the next six figures these same data are analyzed further.

tote is too low. If we try an asymptote of 16, the points from 0.2 to 10 hours, inclusive, fall on a good straight line (Fig. 6-18). But this is unsatisfactory because now the six later observations all lie *below* our assumed asymptote. We have no right to neglect these six points, especially since they show a significant downward trend. It is clearly impossible to fit the data by a single exponential term and a constant asymptote.

As the first step in trying to find a multiple exponential function to fit the data, let us plot y against time on semilog graph paper (Fig. 6-19). Our next step should be to fit a straight line to the last few points. But we are at once confronted by a practical difficulty. How many points belong in the "last few"? Five? Six? Eight? To get an unbiased answer to this question, we might logically begin with the last three or four points and work backwards, including more and more points until the probability that they lie upon a curve is greater than the probability that they do not lie upon a curve.* We would then drop

* More properly speaking, until the probability is less than 0.5 of getting a sum of squares for curvature as large as or larger than the observed sum of squares when a random sample of the same size is drawn from a population in which there is no curvature. In statistics, the accuracy of a statement seems to be inversely proportional to its simplicity!

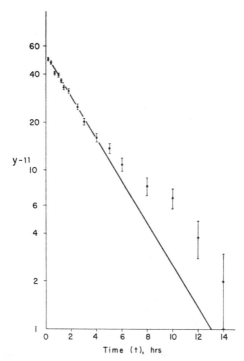

Fig. 6-17. First Attempt to Fit a Single Exponential

If we assume an asymptote of 11, subtract this from each value of y in Table 6-4, and plot the differences against time on semilogarithmic paper, the resulting points for $y - 11$ cannot be fitted by a straight line. In this and the following three figures the *vertical bars* again indicate one standard deviation above and below each point.

the point which had last been added, and find the best straight line for the remaining points. All of this *could* be done—with a wholly inappropriate expenditure of time and effort—by using standard statistical techniques. But we will do well enough fitting by eye if we bear in mind the general philosophy of the statistical procedure. For example, we should realize that an absolute minimum of four points is needed to distinguish a curvilinear from a rectilinear trend unless we have an independent estimate of error. (See the discussion of degrees of freedom in Section 3-11.)

To return to the problem at hand, suppose we decide to fit a straight line to the last *six* points (Fig. 6-19). By subtracting the values read

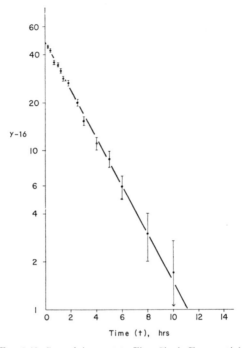

FIG. 6-18. Second Attempt to Fit a Single Exponential

If we assume an asymptote of 16, the function $y - 16$ plotted on semilogarithmic paper against time gives a good straight line at all times up to, and including, 10 hours. But all of the later points must be disregarded. Hence this attempt to fit the points by a single exponential function is no more satisfactory than the one illustrated in Figure 6-17.

from the line from the original values of y, we will obtain a series of differences which give a reasonably good straight line on the semilog paper. From these two lines we obtain the equation

$$y_t = 42.0e^{-0.366t} + 21.5e^{-0.035t}$$

A similar analysis, but with a straight line fitted to the last *eight* points, is shown in Figure 6-20. This gives the equation

$$y_t = 37.7e^{-0.506t} + 28.0e^{-0.049t}$$

Yet the *original* function, about which the values of y were scattered, was

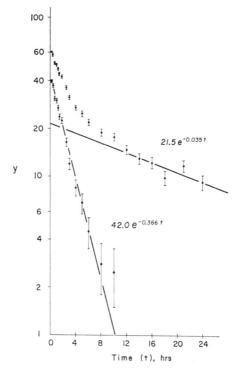

FIG. 6-19. First Attempt to Fit a Double Exponential

The upper series of points represents the original values of y plotted on a logarithmic scale against time on an arithmetic scale. The last six points have been fitted by a straight line which might represent the second or "slower" of the two exponential terms whose sum is perhaps equal to y. The lower series of points represents the series of differences obtained by subtracting from each of the original values of y the simultaneous value of the presumed second term, $21.5e^{-0.0^.5t}$, read from the upper line. This series of differences, plotted on semilogarithmic graph paper, can be fitted by a single straight line, which may accordingly be taken to represent the first or "faster" of the two exponential terms. From this analysis we conclude that the observed points can be fitted reasonably well by the sum of the two exponential functions of time corresponding to the above two straight lines: $y_t = 42.0e^{-0.366t} + 21.5e^{-0.035t}$

$$y_t = 36.0e^{-0.500t} + 22.0e^{-0.100t} + 8.0$$

Which of these equations is "correct"? It is impossible to tell! The standard deviations of the points from the curves are virtually identical: ±1.10, ±1.11, and ±1.12, respectively. To make matters worse,

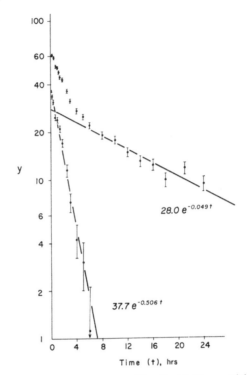

FIG. 6-20. Second Attempt to Fit a Double Exponential

The sequence of steps illustrated in this figure is precisely the same as for Figure 6-19, except that the upper straight line corresponding to the second or "slower" exponential term has been fitted to the last *eight* points. Again the differences between this line and the original values of y for the remaining points can be fitted by a straight line. This shows that the original observations of y can be fitted reasonably well by the equation

$$y_t = 37.7e^{-0.506t} + 28.0e^{-0.049t}$$

it is not difficult to find a *single* exponential, declining toward a straight-line function of time as an asymptote:

$$y_t = 43.5e^{-0.393t} - 0.48t + 20.5$$

which, with a standard deviation of ± 1.08, is just as satisfactory as the other three. *In spite of their very different parameters, all four curves fit the observed values of y equally well* (Fig. 6-21). Yet if these four func-

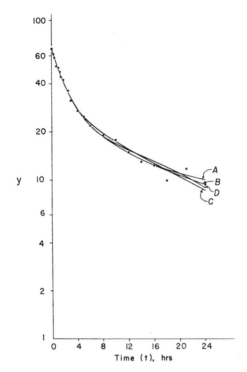

FIG. 6-21. Four Different Functions, Each Providing a Satisfactory
Fit for the Same Points

The original values of y (Table 6-4) were obtained by allowing small random deviations to occur around *Curve A*:

$$y_t = 36.0e^{-0.500t} + 22.0e^{-0.100t} + 8.0$$

But the same points are equally well fitted by *Curve B*:

$$y_t = 42.0e^{-0.366t} + 21.5e^{-0.035t}$$

obtained by the analysis of Figure 6-19, or by *Curve C*:

$$y_t = 37.7e^{-0.506t} + 28.0e^{-0.049t}$$

obtained by the analysis of Figure 6-20, or even by the single exponential function of
Curve D:

$$y_t = 43.5e^{-0.393t} - 0.48t + 20.5$$

for which the asymptote is a straight line. But now look at Figure 6-22.

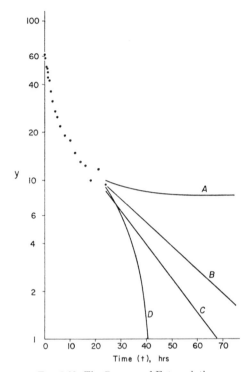

Fɪɢ. 6-22. The Dangers of Extrapolation

The four functions which fitted the points in Figure 6-21 equally well are here extrapolated beyond the period during which y was actually observed. It is obviously hazardous to predict what will happen to a variable beyond the time of last observation merely by extrapolating an empirical equation fitted to the points, however close the fit may appear to be.

tions are extrapolated beyond the actual data to 72 hours, how very different the curves become (Fig. 6-22)!

These examples warn us not to take too seriously any *particular* set of coefficients and rate constants which we may get by plotting data on semilogarithmic paper and ferreting out the exponential terms in the fashion described above. The need for such a warning is all too evident from the preposterously elaborate exponential equations which are sometimes published. The technique of "peeling off" successive terms is so deceptively easy! Fit a straight line, subtract, plot the differences, fit another straight line, subtract, plot the differences. How solid and impressive the resulting sum of exponentials looks! And how remarkably well the curve agrees with the observations. Surely the investigator can be pardoned a certain self-satisfaction for having so clearly identified the individual components which were contributing to the overall change. Yet the examples discussed above show how groundless his satisfaction may be. It is undoubtedly true that the particular sum of exponentials which he happened to pick, plot, and publish fits the points with gratifying accuracy. But so also may other equations of the same general form but with quite different parameters. It is no great trick to have found *one* such equation. Even with a single exponential declining from a known value at time zero toward an unknown constant asymptote there are two parameters—the rate constant and the asymptote—to be fitted to the data. The effect of a considerable change in either may be largely offset by making a compensatory change in the other (see Section 6-13). Add a second exponential term with two more parameters to be estimated from the data, and the number and variety of "closely fitting" equations become truly bewildering. Worst of all, there are no simple statistical measures of the precision with which any of the parameters have been estimated. These considerations do not destroy the value of fitting an exponential equation to experimental data when it is suggested by some underlying theory or when it provides a convenient empirical way of summarizing a group of observations mathematically.* But they make very clear the danger of using an empirical exponential equation to predict what may happen beyond the period actually covered by the observations. It is equally clear that we must be exceedingly skeptical when attempts are made to match the individual terms of an empirical exponential equation with supposedly corresponding processes or regions in the body.

6-15. Calculation of a Logarithmic Mean

Occasionally it is necessary to calculate the average or mean value of a quantity which is changing exponentially between time t_1 and time t_2. For example, suppose we want to estimate the renal plasma clearance of a substance x while the concentration of x in the plasma is decreasing exponentially. We collect the urine excreted between t_1 and t_2, and we divide the total quantity of x in the urine sample by the time, $t_2 - t_1$, during which it was collected to obtain the mean rate of excretion of x during the collection period. Now to calculate the renal plasma clearance, we must divide this *mean* rate of excretion by the corresponding *mean* plasma concentration, *i.e.*, the plasma concentration which, *if it had remained unchanged*

---

* If there is no theoretical reason for choosing an exponential function to fit the data, the choice can be justified *only* by custom and convenience, not by appealing to the law of parsimony; for there are likely to be many other functions, equally obedient to the law, which would fit the data as well.

throughout the period, would have accounted for the observed excretion of
x. If C_P, the plasma concentration of x, were declining from $C_{P.t_1}$ at
t_1 to $C_{P.t_2}$ at t_2 as a straight-line function of time, the mean concentration
would simply be the concentration at the mid-point of the period. But this
would not be true with an exponential decline which follows the equation

$$C_{P.t} = C_{P.t_0}e^{-kt} \qquad (6\text{-}48)$$

However, it is easy to see (Fig. 6-23) that whatever the slope of the concentration curve, the effective *mean* concentration, $\bar{C}_{P.t_{1-2}}$, is equal to the
shaded area, between the curve and the baseline, and between t_1 and t_2,
divided by the length of the period, $t_2 - t_1$. (This "area" has the dimensions
[concentration] [T]. It is an area only in the graphical sense.) The area is
simply the integral of Equation 6-48 between the limits t_1 and t_2:

$$\text{Area} = \int_{t_1}^{t_2} C_{P.t_0}e^{-kt}\,dt = C_{P.t_0}\int_{t_1}^{t_2} e^{-kt}\,dt \qquad (6\text{-}49)$$

In a table of integrals we find

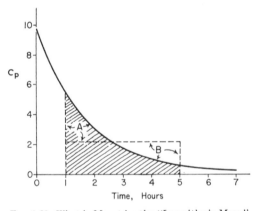

Fig. 6-23. What is Meant by the "Logarithmic Mean"

Let C_p be a plasma concentration which is decreasing toward zero exponentially
(*solid curve*). The logarithmic mean concentration during the period between 1 and 5
hours is the *shaded area* under the curve (measured in units of "concentration-hours")
divided by 4 hours, the length of the period. Graphically, the logarithmic mean concentration is represented by the *horizontal broken line* at a concentration of about 2.2
mg. per liter which forms the top of a rectangle of the same area. Area A, part of the
original shaded area but not part of the *rectangle*, is exactly equal to area B, part of
the *rectangle*, but not part of the *original shaded area*. Notice that the logarithmic
mean concentration is somewhat higher than the concentration at 3 hours, the mid-
point of the period. How to calculate the logarithmic mean is explained in the text.

$$\int e^{ax} \, dx = e^{ax}/a \qquad (6\text{-}50)$$

For our definite integral, this becomes

$$\int_{t_1}^{t_2} e^{-kt} \, dt = (e^{-kt_2}/-k) - (e^{-kt_1}/-k) \qquad (6\text{-}51)$$

Hence, from Equation 6-49 and 6-51:

$$\text{Area} = (C_{P.t_0}e^{-kt_2}/-k) - (C_{P.t_0}e^{-kt_1}/-k) \qquad (6\text{-}52)$$

Combining Equations 6-48 and 6-52,

$$\text{Area} = (C_{P.t_2} - C_{P.t_1})/-k \qquad (6\text{-}53)$$

but $-k$ is the slope of the straight line obtained when $\ln C_P$ is plotted against time, in other words, $\Delta \ln C_P/\Delta t$. In this instance, $\Delta \ln C_P$ is $\ln C_{P.t_2} - \ln C_{P.t_1}$, and Δt is $t_2 - t_1$. Hence,

$$-k = (\ln C_{P.t_2} - \ln C_{P.t_1})/(t_2 - t_1) \qquad (6\text{-}54)$$

Combining Equations 6-53 and 6-54,

$$\text{Area} = (C_{P.t_2} - C_{P.t_1})(t_2 - t_1)/(\ln C_{P.t_2} - \ln C_{P.t_1}) \qquad (6\text{-}55)$$

To obtain the mean plasma concentration, we divide this area by the length of the period, $t_2 - t_1$:

$$\bar{C}_{P.t_{1-2}} = \frac{C_{P.t_2} - C_{P.t_1}}{\ln C_{P.t_2} - \ln C_{P.t_1}} = \frac{C_{P.t_1} - C_{P.t_2}}{\ln (C_{P.t_1}/C_{P.t_2})} \qquad (6\text{-}56)$$

Equation 6-56 is valid, of course, only when the plasma concentration of x is decreasing exponentially toward zero, but it is not difficult to derive a more general equation (see Exercise 8). Unless a very substantial decrease of C_P takes place between t_1 and t_2, the logarithmic mean given by Equation 6-56 will be closely approximated by the arithmetic mean, $(C_{P.t_1} + C_{P.t_2})/2$ (see Appendix C).

6-16. Other Aspects of Exponential Change

It would be a mistake to leave the reader with the impression that exponential change is always a function of time, though time is undoubtedly the commonest independent variable. Exercise 4 gives an example of a change occurring as an exponential function of distance. The mathematical treatment is exactly the same, though one can obviously no longer speak of a "half-time."

The rather simple exponential functions discussed in this chapter are not the only kind which the biologist may encounter. For example, many growth curves have a sigmoid shape which can often be fitted by the *logistic curve* or one if its variants. The *curve of normal distribution* (Gaussian curve) is a still more complex exponential function. A discussion of such curves lies beyond the scope of this book.

Finally, it should not be assumed that exponential functions are the only ones useful for the empirical description of variables which change with time. For example, the long-term retention of certain radioactive elements in bone has been described, purely empirically, by a *power function* of the form

$$y = At^{-b} \qquad (6\text{-}57)$$

or,

$$\log_{10} y = \log_{10} A - b(\log_{10} t) \qquad (6\text{-}58)$$

where y is the proportion of a single dose of the element (corrected for disappearance by radioactive decay) which remains in the body at time t, time zero being the time of administration. The equation is not valid when t is less than unity. From Equation 6-58 it is obvious that one should get a straight line by plotting y against time on log-log graph paper. For further discussion of the advantages and limitations of this kind of equation, the original paper should be consulted (79).

EXERCISES. CHAPTER 6

Exercise 1

You had a particularly prudent and farsighted ancestor who, in the year 1, invested the equivalent of $1.00 in The Second Imperial Bank of Rome, a small but highly reliable and conservative firm paying interest at the rate of 2 per cent per annum, compounded continuously. In spite of war, pestilence, and the barbarian hordes, this bank has managed to survive the centuries, and your ancestor's account has remained untouched at the same interest for the past 1962 years. While sightseeing in Rome, you decide to stop in at the good old Second Imperial just to inquire how the little account is coming along. "We are indeed honored by a visit from our most important depositor!" says the manager of the bank. "According to our latest figures, you now have approximately ———— to your credit. I hope we may have the pleasure of continuing to serve you during the next two millenia."

Exercise 2

By differentiation, prove that when $N_t = N_0 e^{-kt}$, $dN/dt = -kN_t$.

Exercise 3

In each of the six graphs of Figure 6-24, the *solid curve* represents some variable, *y*, which is approaching an asymptote exponentially. The asymptote is represented by the *broken line*. For each *solid line* write the appropriate specific form of the general Equation 6-39a.

Exercise 4

Suppose that $\dot{N}_{\text{phot.0}}/A$ monochromatic and uniformly distributed photons per square centimeter per second enter a uniform absorbing material in a direction normal to a plane surface of the material. Assuming no reflection or scattering of the photons, derive an equation for $\dot{N}_{\text{phot.}x}/A$, the flux of photons at a distance of *x* cm. perpendicularly below the surface. What name is commonly given to this equation? How must the equation be modified to include the concentration of a light-absorbing compound dissolved in a nonabsorbing solvent?

Exercise 5

Let C_D be the concentration of drug *D* in plasma. Suppose that *D* is effective only when C_D exceeds $C_{\min}$ mg./L. of plasma. Suppose further that C_D decreases exponentially toward zero with a half-time of $t_{1/2}$ hours. Finally, suppose that the maximum concentration of the drug, $C_{D.0}$, is achieved at

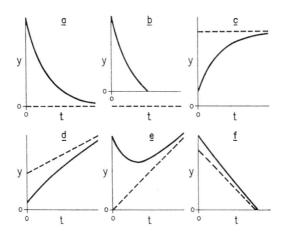

Fig. 6-24. Several Ways in Which a Variable, *y* (*Solid Curves*), Might Approach an Asymptote (*Broken Lines*) Exponentially

See Exercise 3.

time zero, and is directly proportional to the single dose of the drug, Q_D, according to the equation

$$C_{D.0} = aQ_D$$

Find an equation for the number of hours, t_{eff}, during which the drug will remain effective after a single dose, Q_D.

Exercise 6

In the dog, an intravenous dose of 30 mg. of pentobarbital sodium per kilogram of body weight will usually produce surgical anesthesia. Also in the dog, pentobarbital has a biological half-life of about 4.5 hours, due almost entirely to metabolism.

You anesthetize a 14-kg. dog with the above dose of pentobarbital. Two hours later the anesthesia is obviously beginning to lighten and you want to restore the original depth of anesthesia. How many milligrams of pentobarbital sodium should you inject?

Exercise 7

A graduate student, who was interested in ion transport, leached slices of kidney cortex in a potassium-free medium. The slices lost potassium and gained sodium. On subsequent incubation at 37°C. in a potassium-containing medium, the slices lost sodium and gained potassium as indicated by the following mean values:

Time of Incubation	Conc. of K^+ in Slices as % of Normal Conc.
min.	
0	20.5
0.5	43.6
2.0	56.9
6.0	73.5
10.0	81.1
20.0	91.6
40.0	96.1

A. Can potassium accumulation be described as a single exponential process?

B. Can you find an exponential equation to fit the points?

Exercise 8

Generalize Equation 6-56 for the logarithmic mean so as to make it applicable to situations in which the asymptote is not equal to zero.

Exercise 9

Suppose a gas, X, with a solubility in blood, S_B (*i.e.*, an equilibrium blood/air distribution ratio), of 3.0 achieves instantaneous equilibrium between blood and air across the alveolar membrane. Let $\bar{C}_{art} = 0.11$ mM./L. represent the *logarithmic mean* concentration of X in the blood leaving the lungs. Suppose that during a period of breath holding which lasts for 3 sec. the concentration of X in alveolar air decreases exponentially from $C_{A.1}$ to $C_{A.2}$ mM./L., the proportional rate of decrease being 5.0 per cent per second. Neglecting any changes in alveolar volume, and assuming that no X returns to the lungs in the venous blood, calculate the values of $C_{A.1}$ and $C_{A.2}$.

Exercise 10

There is theoretical reason to believe that the following data can be fitted by an exponential equation of the form

$$y_t = Ae^{-k_1 t} + Be^{-k_2 t}$$

From the data, estimate the parameters, A, B, k_1, and k_2.

Time, t	y_t
min.	*mg./L.*
5	50.0
10	41.9
15	36.0
20	32.0
25	28.0
30	24.8
35	22.0
40	19.5
45	17.5
50	16.1
60	13.2

Exercise 11

Figure 3 on page 248 of a paper by Baker *et al.* (7) deserves your critical scrutiny. Note that the authors are trying to fit their experimental points by a multiple exponential, their "H-primes" being intercepts and their "g's" being slopes.

Footnote for page 132: Notice that P, though invariant with time for any particular Δt, becomes smaller and smaller as Δt is taken smaller and smaller. Therefore, in Equation 6-25, P behaves exactly like a differential, and the ratio P/dt can be given a perfectly definite finite value.

7

TRANSFER OF SUBSTANCES BETWEEN BIOLOGI-CAL COMPARTMENTS. SIMPLE DIFFUSION

This chapter and the following two chapters will be concerned chiefly with how solutes get from one place to another in physiological systems. Only the most elementary aspects of the problem will be presented here, for an extended discussion would require us to consider major portions of the fields of physiology, pharmacology, biochemistry, and biophysics.

7-1. Steady States and Equilibrium States

Throughout the subsequent discussion we shall be using the terms, "steady state" and "state of equilibrium" (or simply "equilibrium"). These terms are often confused or used as though they were synonyms. Since they have entirely different meanings, we must begin by distinguish. ing clearly between them. This is most easily done by means of the hy. draulic analogy which is illustrated in Figure 7-1.

W and X are two reservoirs connected by an intervening pipe. Water can flow into W from a faucet, and out of either W or X by a drain. In the various diagrams the direction of flow is indicated by arrows. The reader should have no trouble in following the changes of inflow, flow between W and X, outflow, and amounts of water in the reservoirs, which are depicted in the diagrams. With respect to water, *reservoir W is in equilib-rium with reservoir X when there is no net transfer of water between them, i.e.,* when the flow through the intervening pipe is zero. This may (Fig. 7-1$\underline{d}$) or may not (Fig. 7-1$\underline{b}$) coincide with a steady state. *Reservoir W is in a steady state when the quantity of water in W remains constant, i.e.,* when the flow into W exactly equals the flow out of W. Again, this may (Fig. 7-1$\underline{d}$) or may not (Fig. 7-1$\underline{c}$) coincide with equilibrium between W and X. When it does, we may speak of a *steady state of equilibrium*. Notice that the concept of equilibrium involves at least two regions so connected with each other that transfer can occur between them in both directions, whereas the concept of a steady state can be applied to a single region. Notice, too,

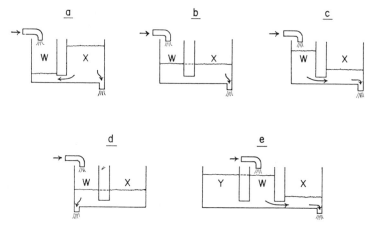

FIG. 7-1. Hydraulic Models to Illustrate the Terms "Steady State" and "Equilibrium"

Model a. Neither a steady state nor an equilibrium. *Model b.* W and X are in equilibrium with each other, but neither is in a steady state. *Model c.* (inflow equal to outflow). Both W and X are in a steady state, but they are not in equilibrium with each other. *Model d* (inflow equal to outflow). W and X are in a steady state of equilibrium. *Model e* (inflow equal to outflow). W, X, and Y are each in a steady state, but only W and Y are in equilibrium with each other.

that in a system with several reservoirs part of the system may be in equilibrium, or even, as in Figure 7-1*e*, a steady state of equilibrium (*Y* and *W*) when another part of the same system (*W* and *X*) is not in equilibrium. Ordinarily, however, when one part of a system is in a steady state so also are all other parts of the same system.

In accordance with the illustration just discussed, we shall now give more general definitions of equilibrium and of steady state.

Let S = the substance which is being transferred

W and X = two interconnected compartments

$(dQ_S/dt)_{W \to X}$ = the rate of transfer of S from W to X at time t

$(dQ_S/dt)_{X \to W}$ = the rate of transfer of S from X to W at time t

$dQ_{S \cdot W}/dt$ = the rate at which the quantity of S in W is changing at time t

$dQ_{S \cdot X}/dt$ = the rate at which the quantity of S in X is changing at time t

Then, W and X are in *equilibrium* with respect to S when there is *no net transfer of S* between them, *i.e.*, when $(dQ_S/dt)_{W \to X} = (dQ_S/dt)_{X \to W}$. W is in a *steady state* with respect to S when the *quantity of S in W remains constant, i.e.,* when $dQ_{s \cdot W}/dt = 0$. In the steady state, the total quantity of S entering W per unit of time must be exactly equal to the quantity of S leaving W per unit of time. Similarly, X is in a steady state when $dQ_{s \cdot X}/dt = 0$.

Note carefully the distinction between $(dQ_s/dt)_{W \to X}$ which means the rate at which S *is being transferred from W to X* at time t, and $dQ_{s \cdot W}/dt$ which means the rate at which the quantity of S *present in W* is changing at time t. In the hydraulic system of Figure 7-1, the substance, S, whose transport was under scrutiny was water. But S might equally well be a solute, a radioactive isotope, or even an electric charge. And in general, W and X are not called reservoirs but rather pools or *compartments* as defined in the next section.

7-2. What Is Meant by the Term "Compartment"?

Suppose we have a rectangular chamber filled with a solution which is divided into two parts by an extremely thin barrier extending across the middle of the chamber (Fig. 7-2). Suppose further that the molecules of a particular solute, S, are able to cross the barrier either by going through holes or "pores" in the barrier or by traversing the substance of the barrier itself. Finally, suppose the solution on each side of the barrier is so well stirred that, when S is added to either side, it immediately reaches a uniform concentration on that side. However, because of the barrier, instantaneous mixing of the fluid on one side with the fluid on the other side does not occur, so that it takes a measurable time for S to approach a final steady state of equilibrium across the barrier. Under these circumstances

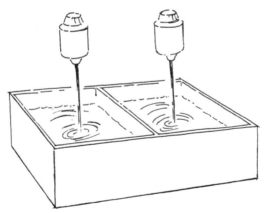

Fig. 7-2. Two Well-stirred Compartments, Separated by a Barrier

the reader will readily agree that the barrier divides the chamber into two distinct compartments. There they are! We could, if we wished, measure them—length, breadth, and depth. Like the compartments in a Pullman car or an egg box, they are real physical entities. Yet it is not their physical reality which identifies them as separate compartments in the sense in which we shall use the term; *it is rather the behavior of S in the system*. For example, suppose that the "barrier" were perforated so freely with big holes that mixing of S across it was complete before we could measure the concentration of S on either side. Then as far as our study of S is concerned, there would be only one compartment. At the other extreme, suppose the barrier were completely impermeable to S. Then also, as far as our study of S is concerned, there would be only one compartment, *i.e.*, the one into which S was originally placed. The other side might just as well not be there at all.* We are justified in talking about two compartments only when we can actually investigate the rate of transfer of S from one side to the other.

Let us take another example. If radioactive potassium ions are added to a well-shaken suspension of red blood cells in an isotonic fluid, the rate at which the radioactive potassium enters the red cells can be studied by removing aliquots from time to time and measuring the radioactivity in the centrifuged cells, or the supernatant fluid, or both. Now there are millions of individual cells in the suspension, each constituting a separate little physical chamber. But the technique of measurement allows us to study only what is going on in the group of red cells *taken as a whole*. There are, therefore, only two compartments—cells and surrounding fluid.

As a final extension of the meaning of the term "compartment" consider the chemical state of iodine in blood plasma. Some of the iodine is inorganic iodide ion, while some—in the thyroid hormone—is organic iodine, chiefly in thyroxine. No iodine is exchanged between these two forms except by complicated processes of hormone synthesis in the thyroid gland or hormone degradation in the tissues. For the mathematical description of iodine metabolism, it is convenient to regard inorganic iodide (everywhere in the body) and hormonal iodine (in all extrathyroidal tissues) as existing in two separate compartments, even though physically both forms of iodine occur together in plasma and in various other body fluids.

In accordance with the broadening of meaning illustrated by these examples, we may define the term "compartment" as follows:

If a substance, S, is present in a biological system in several distinguishable forms or locations, and if S passes from one form or loca-

---

* In the present argument, we neglect any osmotic or electrostatic effects which S may produce across the barrier.

tion to another form or location at a measurable rate, then each form or location constitutes a separate compartment for S.

It may seem strange to lump different chemical forms and different locations together in this definition, but in fact the mathematical description of transformation from one compound to another is so similar to the mathematical description of transportation from one place to another that the apparent incongruity is justified. In most of this chapter, however, we shall be concerned only with the actual movement of substances from one place to another.

7-3. The Importance of Rapid Distribution within a Compartment. The Volume of a Compartment

The definition of a compartment given above is a definition of sheer convenience. It allows us to postulate as many compartments or as few compartments as are required for analyzing a given problem. But there is one important restriction implicit in the definition which limits our freedom to choose what we shall regard as a compartment: The S in one part of a compartment must be able to interchange rapidly enough with the S in all other parts of the same compartment so that for the particular problem at hand we do not have to worry about transport of S within the compartment. For if S were too slowly distributed, we would be forced to postulate not one but two or more different compartments. This does not mean that the concentration of S must necessarily be uniform throughout the compartment. It does mean that if a small increment of S is added to one part of the compartment, the added S must soon permeate the entire compartment so that the concentration of S in all parts will undergo the same proportional increase. For example, when the synthesis of thyroid hormone has been blocked by a drug such as thiourea, the actual concentration of iodide ion in the thyroid gland may be many times higher than in plasma. Yet the iodide ion in the thyroid exchanges so quickly and freely with the iodide ion in the blood stream that both may *usually* be regarded as belonging to the same iodide compartment. (For an example in which this is *not* true, see Exercise 6, Chapter 9.)

Since a single compartment may consist of regions with different solute concentrations, it becomes necessary to choose one of its regional concentrations as a reference standard and to pretend that the entire quantity of S in the compartment is at a uniform concentration equal to the real concentration of S in the reference region. This pretense allows us to define the *volume of the compartment* as the volume it would have if all of the S contained in it were actually distributed at a uniform concentration equal to that in the reference region. For example, suppose that 20 L. of extracellular fluid (represented by plasma) contained inorganic iodide at a concentration of 3 μg. per liter, and that 0.04 L. of thyroid gland contained inorganic iodide ion at a concentration of 300 μg. per liter. The total iodide in the iodide compartment would then be $(3)(20) + (300)(0.04) = 72$ μg. Now 72 μg. would have to be distributed through a volume of 24 L. to make a uniform concentration equal to the actual concentration in plasma, namely, 3 μg. per liter. We would say accordingly that the volume of the iodide compartment is 24 L. To be absolutely specific, we should say that the volume of the iodide compartment, *with respect to the iodide concentration in plasma*, is 24 L.; for in theory it would be equally correct to say that the volume of the compartment is 0.24 L. *with respect to the concentration of inorganic iodide in the thyroid gland.* But it is rarely necessary to make this distinction. When the reference concentration is not specified, it is commonly understood to be the concentration in plasma which can be sampled and analyzed directly. The imaginary compartment volume thus defined is really a "volume of distribution" (see Section 8-5).

Three processes contribute to the rapid distribution of various substances throughout various compartments. First, *stirring and mixing* by currents within a body of fluid. This is obviously important in the blood stream, in the lumen of the intestine, and presumably in certain other hollow viscera which contain fluid. Second, actual *transportation* of the substance by a flowing stream. It is the interconnection of remote regions by the all-pervading blood stream which allows us to regard "extracellular fluid" for many purposes as a single compartment. Both of these processes can distribute substances rapidly over considerable distances. In contrast the third process, *diffusion*, is quite effective in distributing solutes over very short distances, but is practically useless over long distances. To understand why this is true, we must examine Fick's law of diffusion.

7-4. Fick's Law of Diffusion

Consider an unstirred solution of a solute, S, maintained at constant temperature, in which the concentration of S is not uniform throughout the solution but varies from place to place. Consider a very small cubical volume of this solution, measuring dx by dy by dz units of length, at some point where there is a concentration gradient of S (Fig. 7-3). Let the concentration gradient be in the x direction so that one passes from a region of higher to a region of lower concentration as one proceeds in the direction of *increasing* distance along the x axis. Now, because of the random movements of thermal agitation, molecules of S will be entering and leaving the cube on all sides. Since there is no concentration gradient in the y direction, the mean concentration of S at the top face of the cube is equal to the mean concentration at the bottom face, so that on the average as many molecules of S traverse the cube from top to bottom as from bottom to top. There

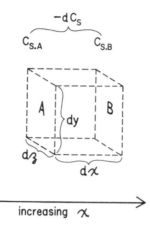

Fig. 7-3. An Infinitesimal Cube through Which Diffusion Is Occurring

The cube is so oriented that the concentration gradient for S, the diffusing solute, lies parallel to the x axis and perpendicular to faces A and B. For the rate of transfer of solute, (dQ_S/dt) to be *positive* in Equation 7-1, distance along the x axis must be measured in the direction of transfer, *i.e.*, from *left* to *right* in the figure (*arrow*). Accordingly, the *distance* at B is *greater* than the distance at A, so that the increment of distance, dx, is positive. But the *concentration* at B is *less* than the concentration at A, so that the increment of concentration, $-dC_S$ (which must be measured in the same direction), is negative.

is, therefore, no *net* movement of S in the y direction. The same is true of the z direction. But in the x direction, more molecules of S will happen to traverse the cube from face A to face B than in the opposite direction from face B to face A. (There is nothing mysterious about this. It is simply that at face A, where the concentration of S is higher than at face B, there are more molecules moving about in *all* directions than there are at face B.) As a consequence, there is a net transfer of S by diffusion in the x direction.

What factors will determine how many moles of S will be transferred from face A to face B per unit time? In other words, upon what will the net rate of transfer of S, (dQ_S/dt), depend? To begin with, it will be directly proportional to the concentration gradient, *i.e.*, to the *decrease* in concentration per unit *increase* in x. For our infinitesimal cube, this is $-dC_S/dx$. Notice that we cannot properly say that the rate of transfer is directly proportional to the concentration difference itself unless we regard the distance, dx, as fixed. It is the difference in concentration per unit of distance (the concentration *gradient*) which determines the rate of movement of S. If we *doubled* the distance over which the *same* concentration difference occurred, the gradient, and hence the rate of transfer of x, would be cut in half. Next, the quantity of S transferred per unit of time is obviously directly proportional to the area through which the transfer is taking place. Double the area of faces A and B and the amount of S transferred from A to B will double. For our infinitesimal cube, the area is $(dy)(dz)$. With these factors identified, we can write an equation for rate of transfer, by defining an appropriate proportionality constant, D_S :

$$(dQ_S/dt) = D_S(dy)(dz)(-dC_S/dx) \qquad (7\text{-}1)$$

Equation 7-1 is a general form of Fick's law of diffusion. It is closely analogous to the corresponding equation for the rate of transfer of heat along a temperature gradient. The proportionality constant, D_S, is called the *coefficient of diffusion* of S, or the *diffusivity* of S. In order to give a more well-defined meaning to diffusivity, let us solve Equation 7-1 for D_S :

$$D_S = \frac{dQ_S/dt}{(dy)(dz)(-dC_S/dx)} \qquad (7\text{-}1a)$$

This makes it clear that D_S is equal to the number of moles of S which would diffuse in unit time across unit area when the concentration gradient is unity. Since the dimensions of D_S are $[L^2T^{-1}]$, its actual numerical value will depend upon what units of measurement are chosen for time and distance.

D_S varies with temperature, with the molecular weight of S, and with forces of interaction between molecules of S and molecules of the solvent which tend to impede the movement of S from one place in the solution to another. D_S also varies somewhat with the concentration of S, particularly when the concentrations are high (61). But the variation with concentration is usually small enough to be neglected in the ranges of concentration with which physiologists are concerned. In gases, diffusion is much more rapid than in liquids because the mean free path of a molecule between collisions is longer and the forces of interaction are weaker. *Graham's law*, which states that diffusivity is inversely proportional to the square root of molecular weight, is followed rather closely in gases but only approximately in liquids (57). For a comment on the relative rates of diffusion in gases and in liquids, see Exercise 2.

The validity of Equation 7-1 is not restricted to any particular pattern of diffusion or geometrical arrangement of concentration gradients, for it describes only what is happening in an infinitesimal volume of solution during an infinitesimal interval of time. In order to calculate the actual changes in concentration which occur through measurable distances and during finite intervals of time, we must use some integrated form of Equation 7-1. The technique of integrating such an equation is complex and need not concern us here, but it is important to realize that different integral

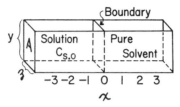

FIG. 7-4. Diffusion in a Finite System with Simple Geometry

The conditions illustrated are for time zero. The solution of Fick's general law of diffusion for this system is discussed in the text.

equations must be used for different geometrical arrangements. An equation which is valid for diffusion from a cylindrical volume cannot be used for diffusion from a spherical volume. Specific solutions for a number of geometrical arrangements have been worked out (25, 56, 61, 93). As an illustration, consider a uniform solution of S, at an initial concentration of $C_{S.0}$, adjacent to pure solvent in a long rectangular tube. At time zero, a sharp boundary between solution and pure solvent extends across the tube at right angles to its long axis at the point where x equals zero. The columns of both solution and solvent must be long enough so that no concentration changes will occur at the extremities of the tube during the time of observation (Fig. 7-4). For these conditions, Equation 7-1 may be simplified somewhat because throughout the entire system the diffusion gradients will remain parallel to the x axis. Accordingly we can replace the infinitesimal area $(dy)(dz)$ by the actual cross-sectional area of the tube, $A = yz$:*

$$(dQ_s/dt)_x = D_sA(-dC_s/dx) \tag{7-2}$$

where $(dQ_s/dt)_x$ is the quantity of solute diffusing per unit of time across the plane (of area A) which lies perpendicular to the diffusion gradient at a distance x units of length away from the initial boundary where $x = 0$.

For the arrangement of Figure 7-4, the integral form of Equation 7-2 is

$$C_{s.x.t} = (C_{s.0}/2)\left[1 - (2/\sqrt{\pi}) \int_0^y e^{-y^2} dy\right] \tag{7-3}$$

where y is not a distance (as above) but is defined as

$$y = x/2\sqrt{D_s t} \tag{7-4}$$

---

* Equation 7-1 had to be written in terms of an infinitesimal area $(dy)(dz)$ because in the general case, the direction of the concentration gradient (which is always normal to a surface of equal concentration) may vary from point to point. For example, if S were diffusing outward from a spherical volume, the concentration gradients would be in the direction of the radii of the sphere which, of course, point in different directions.

and $C_{s.x.t}$ is the concentration of S at point x and at time t. $C_{s.0}$ is the concentration of S in the original solution.

Now the expression

$$(2/\sqrt{\pi}) \int_0^y e^{-y^2} dy$$

is called the error function of y and may be symbolized by "erf(y)." It is closely related to the normal curve of error (normal curve of distribution), a fact which reminds us that the distribution of diffusing molecules is the result of random motions and accordingly follows statistical laws. Values of erf(y) for various values of y may be looked up, for example, in Dwight's *Mathematical Tables* (36). Alternatively, they may be calculated from a table of integrals of the normal curve by a method given in the *Handbook of Chemistry and Physics* (58). We may therefore simplify Equation 7-3 by writing it in the form

$$C_{s.x.t} = (C_{s.0}/2) [1 - \text{erf}(y)] \tag{7-5}$$

Combining Equations 7-4 and 7-5,

$$C_{s.x.t} = (C_{s.0}/2) [1 - \text{erf}(x/2\sqrt{D_s t})] \tag{7-6}$$

Now as t approaches infinity, $x/2\sqrt{D_s t}$ approaches zero, $1 - \text{erf}(x/2\sqrt{D_s t})$ approaches unity, and $C_{s.x.t}$ approaches $C_{s.0}/2$. Hence,

$$C_{s.x.\infty} = C_{s.0}/2 \tag{7-7}$$

Combining Equations 7-6 and 7-7 by eliminating $C_{s.0}$, and rearranging,

$$1 - (C_{s.x.t}/C_{s.x.\infty}) = \text{erf}(x/2\sqrt{D_s t}) \tag{7-8}$$

But the fraction $C_{s.x.t}/C_{s.x.\infty}$ is simply the fraction of the final equilibrium concentration at point x which has been attained at time t. If we symbolize this fraction as F_{eq}, we can rewrite Equation 7-8 as

$$1 - F_{eq} = \text{erf}(x/2\sqrt{D_s t}) \tag{7-9}$$

For any *particular* fraction of equilibrium, the error function will be a constant, and for any constant error function, $x/2\sqrt{D_s t}$ will likewise be a constant. Let us call this latter constant $K_{F_{eq}}$ so that

$$K_{F_{eq}} = x/2\sqrt{D_s t} \tag{7-10}$$

For any particular F_{eq} it is easy to obtain the numerical value for $K_{F_{eq}}$ from a table of the error function. Equation 7-9 shows that we must first locate the value of $1 - F_{eq}$ in the *body* of the table. Then by Equation 7-10 the corresponding *marginal* entry will be $K_{F_{eq}}$. To calculate the time, $t_{F_{eq}}$, needed to attain the specified fraction of equilibrium, we must square both sides of Equation 7-10 and solve for t:

$$t_{F_{eq}} = (x^2/D_S)(1/2K_{F_{eq}})^2 \qquad (7\text{-}11)$$

Equation 7-11 states that the time needed for S to attain any specified fraction of its final equilibrium concentration is inversely proportional to the diffusivity of S and directly proportional to the *square* of the distance through which S must diffuse. It is for this reason that distribution of a substance solely by diffusion occurs rapidly enough over very short distances but is so slow as to be virtually useless over long distances. To get some feeling for the actual times and distances involved, let us work out a specific example.

Example

The diffusivity of nitrogen in water at 20°C. as tabulated by Hitchcock in Höber's *Physical Chemistry of Cells and Tissues* (57) is 2.02 (10^{-5}) cm^2 per second. If nitrogen is allowed to diffuse under conditions for which Equation 7-11 is valid, *i.e.*, conditions similar to those of Figure 7-4, how long will it take to achieve 95 per cent of equilibrium, $F_{eq} = 0.95$, at various distances from the original boundary?

In a table of the error function, we find that $1 - F_{eq} = 0.0500$ in the body of the table corresponds to $K_{F_{eq}} = 0.0443$ in the margin of the table. Substituting this value, and the value for D_{N_2} in Equation 7-11, and performing the indicated arithmetic, we obtain

$$t_{0.95} = 6.31(10^6)x^2$$

where t must be in seconds and x in centimeters because D was expressed in square centimeters per second. From this equation, it is easy to calculate values of $t_{0.95}$ for the values of x given in Table 7-1.

This example shows that by diffusion alone small molecules can achieve practically complete (95 per cent) equilibrium in a matter of seconds at distances of the order of cellular diameters and intercapillary distances. (For a more rigorous discussion, based upon diffusion from a cylindrical capillary, see Kety (64) and Roughton (93).) Indeed, the rate of distribution of essential metabolites by diffusion through cells and tissues is one of the most important factors which determines the optimal size of cells and the optimal spacing of capillaries. At distances of the order of 0.1 mm., 95 per cent of equilibrium is still achieved within a few minutes. But at distances much greater than a millimeter, the time to achieve 95 per cent of equilibrium by diffusion alone is to be reckoned in days. However, for two reasons these exemplary values for 95 per cent of equilibrium tend to overemphasize the ineffectiveness of diffusion as a distributing agency. In

TABLE 7-1

Time required to reach 95 per cent of equilibrium by diffusion through various distances in the system of Figure 7-4

Distance		x^2 (cm.^{2})	Time for 95% of Equilibrium at x			
μ	x (cm.)		$t_{0.95}$ (sec.)	min.	hr.	days
1	0.0001	$1(10^{-8})$	0.063			
3	0.0003	$9(10^{-8})$	0.57			
10	0.001	$1(10^{-6})$	6.3			
30	0.003	$9(10^{-6})$	57	1		
100	0.01	$1(10^{-4})$	630	11		
300	0.03	$9(10^{-4})$	5,700	95	1.6	
1,000	0.1	$1(10^{-2})$	63,000		18	
3,000	0.3	$9(10^{-2})$	570,000		160	6.6
10,000	1.0	1.0	6,310,000			73

the first place, over a considerable range of values of F_{eq} the time calculated for any particular value of D_S and any particular distance is roughly *inversely proportional to the square of the fraction of equilibrium not yet attained*, *i.e.*, inversely proportional to $(1 - F_{eq})^2$ (Exercise 4). For example, the time needed to attain 50 per cent of equilibrium at any given point is roughly 1/100 of the time for 95 per cent of equilibrium. In the second place, the times calculated above are for 95 per cent of equilibrium at a *single point* (or, more properly, a *single plane*) x cm. away from the initial boundary. But clearly the *average* fraction of equilibrium attained throughout the entire volume lying between the initial boundary and the plane at x is greater than 0.95 (Exercise 3).

7-5. Diffusion between Two Different Phases

If a diffusing substance must cross an interface between two different media, the absolute rate of diffusion but not the rate of approach to equilibrium, will be influenced by the *distribution ratio* for the substance between the two phases, *i.e.*, the ratio of concentrations at equilibrium.

Example

Figure 7-5 illustrates the diffusion of two different gases, A and B, from a gas phase into a liquid phase. For simplicity, let A and B have the same constant concentration, 3.0 mM. per liter in the gas phase, let the gas phase be well mixed, and let A and B have the same diffusivity in the liquid phase which is not stirred. Gas A has a solubility

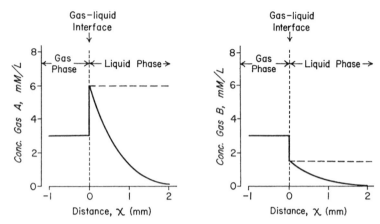

Fig. 7-5. Concentrations of a Gaseous Solute Diffusing Freely between a Gas
Phase and a Liquid Phase

The gas phase is assumed to be well stirred; the liquid phase, unstirred. At the gas-liquid interface (at 0 mm.) it is assumed that equilibrium is established at once, so that the partial pressure of the diffusing gas in the gas phase is equal to its partial pressure in the liquid phase at the interface.

In the graph at the *left*, the solubility of gas A in the liquid is 2, so that at the interface the millimolar concentration suddenly doubles as one passes from gas phase to liquid phase. This provides a relatively high concentration gradient for diffusion through the liquid. In the graph at the *right*, the solubility of gas B in the liquid is 0.5 so that the concentration of gas B in the liquid at the interface is only one fourth as great as for gas A, and the concentration gradient for diffusion is therefore only one fourth as great. Note, however, that the proportion of the final equilibrium concentration in the liquid (*horizontal broken line*) attained by diffusion at a given distance from the interface is the same for both gases. For a given partial pressure in the gas phase, gas A diffuses into the liquid four times as fast as gas B. But the total quantity of gas A which must diffuse to reach equilibrium is also four times as great as for gas B. Hence the rate of approach to equilibrium is the same. (It is assumed in this example that the diffusivity of the two gases in the liquid is the same.)

in the liquid (*i.e.*, a liquid/gas distribution ratio or partition coefficient) of 2.0.* Gas B has a solubility of 0.5.

To begin with, consider the concentrations just *at* the boundary between gas and liquid. If we think of the boundary itself as a plane with no measureable thickness, we will see that equilibrium must be instantaneously established and continuously maintained between the last layer of gas phase before the boundary and the first layer of liquid phase after the boundary. Hence, the concentration of gas in

* Note that for gases the terms *solubility*, *liquid/gas distribution ratio*, and *liquid/gas partition coefficient* are all synonymous.

the first layer of liquid, 6.0 mM. per liter for gas A, and 1.5 mM. per liter for gas B, will be equal to its concentration in the gas phase, here 3.0 mM. per liter, multiplied by its solubility. Thus the concentration gradient in the liquid is directly proportional to the solubility of the gas. By Fick's law, therefore, the moles of gas transferred by diffusion per unit of time must also be directly proportional to the solubility, and in this sense gas A diffuses into the liquid four times as fast as gas B. *Note well*, however, that to achieve its final equilibrium concentration throughout the liquid the *total amount* of gas A which must be transferred is *also* four times as great as for gas B. Consequently, when a gas diffuses into a liquid, the rate of approach to its equilibrium concentration in the liquid is *independent* of its solubility.

Although the rate of approach to equilibrium when a substance diffuses between *two* phases is independent of the distribution ratio, we shall see in the next section that distribution ratios may indeed influence the rate of equilibration when two phases are separated by a third phase.

7-6. Diffusion across Thin Membranes

We have already concluded that diffusion alone can maintain a practically even distribution of small molecules throughout such volumes as are contained within most single cells, although we have no right to assume a uniform concentration of such substances as oxygen and carbon dioxide which are very rapidly consumed by, or produced by, the cell. Diffusion can also account for the rapid distribution of solutes through the extracellular fluid between capillaries and cells. We must now consider the diffusion of substances across the intervening biological membranes such as the plasma membrane of cells. For simplicity the following discussion will be limited to the passive diffusion of uncharged solutes through homogeneous membranes. Whereas the conclusions which we reach from this simple approach will be applicable, with suitable qualifications, to the behavior of certain real membranes, the reader must realize that the whole subject of transport of substances across biological membranes by active and passive processes is exceedingly complex and that we are here deliberately avoiding its most engrossing intricacies.

Let us consider a homogeneous fluid compartment, W, separated from another homogeneous fluid compartment, Z, by a thin homogeneous membrane, M. Suppose first that W, Z, and M are all composed of the same medium, so that they are divided into two separate compartments and an intervening membrane only in our mind's eye (Fig. 7-6a). Then if a solute, S, is introduced into compartment W at the side farthest away from M, it will begin to diffuse without special hindrance across W, M, and Z at a uniform rate determined by a single coefficient of diffusion. Because the

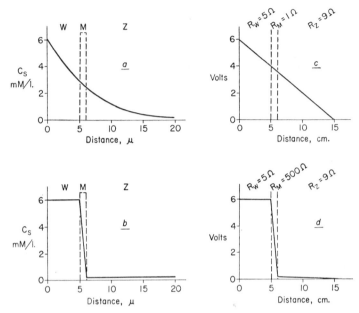

Fig. 7-6. A Membrane Viewed as a "Resistance" to Solute Flow

In *Graph a*, the thin membrane, M, offers no special hindrance to the diffusion of solute, just as in *Graph c* the short resistance, R_M, offers no special hindrance to the flow of current. But in *Graph b*, the thin membrane severely limits the rate of diffusion, just as in *Graph d* the short wire of high resistance limits the rate of current flow. In Graphs *b* and *d*, by far the greater part of the total gradient (of concentration or voltage) is across the rate-limiting segment of the system.

dimensions of the system are small, the concentration gradients will rapidly disappear as S becomes well distributed by diffusion throughout the whole system. Now let M assume more realistic properties as a membrane which considerably slows the diffusion of S, either because M has a limited total area of "pores" through which S can diffuse (83), or because the physicochemical nature of M impedes the progress of S across it. Then the total time needed for transfer of a given amount of S from W to Z will be substantially increased, so that the rate of change of Q_s in both W and Z will be small compared with the rate of distribution of S by diffusion throughout these compartments. As a result, both W and Z behave at all times as if they are well stirred, and the entire concentration gradient is practically confined to M (Fig. 7-6*b*). The situation is somewhat similar to the flow of electrons through a circuit in accordance with a voltage gradient supplied by a battery. When the terminals of the battery are connected by three wires in series whose resistance per unit length is equally low, the voltage gradient is uniform through the whole circuit and electron flow is rapid (Fig. 7-6c). But when the middle wire is replaced by one with a considerably higher resistance, the time needed for transfer of a given number of electrons from one terminal of the battery to the other is much increased, and practically the whole voltage gradient occurs across the high resistance (Fig. 7-6d).

Now it can be shown that if a given difference of concentration is imposed between W and Z across M, and if M is sufficiently thin (say 10 μ or less) the quantity of S *entering* M per unit of time at its interface with W will very rapidly become practically equal to the quantity of S *leaving* M per unit of time at its interface with Z. (For a proof of this statement see Jacobs (61).) But if equal quantities per unit of time pass across both faces of M, the same quantity per unit of time must pass across *every* plane within M which lies parallel to the two interfaces. In other words, $(dQ_S/dt)_x$ is constant everywhere within M. But from Equation 7-2 it is evident that if $(dQ_S/dt)_x$ is constant, the concentration gradient in M, $(-dC_S/dx)_M$, must also be constant throughout M, and must therefore be equal to the total difference in concentration across M, $-\Delta C_{S.M}$, divided by the total thickness of M, Δx_M:

$$(-dC_S/dx)_M = -\Delta C_{S.M}/\Delta x_M = -(C_{S.M.Z} - C_{S.M.W})/\Delta x_M = \\ (C_{S.M.W} - C_{S.M.Z})/\Delta x_M \quad (7\text{-}12)$$

where $C_{S.M.W}$ and $C_{S.M.Z}$ are the concentrations of S in the membrane at its interface with W and at its interface with Z, respectively.

Combining Equations 7-2 and 7-12 with the elimination of $-dC_S/dx$, we obtain the following equation for diffusion from W to Z across a thin membrane:

$$(dQ_S/dt)_M = (dQ_S/dt)_{W \to Z} = D_{S.M}A_M(C_{S.M.W} - C_{S.M.Z})/\Delta x_M \quad (7\text{-}13)$$

Equation 7-13 is a very useful simplification of Fick's law. Notice that it has to do entirely with what happens *in* the membrane. Notice in particular that $D_{S.M}$ is the effective diffusivity of S in the membrane, not in W or Z. If it is known that S diffuses through pores in the membrane which are filled with the same medium as W and Z, comparison of the diffusivity in the membrane, $D_{S.M}$, with the diffusivity in free solution provides a means of estimating what proportion of the total area of the membrane, A_M, is, in effect, available for free diffusion (83). But for the most part the plasma membrane of cells does not behave like a porous membrane, so that substances diffusing from the outside to the inside of cells presumably do so by dissolving in the substance of the membrane. For such membranes, let $R_{S(M/W)}$ be the equilibrium distribution ratio or partition coefficient for S between M and W:

$$R_{S(M/W)} = (C_{S.M}/C_{S.W})_{eq} \tag{7-14}$$

where the subscript eq means "at equilibrium." Similarly,

$$R_{S(M/Z)} = (C_{S.M}/C_{S.Z})_{eq} \tag{7-15}$$

Then if we assume, as we did before, that equilibrium is always present at the interfaces, and that compartments W and Z are well stirred, we can apply Equations 7-14 and 7-15 directly to the interfaces:

$$C_{S.M.W} = R_{S(M/W)}C_{S.W} \tag{7-16}$$

and

$$C_{S.M.Z} = R_{S(M/Z)}C_{S.Z} \tag{7-17}$$

Substituting these values into Equation 7-13,

$$(dQ_S/dt)_{W \to Z} = D_{S.M}A_M(R_{S(M/W)}C_{S.W} - R_{S(M/Z)}C_{S.Z})/\Delta x_M \tag{7-18}$$

If W and Z are both aqueous media, it is likely that $R_{S(M/W)}$ and $R_{S(M/Z)}$ will be equal, so we may define

$$R_{S(M/aq)} = R_{S(M/W)} = R_{S(M/Z)} \tag{7-19}$$

Combining Equations 7-18 and 7-19,

$$(dQ_S/dt)_{W \to Z} = D_{S.M}A_M R_{S(M/aq)}(C_{S.W} - C_{S.Z})/\Delta x_M \tag{7-20}$$

There is evidence from many sources that the plasma membrane consists largely of lipid (31). According to Equation 7-20, the rate of diffusion of S across such a membrane should be directly proportional to the equilibrium distribution ratio for S between the membrane and the adjacent aqueous media. In a general way, this prediction is borne out by experiment. Compounds with a high oil/water partition coefficient (which is not necessarily identical with, but presumably similar to $R_{S(M/aq)}$) usually do penetrate cells more rapidly than compounds of similar molecular weight which are not as soluble in oil (31). This fact is very important, for the distribution of many drugs in the body is strikingly influenced by their relative solubility in aqueous and lipid media (see Section 10-17).

Equation 7-20 is a useful summary of how various important factors influence the rate at which a diffusing solute penetrates a membrane. But Equation 7-20 was derived from very simple, indeed naive, assumptions about the partitioning of a diffusing substance between the membrane and the adjacent aqueous media. By taking account of the successive energies of activation which a diffusing molecule must acquire first to pass from W

to M, then to pass through M, and finally to pass from M to Z, Danielli has derived a considerably more elaborate theory of diffusion across nonporous membranes (28). According to Danielli's analysis, the rate of diffusion is not directly proportional to the distribution ratio, at least for substances which diffuse very slowly across the membrane. However, for present purposes we will neglect this important refinement of theory and we will continue to use Equations 7-18 and 7-20.

7-7. Practical Measures of Membrane Permeability

As it stands, Equation 7-20 is too complex for most biological applications. While it may be possible to measure the concentration of a substance in the two aqueous phases, and the area of the membrane across which diffusion is taking place, usually no reliable estimates can be made of $D_{S.M}$, $R_{S(M/aq)}$, or Δx_M. However, for any particular system, all three of these quantities can be assumed constant, and they may therefore be combined by defining a new *permeability constant*, $(k_{perm})_S$:

$$(k_{perm})_S = D_{S.M} R_{S(M/aq)} / \Delta x_M \qquad (7\text{-}21)$$

Notice that $(k_{perm})_S$ has the dimensions of velocity: (LT^{-1}). We may now write Equation 7-20 in the simplified form:

$$(dQ_S/dt)_{W \to Z} = (k_{perm})_S A_M (C_{S.W} - C_{S.Z}) \qquad (7\text{-}22)$$

$(k_{perm})_S$ can be calculated by measuring all of the other quantities in Equation 7-22. Thus, $(k_{perm})_S$ provides a practical measure of the relative ease with which different solutes can diffuse across a given membrane.

Still other measures of rates of diffusion across membranes are in use, particularly when the substance diffusing is a gas. For example, respiratory physiologists commonly prefer to think in terms of gradients of partial pressure rather than gradients of concentration. Furthermore, they prefer to express the quantity of a gas as its volume at standard temperature and pressure (std T,P), and the concentration of a gas, C_G, as volume of gas (std T,P) per volume of solution. Then according to Equation 2-27

$$\alpha_G P_G = C_G = V_{G(\text{std } T,P)\text{liq}} / V_{\text{liq}} \qquad (7\text{-}23)$$

where α_G is the Bunsen solubility coefficient of gas G (Chap. 2). P_G is the partial pressure of G in atmospheres. Equation 7-2 may then be written

$$(dQ_G/dt)_x = A D_G \alpha_G (-dP_G/dx) \qquad (7\text{-}24)$$

for gases in liquids, Q_G being in units of volume, or

$$(dQ_G/dt)_x = (\text{diffusion constant}) \, A (-dP_G/dx) \qquad (7\text{-}25)$$

where the *diffusion constant* is defined as

$$(\text{diffusion constant}) = \alpha_G D_G \qquad (7\text{-}26)$$

and has the dimensions $[M^{-1}L^3T]$. The diffusion constant is the volume of gas (std T,P) diffusing per unit time across unit area in response to unit gradient of partial pressure. For a further note on this diffusion constant, see Exercise 1.

Finally, the *diffusing capacity* of the lung is used to describe the diffusion of gases between the alveoli and the blood across the alveolar-capillary "membrane" whose thickness and area are both unknown. The diffusing capacity for G is the total volume of G diffusing per unit time when the *mean difference* of partial pressure (*not* the gradient) between alveolar air and capillary blood is unity. The diffusing capacity has the dimensions $[M^{-1}L^4T]$. If the membrane through which G diffuses were simply an aqueous layer which permitted free diffusion, the diffusing capacity could be defined as

$$(\text{diffusing capacity}) = D_{G.\text{aq}} A_M \alpha_{G.\text{aq}} / \Delta x_M \qquad (7\text{-}27)$$

an equation whose sole merit is its dimensional correctness. But it is far better to think of the diffusing capacity as merely an empirical measure of the rate at which a given gas diffuses across a membrane of unknown characteristics in response to unit difference of partial pressure.

This brief discussion of several measures of diffusion only begins to indicate the frustrating confusion of units, dimensions, and definitions which perplexes students in this field!

7-8. The Kinetics of Equilibration by Diffusion of a Solute between Two Compartments

Up to this point we have focused our attention upon the process of diffusion across a membrane. Very often, however, we are not as much concerned with what is happening in the membrane as we are with the resulting changes in the concentration of the diffusing solute in compartments W and Z. If the volume of compartment W, the volume of compartment Z, and the total amount of S in the whole system, $Q_{S.\text{tot}}$, all remain constant, we can easily derive equations for the concentration of S in W and the concentration of S in Z as functions of time. Figure 7-7 illustrates such a two-compartment system, together with the intervening membrane whose thickness, for convenience, has been greatly exaggerated. Actually, we shall assume that the membrane is so thin that it contains a negligible quantity of S. Since we shall be concerned only with a single solute, the subscript S will be omitted from the symbols used in the following derivation.

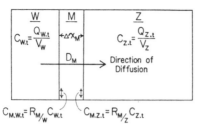

Fig. 7-7. Factors Concerned in the Kinetics of Diffusion of a Solute from W to Z across a Thin Membrane, M

It is helpful to begin by listing all of the *individual factors* which characterize the system.

1. The *independent variable* is time, t.
2. At any time, t, compartment W is fully characterized by
   V_W , the volume of W (constant)
   $Q_{W.t}$, the quantity of S in W (variable with time)
3. At any time, t, compartment Z is fully characterized by
   V_Z , the volume of Z (constant)
   $Q_{Z.t}$, the quantity of S in Z (variable with time)
4. At any time, t, the membrane, M, is fully characterized by
   A_M , the area of M (constant)
   x_M , the thickness of M (constant)
   D_M , the diffusivity of S in M (constant)
   $R_{M/W}$, the equilibrium distribution ratio for S between M and W (constant)
   $R_{M/Z}$, the equilibrium distribution ratio for S between M and Z (constant)

Since the entire system consists of W, M, and Z, it is itself fully characterized at any time, t, by the nine factors listed above.

Next, we should list, as equations, all of the *relationships* which characterize the system. A general equation for diffusion of S across the membrane has already been derived. It is Equation 7-18:

$$(dQ/dt)_{W \to Z} = D_M A_M (R_{M/W} C_{W.t} - R_{M/Z} C_{Z.t})/\Delta x_M \qquad (7\text{-}18)$$

We are assuming that the total quantity of S, Q_{tot}, remains constant. Hence,

$$Q_{W.t} + Q_{Z.t} = Q_{tot} \qquad (7\text{-}28)$$

The concentration of S in compartment W at any time, t, is

$$C_{W.t} = Q_{W.t}/V_W \qquad (7\text{-}29)$$

Similarly, in compartment Z

$$C_{Z.t} = Q_{Z.t}/V_Z \qquad (7\text{-}30)$$

Now the only process which is causing the quantities and concentrations of S in W and Z to change is the diffusion of S across the membrane. We have already assumed (in deriving Equation 7-18) that, with thin membranes, the rate of entry of S into M from W is at all times equal to the rate of exit of S from M into Z. (This assumption is, in fact, equivalent to the assumption stated above that the quantity of S within M is negligible.) Conservation of matter therefore requires

$$-dQ_W/dt = (dQ/dt)_{W \to Z} = dQ_Z/dt \qquad (7\text{-}31)$$

Equation 7-31 simply states that the rate at which W is *losing* (negative sign) S, the rate at which S is *diffusing across* the membrane, and the rate at which Z is *gaining* S, are all equal. Now it should be obvious that at time infinity the gradient for diffusion across the membrane will have disappeared. There will then be no further net transfer of S, and the whole system will be in a steady state of equilibrium. Hence, at time infinity, Equation 7-31 becomes

$$-dQ_{W.\infty}/dt = (dQ/dt)_{W \to Z.\infty} = dQ_{Z.\infty}/dt = 0 \qquad (7\text{-}32)$$

Finally, we must note the initial conditions* for the three variables: When $t = 0$, $Q_W = Q_{W.0}$, and $Q_Z = Q_{Z.0}$. Also, $R_{M/W}C_{W.0} > R_{M/Z}C_{Z.0}$. These boundary conditions show that the gradient for diffusion is from W to Z, and that at time zero there may already be some S present in Z. (Our derivation would be less general if we assumed that when $t = 0$, $Q_W = Q_{tot}$, and $Q_Z = 0$.)

We are now ready to derive an equation† for $C_{W.t}$ as a function of time. For this purpose, we are primarily interested in the changes taking place in compartment W, and consequently, we should like to eliminate, if possible, any dependent variables pertaining to M and Z by replacing them by corresponding variables for W. An obvious first step is to combine Equations 7-31 and 7-18 so as to replace $(dQ/dt)_{W \to Z}$ by its equivalent, $-dQ_W/dt$:

$$-dQ_W/dt = D_M A_M (R_{M/W}C_{W.t} - R_{M/Z}C_{Z.t})/\Delta x_M \qquad (7\text{-}33)$$

Since Equation 7-33 is an equation for the decrease of the *quantity* of S

---

* *Boundary conditions* is the general term for any set of values of the dependent variables which characterize a particular system at specified values of the independent variables and which allow one to apply the solution of the differential equation to that particular system. But when, as in this example, the boundary conditions are given for time zero, they are almost always called *initial conditions*.

† A general method for deriving such an equation is discussed in Chapters 12-14.

in W, we would do well to replace the concentrations in the equation by their equivalents in terms of quantity as given in Equations 7-29 and 7-30:

$$-dQ_W/dt = D_M A_M[(R_{M/W}Q_{W.t}/V_W) - (R_{M/Z}Q_{Z.t}/V_Z)]/\Delta x_M \quad (7\text{-}34)$$

Equation 7-34 still contains a variable, $Q_{Z.t}$, which does not pertain directly to compartment W. This fault is easily remedied by solving Equation 7-28 for $Q_{Z.t}$ and substituting the result in Equation 7-34:

$$-dQ_W/dt = D_M A_M \left[\frac{R_{M/W}Q_{W.t}}{V_W} - \frac{R_{M/Z}(Q_{tot} - Q_{W.t})}{V_Z} \right] \Big/ \Delta x_M \quad (7\text{-}35)$$

Equation 7-35 now contains only $Q_{W.t}$ and t as variables. However, these variables are so surrounded by constants that it is a bit difficult to see what to do next. We may therefore resort to the simple device of defining certain new constants in terms of the old ones, being guided entirely by convenience. To indicate that these new *constants of convenience* have no clearly defined intrinsic meaning, we will employ for them the very colorless notation k_1, k_2, etc. Accordingly let us define

$$k_1 = D_M A_M/\Delta x_M \quad (7\text{-}36)$$

$$k_2 = R_{M/W}/V_W \quad (7\text{-}37)$$

$$k_3 = R_{M/Z}/V_Z \quad (7\text{-}38)$$

$$k_4 = k_3 Q_{tot} \quad (7\text{-}39)$$

Using these constants, we can rewrite Equation 7-35 in the form

$$-dQ_W/dt = k_1(k_2 Q_{W.t} + k_3 Q_{W.t} - k_4) \quad (7\text{-}40)$$

or, by further defining

$$k_5 = k_1(k_2 + k_3) \quad (7\text{-}41)$$

$$k_6 = k_1 k_4 \quad (7\text{-}42)$$

in the still simpler form

$$-dQ_W/dt = k_5 Q_{W.t} - k_6 \quad (7\text{-}43)$$

Now it is very easy to separate the variables in Equation 7-43 so that it can be integrated:

$$\int \frac{dQ_W}{k_6 - k_5 Q_{W.t}} = \int dt \quad (7\text{-}44)$$

The integral on the left can be found in any table of integrals. The integral on the right is elementary:

$$(1/-k_5) \ln (k_6 - k_5 Q_{W.t}) = t + k \quad (7\text{-}45)$$

where k is a constant of integration which can at once be evaluated from the initial conditions. When $t = 0$, Equation 7-45 becomes

$$k = (1/-k_5) \ln (k_6 - k_5 Q_{W.0}) \tag{7-46}$$

Combining Equations 7-45 and 7-46 by eliminating k,

$$(1/-k_5) \ln (k_6 - k_5 Q_{W.t}) = (1/-k_5) \ln (k_6 - k_5 Q_{W.0}) + t \tag{7-47}$$

or, multiplying both sides by $-k_5$,

$$\ln (k_6 - k_5 Q_{W.t}) = \ln (k_6 - k_5 Q_{W.0}) - k_5 t \tag{7-47a}$$

Taking antilogarithms,

$$(k_6 - k_5 Q_{W.t}) = (k_6 - k_5 Q_{W.0}) e^{-k_5 t} \tag{7-48}$$

Dividing both sides by k_5, and changing signs,

$$\left(Q_{W.t} - \frac{k_6}{k_5} \right) = \left(Q_{W.0} - \frac{k_6}{k_5} \right) e^{-k_5 t} \tag{7-48a}$$

Now Equation 7-48a is in a form made very familiar by the discussion in the previous chapter. Evidently $Q_{W.t}$ is decreasing toward an asymptote, $Q_{W.\infty}$, which is equal to k_6/k_5. Furthermore, the difference between $Q_{W.t}$ and the asymptote is decreasing exponentially at k_5 proportion per unit of time. It will now be interesting to replace these "constants of convenience" by the original factors they represent so that we can see which of the factors influence the asymptote and which the rate constant. From Equations 7-37 through 7-42

$$Q_{W.\infty} = k_6/k_5 = \frac{k_1 k_4}{k_1(k_2 + k_3)} = \frac{k_4}{k_2 + k_3} = \frac{k_3 Q_{tot}}{k_2 + k_3} = \frac{Q_{tot}}{1 + (k_2/k_3)}$$
$$= Q_{tot} \bigg/ \left[1 + \left(\frac{R_{M/W} V_z}{R_{M/Z} V_w} \right) \right] = Q_{tot} \bigg/ \left[1 + \left(R_{Z/W} \frac{V_z}{V_w} \right) \right] \tag{7-49}$$

where

$$R_{Z/W} = R_{M/W}/R_{M/Z} = \frac{C_{M.eq}/C_{W.eq}}{C_{M.eq}/C_{Z.eq}} = C_{Z.eq}/C_{W.eq} = C_{Z.\infty}/C_{W.\infty} \tag{7-50}$$

Equation 7-49 shows that the *final equilibrium quantity* of S in W is determined by the total amount of S in the system, by its equilibrium distribution ratio between Z and W, and by the ratio of volumes of Z and W. Note that not a single characteristic of the membrane influences the asymptote, for we have even replaced the two distribution ratios by $R_{Z/W}$, the distribution ratio between Z and W. The lack of influence of the membrane on the final distribution is actually just what we ought to expect, because we have assumed that it contains a negligible volume and a negligible quantity of S. It is not a third compartment; it is merely a barrier which slows the attainment of equilibrium—a kind of negative catalyst!

The rate constant, k_5, may be similarly decomposed into its component factors:

$$k_5 = k_1(k_2 + k_3) = \frac{D_M A_M}{\Delta x_M}\left(\frac{R_{M/W}}{V_W} + \frac{R_{M/Z}}{V_Z}\right) \qquad (7\text{-}51)$$

which has the dimension (T^{-1}), as it should. From Equation 7-51 it is obvious that the *rate of approach to equilibrium* is influenced by *all* of the characteristics of the membrane, and, in addition, by the volumes of W and Z.

We may now substitute the original factors, identified in Equations 7-49 and 7-51, for the k-constants in Equation 7-48a:

$$\left[Q_{W.t} - \frac{Q_{tot}V_W}{V_W + R_{Z/W}V_Z}\right] = \left[Q_{W.0} - \frac{Q_{tot}V_W}{V_W + R_{Z/W}V_Z}\right]$$
$$\cdot\exp\left[-\frac{D_M A_M}{\Delta x_M}\left(\frac{R_{M/W}}{V_W} + \frac{R_{M/Z}}{V_Z}\right)t\right] \qquad (7\text{-}52)$$

But we originally set out to find an expression for the *concentration* of S in W at any time t. Equation 7-52 for the quantity of S may easily be converted to an equation for concentration by dividing both sides by V_W:

$$\left[C_{W.t} - \frac{Q_{tot}}{V_W + R_{Z/W}V_Z}\right] = \left[C_{W.0} - \frac{Q_{tot}}{V_W + R_{Z/W}V_Z}\right]$$
$$\cdot\exp\left[-\frac{D_M A_M}{\Delta x_M}\left(\frac{R_{M/W}}{V_W} + \frac{R_{M/Z}}{V_Z}\right)t\right] \qquad (7\text{-}53)$$

Notice that the expression $Q_{tot}/(V_W + R_{Z/W}V_Z)$ is the equilibrium concentration of S in W, $C_{W.\infty}$, which is being approached asymptotically as time increases:

$$C_{W.\infty} = Q_{tot}/[V_W + R_{Z/W}V_Z)] = Q_{tot}/(V_{dist}) \qquad (7\text{-}54)$$

The denominator, $V_W + (R_{Z/W}V_Z)$, is thus the *volume of distribution* of S (Vdist) calculated with reference to its concentration in compartment W. In other words, $V_W + (R_{Z/W}V_Z)$ is the volume which would contain an amount of S equal to Q_{tot} at a uniform concentration of $C_{W.\infty}$.

Derivation of an explicit equation such as 7-53, in which every symbol has a clearly defined physical meaning, is always instructive and intellectually satisfying. But real membranes are structurally much more complex than the simple homogeneous model here assumed. So even if we had detailed information about their physical properties, we might well find that Equation 7-53 was not really applicable to them. Therefore, in the next chapter we must turn to a less thorough and explicit, but more generally useful, analysis of the kinetics of transfer of substances between biological compartments.

EXERCISES. CHAPTER 7

Exercise 1

In textbooks of physiology it is commonly stated that in aqueous media carbon dioxide diffuses about 20 times as rapidly as oxygen. For example, Carlson, in the text by Ruch and Fulton (22), pages 796 to 797, states: "The intrinsic rate of diffusion of any substance is a function of its solubility, its molecular weight, and the permeability of the medium. Although a larger molecule than O_2, CO_2 is so highly soluble in the body fluids that it diffuses through the tissue 20 to 30 times as rapidly as O_2 does." What is the meaning of this statement?

Exercise 2

Forster (38) states, "Since the diffusion constant in air is about 1 million times that in saline . . . diffusion through 1 million microns (1 meter) of perfectly still gas would only demand a pressure difference equal to that normally associated with gas exchange across the pulmonary membrane." (It is assumed that the pulmonary membrane is $1\ \mu$ thick.) Is this statement justified?

Exercise 3

A. V. Hill (56) calculated the time needed for a sheet of muscle 1.0 mm. thick exposed to a constant concentration of oxygen on one side to attain by diffusion alone various average fractions of the equilibrium concentration of oxygen, the average being taken throughout the entire muscle. It was assumed that no oxygen was consumed by the tissue. For the diffusivity of oxygen in muscle he used the value $D_{O_2} = 4.5 \times 10^{-4}$ cm.^{2} per minute. Hill calculated that under these circumstances it would require 5 min. to reach an average of 53.4 per cent of the equilibrium concentration of oxygen throughout the tissue. Is this of the correct order of magnitude?

Exercise 4

Prove the validity of the statement made in the text (Section 7-4) that "over a considerable range of values of F_{eq} the time calculated for any particular value of D_s and any particular distance is roughly inversely proportional to the square of the fraction of equilibrium not yet attained, *i.e.*, inversely proportional to $(1 - F_{eq})^2$."

8

TRANSFER OF SUBSTANCES BETWEEN
BIOLOGICAL COMPARTMENTS.
GENERAL KINETICS

8-1. The Need for a More General Analysis of Transfer between Compartments

In Chapter 7 we undertook a detailed analysis of a particular mechanism —simple diffusion—by which a solute passes from one compartment to another. But it is often desirable to study the transfer of a drug, or a metabolite, or a radioactive isotope from one compartment to another without being concerned about the precise mechanism of transfer. For example, in the system discussed in Section 7-8 at the end of Chapter 7, if we knew Q_{tot} and if we were to measure the concentration of S in serial samples withdrawn from W and Z during the approach to equilibrium, we would be able to estimate V_W, V_Z, and $R_{Z/W}$. We could also calculate an exponential rate constant which would give us a very useful measure of the rate of approach to equilibrium, but would tell us nothing at all about the characteristics of the membrane. Indeed, the same kind of exponential approach to equilibrium can be caused by many processes other than simple passive diffusion (see Section 8-3). Therefore, it is desirable to undertake a more general analysis of the kinetics of transfer between compartments without reference to any particular mechanism of transfer. This analysis will then be applicable to a wide variety of problems.

8-2. Diagrams and Symbols for the Description of Transfer between Compartments

Figure 8-1 illustrates the kind of diagram and the symbols which will be used in the subsequent discussion. Each separate compartment is designated by a different capital letter and is represented in the diagram by a rectangle.* Each compartment is characterized by its volume V_A,

---

* The use of small rectangles for small compartments, large rectangles for large compartments often makes it easier to visualize the system. An even more elaborate

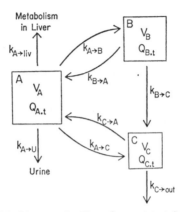

F𝘪ɢ. 8-1. Diagram of a Three-Compartment System

Each compartment is represented by a rectangle. Each pathway for transfer of solute is represented by a unidirectional arrow labeled with the appropriate rate constant. Once such a diagram has been drawn, the differential equations for the rate of change of the quantity of solute in each compartment can be written down by inspection (Section 8-6).

V_B, etc., and by the quantity of solute, $Q_{A.t}$, $Q_{B.t}$, etc., present in it at any instant of time. Each pathway by which the solute, S, moves out of or into a compartment is represented by an *arrow* pointing in the direction of movement. The quantity of S per unit of time which is moving out of a compartment by a particular pathway at any instant is assumed to be proportional to the quantity of S present in the compartment at that time (see Section 8-3). The *proportion* of S lost from the compartment along the pathway per unit of time can therefore be represented by an exponential rate constant, k, of exactly the sort discussed at length in Chapter 6. Each pathway has its own rate constant which is distinguished by a subscript denoting first the origin and then the destination of the pathway. For example, in Figure 8-1 the pathway which represents the renal excretion of some of the S in Compartment A is characterized by the rate constant $k_{A \to U}$; "$k_{A \to U}$" means "the proportion of the S in A which is being transferred *from A to* the urine per unit of time." Similarly, "$k_{B \to A}$" means "the proportion of the S in B which is being transferred *from B to A* per unit

---

pictorial device is to draw a cube in perspective for each compartment. The volume of the cube is made proportional to the volume of the compartment. Moreover, the thickness of an arrow representing a given pathway can be made proportional to the rate of transfer along that pathway. Examples of such diagrams may be found in Reference 88. For ordinary purposes, no such elaborations are needed.

of time"; "$k_{C \to out}$" means "the proportion of the S in C which is being transferred per unit of time *from C to* somewhere outside the system."

Now when the rate of transfer of S from Compartment A to Compartment B is proportional to the quantity of S present in A, the rate of transfer of S *by that particular pathway* will be given by the equation

$$(dQ/dt)_{A \to B} = k_{A \to B} Q_{A,t} \tag{8-1}$$

where

$(dQ/dt)_{A \to B}$ = the rate at which S is being transferred along the pathway from A to B at time t

$Q_{A,t}$ = the quantity of S in Compartment A at time t

8-3. Simplifying Assumptions

To simplify the mathematical description of transfer between compartments we shall make the following assumptions:

1. We shall assume that the size, *i.e.*, the volume, of each compartment remains constant. This means that any equation for the *quantity* of S in a given compartment as a function of time can be converted to an equation for the *concentration* of S in the compartment as a function of time by dividing both sides of the equation by the volume of the compartment.

2. We shall assume that each compartment is well stirred, so that any S entering the compartment is instantaneously distributed throughout the entire compartment. The importance of this assumption has already been discussed (Section 7-3).

3. We shall assume that the proportional rate of transfer along each pathway remains constant so that the resulting change in quantity or concentration is exponential. In other words, we assume that the rate constants, k, are indeed constant. (Occasionally we may want to deal with a pathway through which a *constant quantity* of S per unit time is passing, but we shall characterize such a pathway not by a rate constant k, but by a symbol such as $\dot{Q}_{A \to out}$ which would mean quantity of S transferred out of Compartment A per unit of time.)

When these assumptions are not true for a particular system, the equations to be derived cannot properly be applied to that system. But fortunately, the assumptions are sufficiently valid for many of the biological systems in which we shall be interested. In particular, Assumption 3 is not as restrictive as it might seem, for a remarkable number of processes do transfer solutes at a rate which is proportional to concentration. Exponential change is the rule, not the exception! We have already seen that simple diffusion through a membrane between two well-stirred compartments causes an exponential change in concentration. Similarly, the excretion of a solute solely by filtration through the glomeruli of the kidney will cause exponential disappearance if the glomerular filtration rate remains constant. But it is not such "passive" processes alone which account for exponential changes in concentration. An active process which removes S from a compartment may equally well produce exponential change *provided that the rate of removal is limited by the rate at which S is supplied to the process, and not by the capacity of the process itself.* For example, suppose that 1 out of every 5 molecules of S brought to the liver by a constant hepatic blood flow is destroyed enzymatically by the liver cells regardless of the concentration of S in the blood. Then the process is *supply-limited** and S will disappear exponentially. But if the enzyme system is so fully saturated with S that it is destroying S as fast as it can, then the quantity of S destroyed per unit of time will be constant regardless of the concentration of S in the blood and of the rate of blood flow. Such a process is *capacity-limited.** Many active processes responsible for the transport or removal of solutes become capacity-limited if they are supplied with substrate at a high enough rate. A familiar example is the glucosuria which occurs when the amount of glucose filtered per minute through the glomeruli of the kidney exceeds the reabsorptive capacity, *i.e.*, the "transfer maximum," of the renal tubular cells for glucose.

8-4. The Concept of Clearance

Because so many different processes may be responsible for removing S from a given compartment, it would be convenient to have some very general way of expressing the overall effectiveness of *any* such process. The *absolute rate of removal*, expressed as quantity of S removed per unit of time, would be suitable only when the process is capacity-limited, for otherwise it would be as dependent upon the concentration of S as upon the effectiveness of the process of removal. The *proportional rate of removal*, expressed by the rate constant, k, is also unsuitable as a general term because it is constant only so long as the rate of removal of S from the compartment remains proportional to the quantity of S in the compartment. Furthermore, the magnitude of the rate constant depends as much upon the volume of the compartment as it does upon the effectiveness of the process of removal. In contrast, the *clearance* depends only upon the

---

* Because of their broader meaning, the terms "supply-limited" and "capacity-limited" are used here in preference to the corresponding terms "substrate-limited" and "enzyme-limited." The capacity of some transport mechanisms may well be limited by how much of a nonenzymatic "carrier substance" is available for combination with S rather than by the amount of enzyme needed to form (or to split) the carrier-S complex.

overall effectiveness of removal, and can be used to characterize any process of removal whether it be constant or changing, capacity-limited or supply-limited.

The clearance of S from Compartment A by the pathway leading to B may be defined as the volume of A from which S is, in effect, completely removed, i.e., "cleared," per unit of time via that pathway. It may equally well be defined as the *rate of removal of S from A* via the pathway to B *per unit of concentration in A*:

$$(\dot{V}\text{cl})_{A \to B.t} = (dQ/dt)_{A \to B}/C_{A.t} \qquad (8\text{-}2)$$

where

$(\dot{V}\text{cl})_{A \to B.t} = $ the volume of W being cleared of S per unit of time by the pathway from A to B at time t

Equation 8-2 is a completely general definition of clearance, valid (with appropriate changes in the subscripts) for any substance cleared from any compartment along any pathway at any time. It is important to realize that Equation 8-2 is for clearance via any single *unidirectional pathway*. It will represent *net* removal of S from Compartment A only when there is no movement of S along any return pathway from B to A. But in practice, the term "clearance" is rarely used except for the irreversible removal of S from a compartment by unidirectional pathways of metabolism, storage, or excretion, and it is in this more restricted sense that the term will ordinarily be employed.

Notice that clearance has the dimensions of flow, i.e., of volume per unit of time. It is sometimes called a "virtual flow." Indeed as far as the removal of S from A is concerned, the effect of a clearance of 15 ml. of A per minute is exactly the same as the effect of washing S out of Compartment A by continuous dilution with a real flow of 15 ml. per minute through the compartment (Table 9-1). Now by definition

$$C_{A.t} = Q_{A.t}/V_A \qquad (8\text{-}3)$$

Combining Equations 8-2 and 8-3,

$$(\dot{V}\text{cl})_{A \to B.t} = \frac{(dQ/dt)_{A \to B}}{Q_{A.t}/V_A} \qquad (8\text{-}4)$$

Rearranging terms,

$$(dQ/dt)_{A \to B} = \frac{(\dot{V}\text{cl})_{A \to B.t}}{V_A} Q_{A.t} \qquad (8\text{-}4a)$$

By Assumption 1, Section 8-3, V_A is constant. Suppose that the clearance is also constant. Then the rate constant $k_{A \to B}$ in Equation 8-1 and the constant $(\dot{V}\text{cl})_{A \to B}/V_A$ in Equation 8-4a are identical:

$$k_{A \to B} = (\dot{V}\text{cl})_{A \to B}/V_A \qquad (8\text{-}5)$$

Thus when the volume of the compartment being cleared is constant (Assumption 1), the assumption that the proportional rate of transfer is constant (Assumption 3) is equivalent to assuming that the clearance is constant, the *rate constant being the ratio of the clearance to the compartment volume*. This makes sense. We have learned to think of a rate constant as a proportion per unit of time, and the ratio in Equation 8-5 is simply the proportion of the total volume of Compartment A which is cleared of S per unit of time via the pathway from A to B.

8-5. The Concept of Volume of Distribution

If clearance is a "virtual flow," the volume of distribution of a substance, S, is a "virtual volume." Suppose that the total quantity of S in the body, Q_{tot}, has had time to reach distribution equilibrium throughout all of the compartments which it can enter. Suppose that the concentration of S in some part of one of these compartments (usually blood plasma) can be measured. Let us call the equilibrium concentration in this reference fluid $C_{\text{ref.eq}}$. Then the volume of distribution, (Vdist), of S is simply

$$(V\text{dist}) = Q_{\text{tot}}/C_{\text{ref.eq}} \qquad (8\text{-}6)$$

Notice that the concept of volume of distribution is an *equilibrium* concept.

If the distribution of S is through a *single* compartment, say Compartment A, the term "volume of distribution in Compartment A" is synonymous with the term "volume of Compartment A" as defined in Section 7-3. This is still in keeping with the idea that volume of distribution can be defined only for distribution equilibrium, because the S in any single compartment is *always* supposed to be at distribution equilibrium throughout that compartment (Assumption 2, Section 8-3). Thus the term "initial volume of distribution" simply means the volume of the first compartment through which Q_{tot} is apparently distributed at time zero.

An example of a volume of distribution has already been given in Equation 7-54 which is really a specific example of Equation 8-6. The problem of how to measure the volume of distribution of S throughout a system of compartments will be discussed later (Section 8-11).

8-6. Differential Equations for Transfer between Compartments

Given a system of compartments and pathways such as the one depicted in Figure 8-1, nothing is easier than to write down a differential equation for the rate of change of the quantity of S in each compartment. Equation 8-1 shows that the rate at which S traverses any particular pathway is the product of the rate constant for that pathway and the quantity of S in the compartment from which the pathway is coming. Obviously the *total* change in any compartment is the algebraic sum of all of the individual increments and decrements caused by transfer of S along all pathways leading to and from the compartment. For example, in Compartment A of Figure 8-1 there are two pathways (from B and from C) *adding* S to A, and four pathways (to B, to C, to the liver, and to the kidneys) *subtracting* S from A. At any instant of time, t, the quantity of S in A is therefore changing at the rate given by the following equation:

$$dQ_A/dt = k_{B \to A}Q_{B.t} + k_{C \to A}Q_{C.t} - k_{A \to B}Q_{A.t} - k_{A \to C}Q_{A.t}$$
$$- k_{A \to \mathrm{liv}}Q_{A.t} - k_{A \to U}Q_{A.t} \tag{8-7}$$

Similarly,

$$dQ_B/dt = k_{A \to B}Q_{A.t} - k_{B \to A}Q_{B.t} - k_{B \to C}Q_{B.t} \tag{8-8}$$

$$dQ_C/dt = k_{B \to C}Q_{B.t} + k_{A \to C}Q_{A.t} - k_{C \to A}Q_{C.t} - k_{C \to \mathrm{out}}Q_{C.t} \tag{8-9}$$

Since all rate constants for pathways leading *away* from a given compartment are multiplied by the quantity of S in that compartment, it is convenient to have a single symbol to designate their sum. We will therefore define k_A to mean the sum of all the rate constants for pathways leading *away from* Compartment A. Accordingly, Equations 8-7, 8-8, and 8-9 can be rewritten:

$$dQ_A/dt = k_{B \to A}Q_{B.t} + k_{C \to A}Q_{C.t} - k_AQ_{A.t} \tag{8-10}$$

$$dQ_B/dt = k_{A \to B}Q_{A.t} - k_BQ_{B.t} \tag{8-11}$$

$$dQ_C/dt = k_{A \to C}Q_{A.t} + k_{B \to C}Q_{B.t} - k_CQ_{C.t} \tag{8-12}$$

Equations 8-10, 8-11, and 8-12 are a set of simultaneous differential equations which completely describe the behavior of S in the system at any instant of time. But as usual, the differential equations must be solved, *i.e.*, integrated, before we can use them to find the actual quantity of S in a given compartment at a specific time. Using an analog computer (Section 3-3) to obtain particular solutions of the set of differential equations not only saves much time and effort but may be the only practical way to deal with really complex systems of compartments. Obtaining a general analytical solution even for a two- or three-compartment system requires a knowledge of calculus beyond what is being assumed for this book. Nevertheless, we shall be able to solve one or two elementary problems completely, and we shall also be able to analyze certain more complex systems by using a general solution worked out by others (97). In the following sections, three examples are given in full. Others are suggested

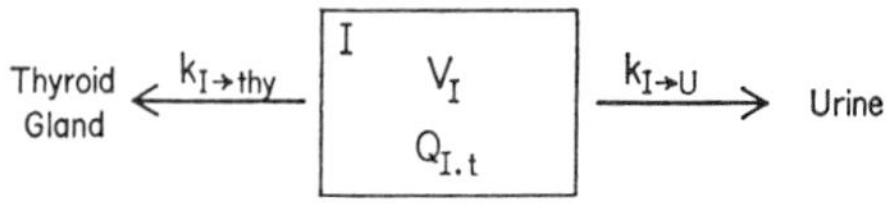

FIG. 8-2. Two Pathways of Loss from a Single Compartment

In the period immediately after the administration of a tracer dose of radioactive iodide, its behavior can often be described adequately by this very simple model (Section 8-7).

as exercises at the end of the chapter. In fact, the reader is strongly advised to work Exercise 1 himself *before* proceeding to the next few paragraphs.

8-7. Irreversible Loss via Several Pathways from a Single Compartment (Fig. 8-2)

Consider the removal of a single tracer dose of radioactive iodide, I^{131}, from the iodide compartment, I, by accumulation in the thyroid gland, thy, and by excretion in the urine, U. We will assume that during the period of interest all of the iodide which enters the thyroid gland is stored there as organic iodine and that none is returned to the iodide compartment or secreted as hormone. (This assumption is often, but by no means always, justified by the actual behavior of a tracer dose of I^{131}.)

Initial conditions: at $t = 0$, $Q_{I.0} = Q_{tot}$, $Q_{thy.0} = 0$, $Q_{U.0} = 0$

By inspection of Figure 8-2, the differential equation for Q_I is

$$dQ_I/dt = -k_{I \to thy}Q_{I.t} - k_{I \to U}Q_{I.t}$$
$$= -(k_{I \to thy} + k_{I \to U})Q_{I.t} = -k_I Q_{I.t} \qquad (8\text{-}13)$$

Integrating Equation 8-13,

$$Q_{I.t} = Q_{I.0}e^{-k_I t} = Q_{tot}e^{-k_I t} \qquad (8\text{-}14)$$

Equation 8-14 shows that even though the I^{131} is leaving Compartment I by two separate pathways, its rate of disappearance from I is controlled by a single exponential term for which the rate constant is the *sum* of the rate constants of the two efferent pathways. Two leaks "are equal to one leak of larger size" (101).

Equation 8-14 enables us to calculate how much of the tracer dose remains in the iodide compartment at any time. According to Figure 8-2, the iodide which has left the iodide compartment must either have accumulated in the thyroid gland or have been excreted in the urine. But how much goes to each? It is obvious from Equation 8-13 that at any particular *instant* of time, the fraction of the total change due to thyroid accumulation, $F_{thy/tot}$, is

$$F_{thy/tot} = k_{I \to thy}Q_{I.t}/k_I Q_{I.t} = k_{I \to thy}/k_I \qquad (8\text{-}15)$$

But since the k's are constant, this fraction is also constant and represents the fraction of $Q_{tot} - Q_{I.t}$ (*i.e.*, the quantity lost from I between time zero and time t) which has accumulated in the thyroid gland. Hence,

$$Q_{thy.t} = F_{thy/tot}(Q_{tot} - Q_{I.t}) = \frac{k_{I \to thy}}{k_I}(Q_{tot} - Q_{I.t}) \qquad (8\text{-}16)$$

Substituting in Equation 8-16 the value of $Q_{I.t}$ from Equation 8-14,

$$Q_{thy.t} = \frac{k_{I \to thy}}{k_I}Q_{tot}(1 - e^{-k_I t}) \qquad (8\text{-}17)$$

Similarly, if $Q_{U.t}$ represents the cumulative excretion of I^{131} in the urine between time zero and time t,

$$Q_{U.t} = \frac{k_{I \to U}}{k_I}Q_{tot}(1 - e^{-k_I t}) \qquad (8\text{-}18)$$

By combining Equations 8-15 and 8-17 we obtain

$$(Q_{thy.t} - F_{thy/tot}Q_{tot}) = (Q_{thy.0} - F_{thy/tot}Q_{tot})e^{-k_I t} \qquad (8\text{-}19)$$

where, according to the initial conditions, $Q_{thy.0} = 0$. Equation 8-19 may be rearranged in like manner. Equation 8-19 makes it clear that the quantity of I^{131} in the thyroid gland is rising toward the asymptote $F_{thy/tot}Q_{tot}$ at a rate determined by k_I, the *sum* of the rate constants for transfer of I^{131} to thyroid *and* to urine. Similarly the rate at which the cumulative excretion of I^{131} in the urine approaches its asymptote depends not just on the rate constant $k_{I \to U}$ for urinary excretion but upon the *sum* of the two rate constants.

This important conclusion may be generalized as follows: Suppose that an amount Q_{tot} of S is placed into Compartment A at time zero. Suppose that S is then *irreversibly* lost from A by several routes (to B, to C, to D, etc.), each characterized by its own rate constant. Then *the cumulative loss by any single route, say A to B, approaches an asymptote equal to* $(k_{A \to B}/k_A)Q_{tot}$ *with a half-time which is determined not by* $k_{A \to B}$ *alone but by* k_A, *the sum of all of the rate constants.* This means that the straight lines obtained by plotting $\ln Q_{A.t}$ against time, $\ln (Q_{B.asymp} - Q_{B.t})$ against time $\ln (Q_{C.asymp} - Q_{C.t})$ against time, etc., will all be parallel with a slope of $-k_A = -(k_{A \to B} + k_{A \to C} + \cdots)$ (see Exercise 2).

8-8. Equilibration by Exchange in a Closed Two-Compartment System (Fig. 8-3)

A specific example of how equilibrium was approached by diffusion between two compartments was considered in detail in Chapter 7. We shall now derive more general equations for equilibration between two compartments, equations which will not depend upon the assumption of

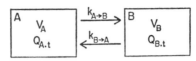

FIG. 8-3. A Closed Two-Compartment System

If a solute is introduced at time zero into either Compartment A or Compartment B, it approaches equilibrium throughout the entire system at the single exponential rate specified by the sum of the two rate constants (Section 8-8).

any particular mechanism of transfer. In the diffusion problem, we dealt only with *net* transfer. In the general case, it is easier to consider the transfer of solute in both directions.

Consider the approach to equilibrium when a solute, S, is added to Compartment A of the system depicted in Figure 8-3. By setting appropriate boundary conditions we will take into account the possibility that there may already be some S present in the two compartments at time zero.

Initial conditions: at $t = 0$, $Q_A = Q_{A.0}$, $Q_B = Q_{B.0}$

The differential equation for Compartment A is

$$dQ_A/dt = k_{B \to A}Q_{B.t} - k_{A \to B}Q_{A.t} \qquad (8\text{-}20)$$

The differential equation for Compartment B is similar but with opposite signs. Since the system is a closed system (*i.e.*, one with no outlets), conservation of S requires that

$$Q_{tot} = Q_{A.t} + Q_{B.t} \qquad (8\text{-}21)$$

Combining Equations 8-20 and 8-21 with the elimination of $Q_{B.t}$,

$$dQ_A/dt = k_{B \to A}Q_{tot} - (k_{A \to B} + k_{B \to A})Q_{A.t} \qquad (8\text{-}22)$$

or, separating the variables prior to integration,

$$dQ_A/[k_{B \to A}Q_{tot} - (k_{A \to B} + k_{B \to A})Q_{A.t}] = dt \qquad (8\text{-}22a)$$

Equation 8-22a may now be integrated in exactly the same way as Equation 7-44, $k_{B \to A}Q_{tot}$ corresponding to k_6 and $(k_{A \to B} + k_{B \to A})$ corresponding to k_5. Therefore, we need not repeat the intermediate steps but can simply copy the result given in Equation 7-48a:

$$\left(Q_{A.t} - \frac{k_{B \to A}}{k_{A \to B} + k_{B \to A}} Q_{tot} \right) = \left(Q_{A.0} - \frac{k_{B \to A}}{k_{A \to B} + k_{B \to A}} Q_{tot} \right)$$
$$\exp\left[-(k_{A \to B} + k_{B \to A})t\right] \qquad (8\text{-}23)$$

From Equation 8-23 it is evident that $Q_{A.t}$ is approaching its asymptote,

$k_{B \to A} Q_{tot} / (k_{A \to B} + k_{B \to A})$, at a rate determined by the sum of the two rate constants. We can easily prove that this asymptote is indeed $Q_{A.\infty}$, the equilibrium value which is being approached as time approaches infinity. For, by definition, at equilibrium as much S must move per unit of time from A to B as from B to A so that the net change in Q_A as given by Equation 8-22 will be zero. Then Equation 8-22 becomes

$$(k_{A \to B} + k_{B \to A}) Q_{A.\infty} = k_{B \to A} Q_{tot} \tag{8-24}$$

Solving Equation 8-24 for $Q_{A.\infty}$,

$$Q_{A.\infty} = k_{B \to A} Q_{tot} / (k_{A \to B} + k_{B \to A}) \tag{8-24a}$$

which is indeed the asymptote of $Q_{A.t}$ in Equation 8-23.

Since Equation 8-23 is a general equation for equilibration in a closed two-compartment system, it should include Equation 7-52 for equilibration by diffusion as a special case. Notice that in Equation 7-52 the asymptote is the product of Q_{tot} and a ratio of volumes, whereas in Equation 8-23 the asymptote is the product of Q_{tot} and a ratio of rate constants. By using Equation 8-5 it is not difficult to prove that these ratios are equal. It is also possible to show that the exponent in Equation 7-52 corresponds to the exponent in Equation 8-23, though the reasoning is a bit more subtle (see Exercise 8).

8-9. A General Solution of the Two-Compartment Problem

The two previous examples have been chosen for their simplicity. Because of such special constraints as unidirectional transport, or absence of any pathways leading out of the system, the solutions contained only a single exponential term. But where no such constraints are imposed, general solutions soon become uncomfortably complex as the number of compartments increases. For the sake of simplicity, let us assume that at time zero a known amount, Q_{tot}, of S is given as a single dose into Compartment A, and that there is initially no S in any of the other compartments. Suppose that we have a system of N compartments, each having pathways both to and from every other compartment as well as a pathway leading out of the system. Then there would be altogether N^2 different pathways, each with its own rate constant. Each compartment will also have a volume, so that such a system of N compartments might have as many as $N^2 + N$ arbitrary constants or parameters. If we assume that all of these parameters are known, it will still be necessary to solve a set of N simultaneous differential equations, one for each compartment. Even with a three-compartment system, the resulting integrated equations are too complex to discuss here, although an explicit solution for the three-compartment system can be obtained (97). However, the explicit equations

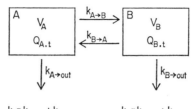

$$k_A = k_{A \to B} + k_{A \to out} \qquad k_B = k_{B \to A} + k_{B \to out}$$

Fig. 8-4. An Open Two-Compartment System

A general solution of the differential equations for this system leads to somewhat complicated expressions which contain two exponential terms (Section 8-9).

for the generalized two-compartment system of Figure 8-4 are not as formidable, and will be found useful for the solution of a number of different problems. These equations have been obtained by simplification of the general solution for the three-compartment system published by Skinner *et al.* (97). The equations, which will be given here without derivation, are for the fraction of the total dose in Compartment A, $F_{A.t}$, and the fraction of the total dose in Compartment B, $F_{B.t}$, at any time, t:

$$F_{A.t} = \left\{ \left(\frac{k_A - k_B + Z}{2Z} \right) \exp \left[-\frac{1}{2}(k_A + k_B + Z)t \right] \right\} \\ + \left\{ \left(1 - \frac{k_A - k_B + Z}{2Z} \right) \exp \left[-\frac{1}{2}(k_A + k_B - Z)t \right] \right\} \tag{8-25}$$

$$F_{B.t} = \frac{k_{A \to B}}{Z} \left\{ \exp \left[-\frac{1}{2}(k_A + k_B - Z)t \right] \\ - \exp \left[-\frac{1}{2}(k_A + k_B + Z)t \right] \right\} \tag{8-26}$$

where

$$Z = \sqrt{(k_A - k_B)^2 + 4k_{A \to B}k_{B \to A}} \tag{8-27}$$

Initial conditions: at $t = 0$, $\qquad F_{A.0} = 1$, $\qquad F_{B.0} = 0$

These equations can readily be converted into equations for the *concentration* of S in the two compartments at any time, t, by multiplying the fraction by the total dose divided by the volume of the compartment:

$$C_{A.t} = Q_{A.t}/V_A = Q_{tot}F_{A.t}/V_A \tag{8-28}$$

and

$$C_{B.t} = Q_{B.t}/V_B = Q_{tot}F_{B.t}/V_B \tag{8-29}$$

It is well worth noting that *both* of the exponential terms in Equation

8-25 are influenced by *all* of the rate constants, and hence by all of the individual processes of transfer illustrated in Figure 8-4. We are thus warned once more that when a multiple exponential equation is fitted empirically to biological data, in general we have no right to identify a particular term with a particular process occurring at the corresponding exponential rate.

8-10. Calculation of Rate Constants and Compartment Volumes from Experimental Observations

In the discussion thus far it has been assumed that the rate constants and the compartment volumes were known, so that the problem was to find a set of equations which would enable one to predict the distribution of S in the system as a function of time. But in practice, it is far more common to observe the concentration of S in one or more compartments as a function of time, and to try to deduce the parameters of the system from the observed behavior of S. For this purpose it is of the utmost value to have some *a priori* knowledge about how many compartments there are and what pathways interlink them. Otherwise, the calculated values are not likely to mean very much physiologically.

Example

There is substantial evidence that in man and in the dog, creatinine is distributed throughout two compartments, of which the first (which we shall call A) includes blood plasma, whereas the second (which we shall call B) is not clearly identified but presumably includes at least some "intracellular fluid." It is not possible to obtain samples for analysis from Compartment B. Exogenous creatinine leaves the body only via the urine. The behavior of creatinine may accordingly be represented by the model shown in Figure 8-5. The parameters in the following example have been taken from the discussion by Dominguez in *Medical Physics* (33).

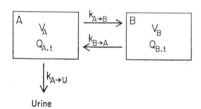

Urine

FIG. 8-5. A Model System for the Behavior of Creatinine in the Mammalian Body

The several parameters of this model system can be estimated from experimental observations as described in Section 8-10.

Ten grams of creatinine were administered intravenously at time zero to a human subject. Thereafter, the concentration of creatinine in plasma, corrected for endogenous creatinine, was measured at various times. The concentration was plotted against time on semilog graph paper, and the data were analyzed as described in Section 6-14. The points could be fitted by the following equation:

$$C_{A.t} = 0.38e^{-1.65t} + 0.18e^{-0.182t} \tag{8-30}$$

where

$C_{A.t}$ = the concentration of creatinine in plasma (regarded as a sample of Compartment A) in grams per liter

t = the time in hours

At time zero, Equation 8-30 gives

$$C_{A.0} = 0.38 + 0.18 = 0.56 \text{ gm./L.} \tag{8-31}$$

Since the dose administered, Q_{tot}, was 10 gm., the initial volume of distribution (taken as a measure of the volume of Compartment A) was

$$V_A = Q_{tot}/C_{A.0} = 10.0/0.56 = 17.9 \text{ L.} \tag{8-32}$$

Multiplying both sides of Equation 8-30 by $V_A/Q_{tot} = 1.79$ to convert it to $F_{A.t}$,

$$F_{A.t} = 0.68e^{-1.65t} + 0.32e^{-0.182t} \tag{8-33}$$

Equation 8-33 is a particular example of Equation 8-25. We can therefore equate the *numerical* values in Equation 8-33 to the corresponding *algebraic* values of Equation 8-25:

$$\tfrac{1}{2}(k_A + k_B + Z) = 1.65 \tag{8-34}$$

$$\tfrac{1}{2}(k_A + k_B - Z) = 0.182 \tag{8-35}$$

$$(k_A - k_B + Z)/2Z = 0.68 \tag{8-36}$$

Solving Equations 8-34 and 8-35 simultaneously for Z, we get $Z = 1.468$. Placing this value in Equation 8-36 and either 8-34 or 8-35 and solving the resulting equations simultaneously, we obtain $k_A = 1.180$ and $k_B = k_{B \to A} = 0.652$. Now by Equation 8-27

$$Z^2 = (k_A - k_B)^2 + 4k_{A \to B}k_{B \to A} \tag{8-27a}$$

Substituting our known values in Equation 8-27a, we find $k_{A \to B} = 0.719$, and, by subtracting this value from k_A, $k_{A \to U} = 0.461$. We have thus been able to calculate values for all of the rate constants as well as for the volume of Compartment A:

$$k_{A \to B} = 0.719 \text{ per hour}$$

$$k_{A \to U} = 0.461 \text{ per hour}$$

$$k_{B \to A} = 0.652 \text{ per hour}$$

$$V_A = 17.9 \text{ L.}$$

We can therefore write a specific equation for $F_{B.t}$ in the form of Equation 8-26:

$$F_{B.t} = 0.49(e^{-0.182t} - e^{-1.65t}) \tag{8-37}$$

We can also multiply Equation 8-37 by Q_{tot} to obtain an equation for the quantity of creatinine in B at any time. But to complete our characterization of the system what we really need is an estimate of V_B, the volume of Compartment B. If we could get even a single sample of B for analysis, we could measure $C_{B.t}$. Since we can calculate $Q_{B.t}$, we could then estimate V_B as $Q_{B.t}/C_{B.t}$. But B is not open to sampling, and we therefore seem to be stuck!

Actually, the problem is not as hopeless as it seems. It is true that we cannot calculate the volume of B with reference to the concentration of creatinine in some portion of B itself. But it is quite easy to calculate the volume of distribution of creatinine in B with reference to its concentration in A as measured in samples of plasma. By definition (Equation 8-6), this will be

$$(V\text{dist})_{B_A} = Q_{B.eq}/C_{A.eq} \tag{8-38}$$

where

$(V\text{dist})_{B_A}$ = the volume of distribution in B with reference to the concentration in A

Also by definition (Equation 7-50),

$$R_{B/A} = C_{B.eq}/C_{A.eq} \tag{8-39}$$

Combining Equations 8-38 and 8-39 with the elimination of $C_{A.eq}$,

$$(V\text{dist})_{B_A} = R_{B/A}Q_{B.eq}/C_{B.eq} \tag{8-40}$$

or, since $Q_{B.eq}/C_{B.eq} = V_B$,

$$(V\text{dist})_{B_A} = R_{B/A}V_B \tag{8-41}$$

Now at equilibrium, the rate of transfer from A to B must be exactly equal to the rate of transfer from B to A. Hence,

$$k_{A \to B}Q_{A.eq} = k_{B \to A}Q_{B.eq} \tag{8-42}$$

or, since $Q_{A.eq} = C_{A.eq}V_A$, and $Q_{B.eq} = C_{B.eq}V_B$,

$$k_{A \to B}C_{A.eq}V_A = k_{B \to A}C_{B.eq}V_B \qquad (8\text{-}43)$$

Combining Equations 8-39 and 8-43 with the elimination of $C_{B.eq}$, solving for $V_BR_{B/A}$, and including Equation 8-41 in the result,

$$(V\text{dist})_{B_A} = V_BR_{B/A} = (k_{A \to B}/k_{B \to A})V_A \qquad (8\text{-}44)$$

Since the three quantities on the right of Equation 8-44 have already been evaluated, $(V\text{dist})_{B_A}$ may be calculated as 19.7 L. The total volume of distribution of creatinine in the entire system, calculated with reference to its concentration in A, is simply the sum of V_A and $(V\text{dist})_{B_A}$ as given by Equation 8-44:

$$(V\text{dist}) = V_A + (V\text{dist})_{B_A} = V_A \left(\frac{k_{A \to B} + k_{B \to A}}{k_{B \to A}}\right) \qquad (8\text{-}45)$$

For this example, the total volume of distribution is 37.6 L.

The example just given shows that in a two-compartment system whose general arrangement is known, careful analysis of the changes in concentration in one of the two compartments coupled with knowledge of the boundary conditions allows one to estimate the rate constants and volumes which characterize the system. Skinner and his collaborators (97) discuss methods by which the parameters of even more complex three-compartment systems can be calculated. However, in interpreting the results of any such analysis one must bear clearly in mind the considerable uncertainties of fitting observed data by exponential equations (see Chap. 6). Whenever possible the calculations should be checked, supplemented, or even partially replaced by additional methods of studying the system. For instance, in the preceding example it would be highly desirable to collect samples of urine as well as of blood plasma and to calculate the renal plasma clearance of creatinine therefrom in the time-honored manner:

$$(\dot{V}\text{cl})_{A \to U} = \bar{Q}_{U.t_{1-2}}/\bar{C}_{A.t_{1-2}} \qquad (8\text{-}46)$$

where

$\bar{Q}_{U.t_{1-2}}$ = the mean rate of excretion of creatinine in the urine between time 1 and time 2

$\bar{C}_{A.t_{1-2}}$ = the mean concentration (properly, the logarithmic mean concentration) of creatinine in the plasma between time 1 and time 2

This value for plasma clearance could then be compared with the value indirectly calculated by the relationship given in Equation 8-5:

$$(\dot{V}\text{cl})_{A \to U} = k_{A \to U}V_A \qquad (8\text{-}47)$$

For the preceding example, $k_{A \to U} V_A = 8.25$ L. of plasma per hour, or 138 ml. of plasma per minute, a value which is in reasonably good agreement with values for the renal plasma clearance of creatinine obtained by direct measurement in man.

8-11. The Measurement of Volume of Distribution

The term "volume of distribution of S" has already been defined as the volume of solution which, if it had a uniform concentration equal to $C_{ref.eq}$, would contain the same total amount of S as is distributed about the entire system at equilibrium, *i.e.*, when there is no net transfer of S between compartments. $C_{ref.eq}$ is the equilibrium concentration in some well-defined portion of the system, usually blood plasma. We must now see how the volume of distribution of S can be measured.

In a closed system where S is neither metabolized, nor excreted, nor hidden away in some storage depot, the problem is simple. A known dose of S, Q_{tot}, is administered at time zero, and the concentration of S in plasma is studied as a function of time until enough data have been gathered to define the equilibrium concentration being approached asymptotically. The volume of distribution of S in the whole system, (Vdist), with reference to its equilibrium concentration in plasma, $C_{P.eq}$, is then given by the following variant of Equation 7-54:

$$(V\text{dist}) = Q_{tot.eq}/C_{P.eq} \qquad (8\text{-}48)$$

where

$Q_{tot.eq}$ = the quantity distributed throughout the entire system at equilibrium (in this instance, equal to Q_{tot})

Usually, however, S is lost more or less rapidly by irreversible pathways of metabolism, excretion, and storage, so that a steady state of equilibrium is not approached with the passage of time after a single dose. Nevertheless, it may be possible to approximate a steady state of equilibrium by infusing S intravenously at a constant rate for a long period of time. For this method to be valid, all of the irreversible loss of S must be from the same compartment as the one into which S is being infused. Consider the system in Figure 8-6. Compartments B, C, and D have no outlet except via Compartment A. During continuous infusion of S at a constant rate of $\dot{Q}_{in \to A}$, S will accumulate in the several compartments of the system until the rate of loss, $k_{A \to out} Q_A$, equals the rate of infusion. In theory, this would occur only at infinite time. But in practice, if interchange between A, B, C, and D is reasonably rapid, equilibrium will be approached sufficiently closely in a finite time, *e.g.*, a few hours. At equilibrium Equation 8-48 can be used if $Q_{tot.eq}$ can be estimated, for example, by measuring S in serial samples of urine collected after abruptly discontinuing the in-

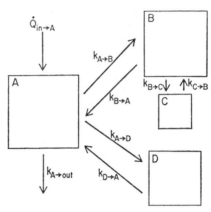

Fɪɢ. 8-6. A System of Compartments Which *Will* Approach Equilibrium with Each Other during Continuous Infusion of the Solute at a Constant Rate into Compartment *A*

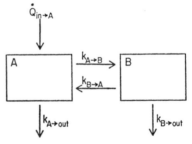

Fɪɢ. 8-7. A System of Compartments Which *Will Not* Approach Equilibrium with Each Other during Continuous Infusion of the Solute at a Constant Rate into Compartment *A*

fusion. This method can *not* be used for any system such as the one depicted in Figure 8-7 because part of the S infused into A is being lost from B. If S is continuously infused at a rate of $\dot{Q}_{in \to A}$, this system, like the previous one, will approach a steady state at which the rate of infusion is just equaled by the total rate of loss, $k_{A \to out}Q_A + k_{B \to out}Q_B$. At this steady state, it is obvious that $k_{A \to B}Q_A$ must *exceed* $k_{B \to A}Q_B$ by an amount equal to $k_{B \to out}Q_B$, whereas distribution equilibrium between A and B can be present only when $k_{A \to B}Q_A$ equals $k_{B \to A} Q_B$. Hence, even though the system approaches a *steady state*, it does not approach a state of distribution *equilibrium*, and Equation 8-48 cannot properly be applied.

A third method of estimating the volume of distribution is to add up the volumes calculated individually for the different compartments. A detailed example for a two-compartment system was discussed above (Equation 8-45).

8-12. Biased Methods of Estimating Volume of Distribution

There are still other methods, which must be employed with great caution because they are biased in the direction of overestimating the volume of distribution. They may therefore give entirely erroneous results if used without proper appreciation of their limitations. Consider again the system in Figure 8-5, in which the only route of loss is from Compartment A to the urine, U. Suppose first that $k_{A \to U}$ is zero. Then the system would be simply a closed two-compartment system approaching both diffusion equilibrium and a steady state with the passage of time. Now suppose that $k_{A \to U}$ is greater than zero, but very much less than $k_{A \to B}$ and $k_{B \to A}$ which control the rate of distribution between A and B. Then the approach of A and B to equilibrium with each other will be scarcely influenced by the slow rate of loss of S in the urine. Therefore, after enough time for distribution has elapsed, $i.e.$, after the first exponential term in Equation 8-25 has become negligibly small, A and B will behave practically like a single compartment, which we shall call $(A + B)$, with a volume equal to $V_A + (V\text{dist})_{B_A}$ and a concentration equal to $C_{A.t}$. We can derive an equation for the concentration in this quasi-singular compartment by dropping the now negligibly small first term from Equation 8-25, and rewriting the equation for concentration as suggested by Equation 8-28:

$$C_{(A+B).t} = (Q_{\text{tot}}/V_A) \left(\frac{Z + k_B - k_A}{2Z} \right) \exp\left[-\frac{1}{2}(k_A + k_B - Z)t\right] \quad (8\text{-}49)$$

By Equation 8-49 the concentration assumed to be present in the "$A + B$" compartment at time zero will be

$$C_{(A+B).0} = (Q_{\text{tot}}/V_A) \left(\frac{Z + k_B - k_A}{2Z} \right) \quad (8\text{-}50)$$

This concentration can be estimated by plotting the observed concentration in Compartment A against time on semilogarithmic graph paper, and extrapolating the straight line for the second exponential back to its intercept at time zero. Then by the relation given in Equation 8-32,

$$V_{(A+B)} = (V\text{dist})_{\text{intercept}} = Q_{\text{tot}}/C_{(A+B).0} \quad (8\text{-}51)$$

where

$(V\text{dist})_{\text{intercept}}$ = the volume of distribution estimated from the *intercept* at time zero

Since both Q_{tot} and $C_{(A+B).0}$ are known, $(V\text{dist})_{\text{intercept}}$ can be calculated.

It is important for us to know how much this estimate differs from the true volume of distribution $(V\text{dist})$ which, by Equation 8-45, is $V_A(k_{A \to B} + k_{B \to A})/k_{B \to A}$. Combining Equations 8-50 and 8-51 with the elimination of $C_{(A+B).0}$,

$$(V\text{dist})_{\text{intercept}} = 2ZV_A/(Z + k_B - k_A) \qquad (8\text{-}52)$$

The ratio of $(V\text{dist})_{\text{intercept}}$ to $(V\text{dist})$ is

$$\frac{(V\text{dist})_{\text{intercept}}}{V\text{dist}} = \frac{2Zk_{B \to A}}{(Z + k_B - k_A)(k_{A \to B} + k_{B \to A})} \qquad (8\text{-}53)$$

If $k_{A \to U}$ is so very small that we can assume $k_A = k_{A \to B}$, this ratio reduces to unity, i.e., $(V\text{dist})_{\text{intercept}}$ equals the true $(V\text{dist})$. But when $k_{A \to U}$ amounts to an appreciable fraction of $k_{A \to B}$, the ratio exceeds unity. $(V\text{dist})_{\text{intercept}}$ then *overestimates the true volume of distribution by an amount which is impossible to calculate unless we already know all of the rate constants* characterizing the system! But if we know the rate constants we should use Equation 8-45. In Figure 8-8, a particular set of physiologically reasonable values has been used to illustrate how preposterously large $(V\text{dist})_{\text{intercept}}$ can be when the renal plasma clearance is even moderately rapid.

A second method, which also overestimates $(V\text{dist})$ when $k_{A \to U}$ is appreciably large, depends upon the relationship expressed by Equation 8-5; namely, that the rate constant for a given pathway is equal to the clearance by that pathway divided by the volume of the compartment being cleared. At present we are assuming that $(A + B)$ behaves as a single compartment being cleared by a single pathway. Hence,

$$k_{(A+B) \to U} = (\dot{V}\text{cl})/(V\text{dist})_{\text{slope}} \qquad (8\text{-}54)$$

or,

$$(V\text{dist})_{\text{slope}} = (\dot{V}\text{cl})/k_{(A+B) \to U}$$

where

$(V\text{dist})_{\text{slope}}$ = the volume of distribution calculated from the rate constant for loss of S in urine and from the renal clearance of S

$k_{(A+B) \to U}$ = the rate constant for the disappearance of S from $(A + B)$ estimated from the *slope* of the straight line for the second exponential obtained when $C_{A.t}$ is plotted on semilog graph paper against time

Here, too, $(V\text{dist})_{\text{slope}}$ is most nearly correct when $k_{A \to U}$ is very small. Unfortunately, when $k_{A \to U}$ is negligibly small, so also is $k_{(A+B) \to U}$! Hence,

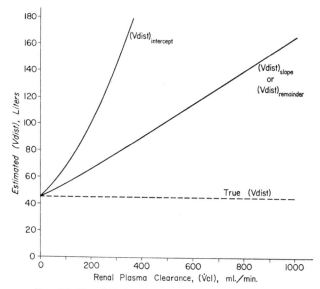

Fig. 8-8. Biased Estimates of the Volume of Distribution

The three biased estimates described in the text have been calculated for various rates of renal clearance with the following parameters for the model system of Figure 8-5: $V_A = 15$ L., $V_B = 30$ L., $k_{A \rightarrow B} = 0.9$ per hour, $k_{B \rightarrow A} = 0.45$ per hour. Except when the renal plasma clearance is very small (so that the two compartments behave practically as a single compartment after distribution between them is complete) all three methods grossly overestimate the true volume of distribution, estimation from the intercept (Equation 8-51) being particularly misleading (see also Figs. 8-9 and 8-10).

even when it is possible to measure a small renal plasma clearance accurately, it will probably not be possible to measure $k_{(A+B) \rightarrow U}$ with any precision. The method therefore suffers from the curious defect of being least accurate when least biased. This is too bad because $(V\text{dist})_{\text{slope}}$ is a less biased estimate of the true volume of distribution than is $(V\text{dist})_{\text{intercept}}$ (Fig. 8-8).

A third method of approximation yields results which are theoretically (i.e., aside from errors of measurement) identical with the results given by the previous method. Suppose that at some time, t, the first exponential of Equation 8-25 has become negligibly small. Then

$$(V\text{dist})_{\text{remainder}} = (Q_{\text{tot}} - Q_{\dot{v} \cdot t})/C_{A \cdot t} \qquad (8\text{-}55)$$

where

$(V\text{dist})_{\text{remainder}}$ = the volume of distribution calculated from the quantity of S *remaining* in Compartments A and B at time t

$Q_{U.t}$ = the quantity of S which has been excreted between time zero and time t

Since this method does not depend upon estimating a very small slope, it is likely to give much more accurate results than the previous method, and should be, in practice, the least objectionable of the three approximate methods described above. But none of these methods has much to recommend it, and, in the author's opinion, there is little justification for regarding the volumes so calculated as equivalent to true volumes of distribution unless the final rate of decrease of the concentration of S in plasma is very small. The fallacy of treating two separate compartments as a single compartment when loss of S is rapid is further illustrated by Figures 8-9 and 8-10 which are worth careful study.

8-13. After Intravenous Injection, Why Is Not the Initial Volume of Distribution Always Equal to the Plasma Volume?

When a single dose of S is injected rapidly into a vein, surely the first "compartment" which it enters is the blood plasma. If samples of blood are taken early enough, and frequently enough, and the concentration of S in plasma is plotted against time, should not the intercept of the curve at time zero always indicate plasma volume? The answer to this perfectly logical question is simply that for most small molecules the blood plasma does not behave like a separate compartment. Mixing in plasma is not instantaneous. Indeed, when a dye which binds almost completely to plasma protein is injected intravenously it appears as a "hump" of concentration in the arterial blood, thus providing one method of measuring cardiac output (Section 9-5). There are often smaller subsequent "humps" during the first few recirculations of blood before mixing is complete. Moreover, with small molecules, filtration and diffusion out of capillary beds into the surrounding tissue fluids is extremely rapid (83), so that by the time concentration in plasma has become reasonably uniform, the solute has already penetrated into a much larger volume. It is this larger volume, including blood plasma, which constitutes the apparent initial volume of distribution for many small solutes such as creatinine. Only with large molecules, or substances firmly bound to large molecules, can one identify a separate plasma compartment by analyzing the early part of the curve of plasma concentration *versus* time. But since both molecular size and extent of binding to plasma protein vary from substance to substance over very wide ranges, we must expect to find some intermediate

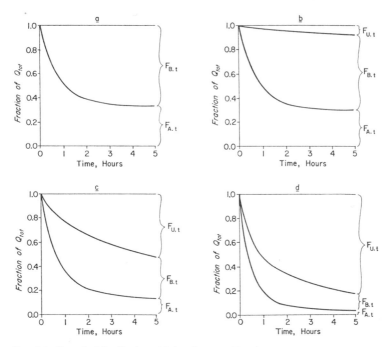

Fig. 8-9. How the Distribution of Solute between Two Compartments Is Influenced by Rate of Clearance

At time zero, Q_{tot} of solute was injected as a single dose into Compartment A of the system of Figure 8-5 with the same parameters as for Figure 8-8. The fractions of this total quantity remaining in Compartment A, $F_{A.t}$, remaining in Compartment B, $F_{B.t}$, or already excreted in the urine, $F_{U.t}$, are plotted as functions of time for clearances of 0 ml. per minute (Graph a), 10 ml. per minute (Graph b), 100 ml. per minute (Graph c) and 300 ml. per minute (Graph d).

With zero clearance (Graph a), the system is, in fact, a closed two-compartment system approaching a steady state of equilibrium with $F_{A.t} = \frac{1}{3}$ and $F_{B.t} = \frac{2}{3}$ at a proportional rate of $k_{A \to B} + k_{B \to A} = 1.35$ per hour, corresponding to a half-time of 0.513 hr. Equilibrium has been achieved, for all practical purposes, by the end of 4 or 5 hr. With a clearance of only 10 ml. per minute, Compartments A and B still come close to equilibrium with each other (Graph b), but with a clearance of 100 ml. per minute (Graph c) or 300 ml per minute (Graph d) a steady state of diffusion equilibrium between A and B is never approached. For example, when the rate of clearance is 300 ml. per minute, less than $\frac{1}{5}$ of the total solute remaining in the body at 5 hr. is present in Compartment A instead of the $\frac{1}{3}$ which would be present at distribution equilibrium (see also Fig. 8-10).

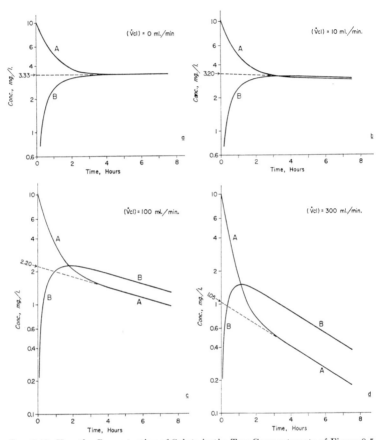

Fig. 8-10. How the Concentration of Solute in the Two Compartments of Figure 8-5 Is Influenced by Rate of Clearance

The parameters of the system are the ones listed in the legend for Figure 8-8 and also assumed for Figure 8-9. A single dose of 150 mg. of solute was placed in Compartment A at time zero. The four rates of clearance are the same as for Figure 8-9. With zero clearance (*top left*) the concentration in A falls, and the concentration in B rises, so that both approach the same equilibrium concentration, 3.33 mg. per liter, with a half-time of 0.513 hr. The true volume of distribution is thus equal to the total dose, 150 mg., divided by the equilibrium concentration, 3.33 mg. per liter, *i.e.*, 45 L. But when solute is being cleared from Compartment A at an appreciable rate, a steady state of equilibrium between A and B is not approached with the passage of time after a single dose. Notice that the concentration in B keeps increasing as long as there is a concentration gradient for diffusion of solute from A to B, *i.e.*, as long as the concentration in A exceeds the concentration in B. Notice also that at some instant of time A and B are momentarily in equilibrium with each other compounds whose apparent initial volume of distribution corresponds neither to plasma volume, nor to "extracellular fluid volume," nor to any other recognizable entity. Indeed the greatest circumspection must be exercised in attempting to identify the volume of distribution of any substance with a particular body fluid.

An extensive table of volumes of distribution is given by Dominguez (33).

EXERCISES. CHAPTER 8

Exercise 1

In each of the following problems, write a differential equation for the rate of change of the quantity of X in Compartment A, dQ_A/dt. Then solve, i.e., integrate, this equation so as to obtain an explicit expression for $Q_{A.t}$, the quantity of X in A at any time t. You will find it helpful to draw a diagram of compartments and pathways for each problem. You should also check each equation for $Q_{A.t}$ by seeing whether it makes sense when $t = 0$ and as $t \to \infty$.

A. Loss of X at a constant absolute rate of $\dot{Q}_{A \to \text{out}}$ mg. per minute from a single compartment, A.

B. Loss of X at a constant proportional rate of $k_{A \to \text{out}}$ per minute from a single compartment, A.

C. Continuous infusion of X into A at a constant rate $\dot{Q}_{\text{in} \to A}$ starting at time zero (no X present in A at time zero).

D. Same as "C" above but with the addition of a pathway for loss of X from A into the urine at a constant proportional rate, $k_{A \to U}$.

E. Same as "D" above, but with the addition of another route of loss at a constant proportional rate, $k_{A \to \text{liv}}$, representing metabolic transformation of X in the liver.

F. Same as "D" above, but with the addition of another route of loss at a constant absolute rate, $\dot{Q}_{A \to \text{out}}$.

(equal concentrations) so that there is no net diffusion between them. It is precisely at this point that the concentration in B achieves its maximum. (See the discussion of precursor-product relationships in Section 9-8.) As clearance of solute from Compartment A continues, the diffusion gradient is reversed, so that the concentration in B, though now decreasing, becomes, and remains, higher than in A. Instead of approaching a steady state of equilibrium, A and B approach a state in which their concentrations decline at the same exponential rate (*parallel straight lines* on the semilog plot of this figure). But as the clearance increases, the intercept of the straight line for Compartment A on the time-zero axis becomes more and more misleading as an index of the volume of distribution. For example, when the clearance is 300 ml. per minute (*lower right*) the "volume of distribution" calculated from the intercept for Compartment A at time zero is 150/1.05 = 143 L. whereas the true value is only 45 L. (Fig. 8-8).

G. At time zero, Q_{tot} of X is present in Compartment B, and no X is in Compartment A. X is transferred *irreversibly* from B to A at a constant proportional rate, $k_{B \to A}$.

Exercise 2

In Section 8-7, it was assumed that the initial behavior of a tracer dose of radioactive iodide (I^{131}) given to a normal subject could be analyzed according to the model shown in Figure 8-2. Accepting this assumption, estimate from the data tabulated below:

A. The half-time for disappearance of iodide from the plasma.
B. The proportion of the tracer dose ultimately accumulated by the thyroid gland.
C. The plasma clearance of iodide by the thyroid gland if the renal plasma clearance was 34 ml. per minute.

(Time zero is the time of administration of the tracer dose of I^{131}.)

Urine Sample Collected between			Per Cent of Tracer Dose in Sample	Cumulative Per Cent
0.0	and	2.13 hr.	17.9	17.9
2.13		4.22	13.5	31.4
4.22		6.25	11.1	42.5
6.25		8.13	7.7	50.2
8.13		24.00	31.6	81.8
24.00		48.00	8.3	90.1

Exercise 3

Consider a closed system consisting of the two compartments, A and B. At time zero, Q_{tot} mg. of S was instantaneously dissolved in Compartment A. The quantity of S in A, $Q_{A.t}$, decreased with time as indicated below:

Time (min.)	10	20	30	40	50	70
$Q_{A.t}$ (mg.)	59	43	35	29	25	22

From these data, estimate

Q_{tot} , $Q_{A.\infty}$, $Q_{B.\infty}$, $k_{A \to B}$, $k_{B \to A}$,   and

the half-time for equilibration.

Exercise 4. Accumulation of a drug given repeatedly

Suppose that a single intravenous dose, Q_D , of a drug is instantaneously distributed throughout a single compartment, producing an immediate peak plasma concentration of C_0. After t hours, this concentration has declined exponentially to C_t. Let F be the fraction remaining at time t, so that (assuming the volume of the compartment remains constant) $F = C_t/C_0$. What will the *maximum* peak plasma concentration just after a single dose ultimately be if the same single dose is administered every t hours for a very long (theoretically, infinite) time? Express the maximum peak concentration, C_{max}, in terms of C_0 and F.

Exercise 5

A substance not metabolized in the body is infused at a constant rate of 85 mg. per minute. The amount of the substance excreted in the urine per minute at various times after starting the continuous infusion is given by the following tabulation:

Time of Infusion	Urinary Excretion
hr.	*mg./min.*
0	0
0.5	25
1.0	40
1.5	41
2.0	46
2.5	52.7
3.0	56.8
3.5	56.5
4.0	58.5

At 4.0 hr. the blood plasma contained 83.6 mg. per cent of the substance.
Estimate: the biological half-life of the substance.
the volume of distribution of the substance.
the renal plasma clearance of the substance.
the extrarenal plasma clearance of the substance.

Exercise 6

Dominguez *et al.* (34) studied the fate of exogenous creatinine in the dog. The following data are taken from an experiment in which 6.66 gm. of creatinine were injected intravenously into a 20.2-kg. dog at time zero, and the concentration of creatinine in plasma was subsequently determined. Evidence was obtained that practically all of the injected creatinine was excreted unchanged by the kidneys. The data have been corrected for the small amount of endogenous creatinine present, so that the values given are for the exogenous creatinine only.

Time of Blood Collection (t)	Plasma Creatinine ($C_{A.t}$)
min.	*mg./100 ml.*
12	81.2
$17\frac{1}{2}$	70.2
34	42.2
47	41.1
$61\frac{1}{2}$	31.6
$92\frac{1}{2}$	24.2
$122\frac{1}{2}$	20.5
$182\frac{1}{2}$	16.9
242	12.5
302	8.6
$363\frac{1}{2}$	7.7
$422\frac{1}{2}$	6.6
$484\frac{1}{2}$	5.0

Assume that these data are in accord with the two-compartment model depicted in Figure 8-5, and that they may therefore be fitted by a double exponential of the general form:

$$C_{A.t} = G_1 e^{-k_1 t} + G_2 e^{-k_2 t}$$

where

$C_{A.t}$ = the concentration of creatinine in plasma at time t

G_1, G_2, k_1 and k_2 are parameters to be estimated from the data.

A. Estimate the parameters of the double exponential equation.

B. Estimate $k_{A \to B}$, $k_{A \to U}$, $k_{B \to A}$, V_A, $(V\text{dist})_{B_A}$, $(V\text{dist})$, and $(\dot{V}\text{cl})$.

Exercise 7

Prove that when both $k_{A \to \text{out}}$ and $k_{B \to \text{out}}$ are zero, so that the two-compartment system of Figure 8-4 is closed, Equation 8-25 (multiplied by Q_{tot} to convert it to an equation for the quantity of solute in A at time t) reduces to Equation 8-23.

Exercise 8

Prove that Equation 7-52 for equilibration by diffusion in a closed two-compartment system is a special case of Equation 8-23, the general equation for equilibration in a closed two-compartment system.

9

FURTHER KINETIC PROBLEMS. FLUID FLOW, METABOLIC TRANSFORMATIONS

9-1. Transport by Fluid Flow

In a number of important problems a solute is carried into or out of a compartment by an actual flow of fluid through the compartment. Provided the underlying assumptions remain valid, the equations previously developed for the kinetics of distribution can be applied equally well to problems of actual flow merely by substituting flow, $\dot{V}$, for clearance, ($\dot{V}cl$). For example, Equation 8-5 which gives the fundamental relationship between rate constant, compartment volume, and clearance becomes

$$k_{W \to X} = \dot{V}_{W \to X} / V_W \qquad (9\text{-}1)$$

where

$\dot{V}_{W \to X}$ = the flow (*i.e.*, volume of fluid per unit time) from W to X

Briefly, the underlying assumptions are that the compartment volumes remain constant, that mixing within each compartment is instantaneous, and that the clearance, now the actual fluid flow, by each pathway remains constant. In addition, we have usually assumed hitherto that a single dose, Q_{tot}, of a solute, S, was introduced at time zero into Compartment W. We shall now see that by restricting consideration to a single compartment we can deal with certain problems in which the concentration of S in the inflow is variable and in which mixing is not necessarily instantaneous. However, we shall retain the assumptions that the compartment volumes and the flows are constant.

In the majority of physiological problems, the fluid which is flowing is blood. Very often the objective is to estimate blood flow from observations of how the concentration of some solute in the blood going to and coming from a particular region varies with time as the region adds solute to or removes solute from the bloodstream. The solution of such problems de-

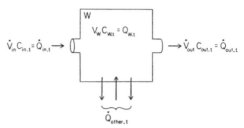

F<small>IG</small>. 9-1. Model of a Compartment Being Washed through by a Stream of Fluid

Conservation of matter (the Fick principle) requires that during any interval of time the quantity of solute entering the compartment be equal to the quantity of solute leaving the compartment by all routes, plus the quantity accumulated in the compartment.

pends upon the application of the Fick principle (not to be confused with Fick's law of diffusion) which we shall now formulate.

9-2. The Fick Principle

The Fick principle is really no more than a statement of the conservation of matter: *During any interval of time, the quantity of S entering a given compartment in the inflowing blood must be equal to the quantity of S being removed from the blood by the compartment plus the quantity of S leaving in the outflowing blood.* This statement must be true whether the compartment is simply accumulating S from the bloodstream, or destroying S by metabolism, or excreting S, or putting S into another fluid such as lymph, or doing more than one of these things at the same time. Furthermore, with proper regard to signs the statement still holds true when the compartment is *adding* S to the bloodstream, for addition of S is equivalent to *negative* removal. We will begin by deriving a fairly general equation for the Fick principle.

Consider the system illustrated in Figure 9-1. W is a region supplied by an inflow of blood, $\dot{V}_{in}$, containing S at a concentration of $C_{in.t}$. The outflow of blood is $\dot{V}_{out}$ with a concentration $C_{out.t}$. As indicated in the figure, S may be entering or leaving W by other routes also, at a constant absolute rate, or a constant proportional rate, or both. We shall designate all such transfers by $\dot{Q}_{other.t}$, taking it as positive for transfer of S out of W. Furthermore, the quantity of S in W, $Q_{w.t}$, may be increasing, remaining constant, or decreasing. However, for simplicity we shall assume that the only actual flow is of blood, so that inflow and outflow are equal.* We may therefore replace both $\dot{V}_{in}$ and $\dot{V}_{out}$ by $\dot{V}$.

---

* When this is *not* true (*e.g.*, in the kidney), we must make sure that we conserve *volume of fluid* as well as *quantity of solute.*

Now let us see what will happen in this system during an infinitesimal interval of time, dt. The quantity of S *entering* W in the inflowing blood is equal to the infinitesimal volume of blood, $\dot{V}dt$, entering W multiplied by the concentration of S in the inflowing blood:

$$dQ_{in} = \dot{V}dtC_{in.t} \qquad (9\text{-}2)$$

Similarly, the quantity of S *leaving* W in the outflowing blood is

$$dQ_{out} = \dot{V}dtC_{out.t} \qquad (9\text{-}3)$$

The quantity of S *removed from the blood* by W will equal the infinitesimal increase in the quantity in W, dQ_W, or, (V_W being constant) $V_W dC_W$, plus the quantity of S lost from W by all other routes, $\dot{Q}_{other.}dt$:

$$dQ_{removed} = V_W dC_W + \dot{Q}_{other.}dt \qquad (9\text{-}4)$$

Now the Fick principle states that during any interval of time, dt,

$$dQ_{in} = dQ_{removed} + dQ_{out} \qquad (9\text{-}5)$$

Substituting in Equation 9-5 the values given by Equations 9-2, 9-3, and 9-4,

$$\dot{V}dtC_{in.t} = V_W dC_W + \dot{Q}_{other.}dt + \dot{V}dtC_{out.t} \qquad (9\text{-}6)$$

which may be solved, if you wish, for $dQ_W/dt = V_W(dC_W/d\,')$:

$$dQ_W/dt = V_W(dC_W/dt) = \dot{V}(C_{in} - C_{out})_t - \dot{Q}_{other.t} \qquad (9\text{-}6a)$$

Equation 9-6 is the desired general equation for the Fick principle. We shall now apply it to a number of specific problems.

9-3. Calculation of Cardiac Output from Oxygen Consumption in the Steady State. The "Direct Fick" Method

Let W be the entire body, and let S be oxygen. In a steady state of metabolism, the rate of oxygen consumption by the body, $\dot{Q}_{O_2.tot}$, is constant. If cardiac output, $\dot{V}_{tot}$, is constant, so also must be the arteriovenous oxygen difference, $C_{O_2.art} - C_{O_2.ven.mix}$, and the quantity of oxygen present in the tissues. Hence $dQ_{O_2.W}/dt$ is zero, and Equation 9-6a becomes

$$\dot{V}_{tot}(C_{O_2.art} - C_{O_2.ven.mix}) - \dot{Q}_{O_2.tot} = 0 \qquad (9\text{-}7)$$

where

$C_{O_2.ven.mix}$ = the concentration of oxygen in the mixed venous blood from the right heart

Note that the specific term "$\dot{Q}_{O_2.tot}$" replaces the general term "$\dot{Q}_{other.t}$"

in Equation 9-6a. Solving Equation 9-7 for $\dot{V}_{tot}$,

$$\dot{V}_{tot} = \dot{Q}_{O_2.tot}/(C_{O_2.art} - C_{O_2.ven.mix}) \quad (9\text{-}7a)$$

Since all of the quantities on the right of Equation 9-7a can be measured experimentally, the cardiac output, $\dot{V}_{tot}$, can be calculated.

9-4. Accumulation of S in a Tissue from a Constant Concentration in Arterial Blood

Consider a uniform tissue, T, perfused by a constant blood flow. S is a solute whose concentration in arterial blood remains constant. S is not metabolized in, nor excreted by T, but enters T in accordance with whatever diffusion gradients exist between blood and tissue. Now it might seem that the blood flowing through the tissue, and the tissue itself ought to be treated as two separate compartments. But fortunately, in most tissues diffusion between blood and tissue is so rapid that, at least for certain solutes, we can assume that distribution equilibrium has been achieved by the time the blood leaves the capillary bed. Hence, the entire volume of tissue-plus-blood can be treated as a single compartment, T, which is being washed through by the blood flow. When this is true,

$$C_{ven.t} = C_{T.t}/R_{T/B} \quad (9\text{-}8)$$

where

$C_{T.t}$ = the concentration of S in tissue T at time t

$R_{T/B}$ = the equilibrium distribution ratio for S between tissue T and blood

Combining Equations 9-6 and 9-8 with appropriate revision of the subscripts and solving for dC_T/dt,

$$dC_T/dt = (\dot{V}/V_T)C_{art} - (\dot{V}/V_T R_{T/B})C_{T.t} \quad (9\text{-}9)$$

Notice that according to the present assumptions, $\dot{Q}_{other.t}$ is zero and does not appear in the equation. The only variables in Equation 9-9 are t and C_T; it is easy to separate them and integrate in the manner previously shown in detail for Equation 7-43:

$$(C_{art}R_{T/B} - C_{T.t}) = (C_{art}R_{T/B} - C_{T.0}) \exp[-(\dot{V}/V_T R_{T/B})t] \quad (9\text{-}10)$$

or, if $C_{T.0}$ is zero,

$$C_{T.t} = C_{art}R_{T/B}\{1 - \exp[-(\dot{V}/V_T R_{T/B})t]\} \quad (9\text{-}11)$$

Equation 9-11 will be used in Section 13-4 as the starting point for an analysis of the accumulation of inert gases in various tissues.

9-5. Calculation of Cardiac Output by the Dye-Dilution Technique

Suppose that Compartment W consisted of all of the blood lying between a peripheral vein and some part of the arterial tree from which blood samples can be removed. W would thus include the blood in the heart and lungs and in certain portions of the larger blood vessels. The blood flow through this compartment would be the cardiac output, $\dot{V}_{tot}$. Let S be a dye which becomes so firmly bound to plasma protein that it leaves the blood stream very slowly. At time zero, a known amount of the dye, Q_{tot}, is suddenly injected into the vein, and its concentration in the arterial blood coming out of the compartment is recorded thereafter as a continuous function of time. *If* distribution throughout W were actually instantaneous, and *if* there were no recirculation of the dye, its concentration in W and in the arterial blood would at all times be equal:

$$C_{W.t} = C_{art.t} \qquad (9\text{-}12)$$

Under these circumstances, both $C_{in.t}$ and $\dot{Q}_{other.t}$ in Equation 9-6 would be zero. Combining Equations 9-6 and 9-12 with appropriate revision of the subscripts, and solving for dC_W/dt,

$$dC_W/dt = dC_{art}/dt = -(\dot{V}_{tot}/V_W)C_{W.t} \qquad (9\text{-}13)$$

On integration, Equation 9-13 takes the familiar form:

$$C_{W.t} = C_{art.t} = C_{W.0} \exp[-(\dot{V}_{tot}/V_W)t] \qquad (9\text{-}14)$$

or, since $C_{W.0} = Q_{tot}/V_W$,

$$C_{W.t} = C_{art.t} = (Q_{tot}/V_W) \exp[-(\dot{V}_{tot}/V_W)t] \qquad (9\text{-}15)$$

From this equation, it is obvious that if we plotted $C_{art.t}$ against time in the usual manner on semilogarithmic paper, we would get a straight line from whose slope we could calculate the rate constant, $(\dot{V}_{tot}/V_W)$, and from whose intercept we could calculate (Q_{tot}/V_W). Knowing Q_{tot}, we could then obtain both $\dot{V}_{tot}$ and V_W.

Unfortunately, however, Equation 9-15 is not applicable to the present problem, because, in fact, distribution of the dye is *not* instantaneous and dye *does* recirculate. Both of these difficulties are illustrated in Figure 9-2 where the actual concentration of a dye, Evans blue, in arterial blood is plotted against time, time zero being the moment at which the dye was injected intravenously. Far from appearing immediately at its maximum concentration, the dye failed to emerge from the compartment at all for about 12 sec., and when it did appear, the concentration rose gradually, not reaching a peak for several more seconds. Thereafter it declined, but not toward zero as it would if it were following a simple exponential curve.

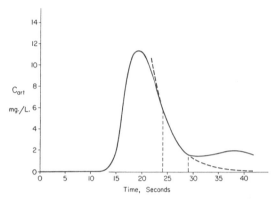

FIG. 9-2. The Indicator Dilution Method for Estimating Cardiac Output

At time zero, a known amount of an indicator (dye or radioactive isotope) is injected intravenously. The *solid curve* shows the actual arterial concentration of the indicator as a function of time. To avoid the complications presented by recirculation of dye (low peak at about 38 sec.) it is assumed that the terminal portion of the curve for the *first* circulation declines exponentially. (*Broken curve* fitted to actual curve between about 24 and about 29 sec. See also Fig. 9-3.) Adapted from Fox and Wood (39).

At about 29 sec., dye which had already passed through the shortest circuits in the greater circulation and had entered the central "Compartment W" a second time began to reappear at the site of arterial sampling. Recirculation of dye thus accounts for the second peak at about 38 sec. The recirculating dye is a nuisance, for we are interested only in the rate at which the *original* dye is diluted out of the compartment. To correct for recirculation, let us replot the declining concentrations against time on semilogarithmic graph paper (Fig. 9-3). A straight line can be fitted to a portion of the resulting curve, in this example from about 24 to about 29 sec. Let us *assume* that if there were no recirculation of dye, the concentration would continue to decline exponentially at the same rate, as indicated by downward extrapolation of the straight line. We can then transfer the extrapolated values back to the original curve, and substitute the *assumed* exponential decrease for the lower part of the *observed* curve (*broken line*, Fig. 9-2). We thus overcome the difficulty caused by recirculation.

To dispose of the second difficulty—that the so-called "central blood volume," here termed "Compartment W," does not behave at all like a well-stirred uniform compartment—we must look again at Equation 9-6, our general equation for the Fick principle. Since this equation was derived from the simple fact that matter is conserved, it should be valid whether

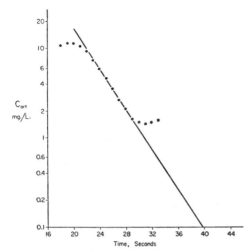

Fig. 9-3. Portion of the Arterial Concentration Curve of Figure 9-2 Plotted on Semilogarithmic Paper

A portion of the declining curve for arterial concentration in Figure 9-2 can be fitted by a straight line when points taken 1 sec. apart are plotted against time on semilogarithmic paper. The *broken curve* in Figure 9-2 was plotted according to this straight line.

mixing is instantaneous or not. But before it can be applied to this problem, or to any similar problem, Equation 9-6 must be integrated. Let us re-arrange it for integration between any two times, t_1 and t_2, by multiplying both sides by dt and by putting in the appropriate integral signs:

$$\int_{Q_{W.1}}^{Q_{W.2}} dQ_{W.t} = V_W \int_{C_{W.1}}^{C_{W.2}} dC_{W.t} = \dot{V} \int_{t_1}^{t_2} (C_{in} - C_{out})_t \, dt - \int_{t_1}^{t_2} \dot{Q}_{other.t} \, dt \quad (9\text{-}16)$$

Part of the indicated integration is simple:

$$Q_{W.2} - Q_{W.1} = V_W(C_{W.2} - C_{W.1}) = \dot{V} \int_{t_1}^{t_2} (C_{in} - C_{out})_t \, dt - \int_{t_1}^{t_2} \dot{Q}_{other.t} \, dt \quad (9\text{-}17)$$

but the remaining integrals can be evaluated mathematically only if we have explicit equations for $(C_{in} - C_{out})_t$ and $\dot{Q}_{other.t}$ as functions of time. But even without an explicit equation, if we can observe $(C_{in} - C_{out})_t$

and plot the observations against time, we can obtain the integral graphically as the area under the curve between t_1 and t_2. If necessary, we could deal with $\dot{Q}_{\text{other} \cdot t}$ in the same way. Let us now apply this idea to the present problem.

Let t_1 be zero, and let t_2 be infinity. Since C_{in} and $\dot{Q}_{\text{other} \cdot t}$ are both zero, Equation 9-17 becomes

$$Q_{W \cdot \infty} - Q_{W \cdot 0} = \dot{V}_{\text{tot}} \int_0^\infty -C_{\text{art} \cdot t} \, dt \qquad (9\text{-}18)$$

But $Q_{W \cdot \infty}$ is zero, and $Q_{W \cdot 0} = Q_{\text{tot}}$. Putting these values into Equation 9-18 and changing signs on both sides,

$$Q_{\text{tot}} = \dot{V}_{\text{tot}} \int_0^\infty C_{\text{art} \cdot t} \, dt \qquad (9\text{-}19)$$

Now the integral in Equation 9-19 can be evaluated by measuring the area under the curve (corrected for recirculation) in Figure 9-2 between the limits of zero and infinity, $i.e.$, the total area. But this poses a problem. Since the descending limb of the revised curve approaches zero asymptotically, the right-hand "tail" of the area to be measured is, at least in theory, infinitely long. However, in practice the curve soon approaches so close to zero that we can neglect the very small area lying beyond a time when the concentration has fallen to some arbitrarily chosen small fraction (say 1 per cent) of its maximum value. Alternatively, the total area beneath the exponential portion of the curve, infinitely long tail and all, can easily be calculated and added to the measured area under the first part of the curve (Exercise 5). This area can be measured with a planimeter, or calculated by adding up the values for C_{art} taken every N seconds along the curve and multiplying the sum by N. (It is usually satisfactory to let $N = 1$.) By Equation 9-19 the cardiac output, $\dot{V}_{\text{tot}}$, is then equal to the amount of dye injected, Q_{tot}, divided by the total area.

9-6. Calculation of Blood Flow per Unit Volume of Brain by the Nitrous Oxide Method of Kety and Schmidt

As a final illustration of the Fick principle let us see how it may be useful even if nothing is known about the actual quantity of the solute whose distribution is being studied.

When a constant concentration of an inert gas such as nitrous oxide is inhaled, a large number of factors influence its concentration in blood (Chap. 13). However, if the concentration of the gas in the arterial blood entering and the venous blood leaving a given tissue can be measured at frequent intervals, the arteriovenous difference can be plotted against time and the area under the curve can be used to estimate *blood flow per unit volume of tissue* by means of the Fick principle. Kety and Schmidt

(65) have applied this method to the study of cerebral blood flow. Inhalation of a constant concentration of nitrous oxide is started at time zero. The concentrations of nitrous oxide in arterial blood and in blood from the internal jugular vein (*assumed* to represent venous outflow from the brain only) are measured from time to time until they become practically equal. It is then *assumed* that distribution equilibrium has been achieved between blood and brain. It is further *assumed* (and this assumption is supported by experimental evidence) that the equilibrium distribution ratio for nitrous oxide between brain and blood is unity. From the last two assumptions

$$C_{\text{ven.brain.eq}} = C_{\text{brain.eq}} = C_{\text{art.eq}} \qquad (9\text{-}20)$$

Strictly speaking, of course, equilibrium is not actually attained until time infinity. For the present problem, $C_{W.0}$ and $\dot{Q}_{\text{other.}t}$ are zero, and Equation 9-17 becomes

$$V_{\text{brain}}(C_{\text{ven.brain.eq}}) = \dot{V}_{\text{brain}} \int_0^\infty C_{\text{art}} - C_{\text{ven.brain}})_t \, dt \qquad (9\text{-}21)$$

Dividing Equation 9-21 by $\dot{V}_{\text{brain}}$ and solving for $\dot{V}_{\text{brain}}/V_{\text{brain}}$,

$$\dot{V}_{\text{brain}}/V_{\text{brain}} = \frac{C_{\text{ven.brain.eq}}}{\displaystyle\int_0^\infty (C_{\text{art}} - C_{\text{ven.brain}})_t \, dt} \qquad (9\text{-}21a)$$

From this equation the blood flow per unit volume of brain can be calculated because both numerator and denominator on the right can be evaluated. The denominator is the total area beneath the curve of arteriovenous difference plotted against time, or (which is the same thing) the total area between the curve for $C_{\text{art.}t}$ and the curve for $C_{\text{ven brain.}t}$ (Fig. 9-4). As in

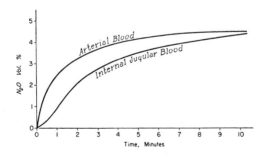

FIG. 9-4. The Concentrations of Nitrous Oxide in Arterial and in Internal Jugular Vein Blood as a Function of Time after Commencing Inhalation of Nitrous Oxide

Adapted from Kety and Schmidt (65).

the preceding problem, the infinitely long "tail" of this area is in practice neglected.

Notice that Equation 9-21a does not give the actual blood flow, but rather the blood flow *per unit volume of brain*, which would be the proportional rate constant if the brain were a uniform compartment washed through by the perfusing blood (*cf.* Equation 9-1). However, there is evidence that inert gas exchange in the brain is governed by two different exchange rates, presumably representing two different compartments with different blood flows (63). Furthermore, as Sapirstein and Ogden (94) have pointed out, if a volume of poorly perfused tissue still far from equilibrium were contributing to the venous outflow being sampled, it might not prevent C_{art} and C_{ven} from approaching each other so closely as to suggest that equilibrium throughout the entire region had been practically achieved. Yet if accurate measurements could be continued for a much longer time, it would be found that the "tail" of the arteriovenous difference was not actually negligible because, though very thin, it was long enough to contain a considerable proportion of the total area. (For a concrete illustration of this, see Figure 13-2.) But in spite of the criticisms which have been directed at the several assumptions underlying this method, it has yielded valuable information about cerebral blood flow.

9-7. The Use of Isotopes as Tracers in Studying Metabolic Pathways

The foregoing concepts and equations describe the behavior of a solute, S, in one or more well-stirred compartments of constant volume. With a mere change in notation, we shall be able to apply precisely the same concepts and equations to the behavior of an isotopic tracer, S^*, in one or more well-mixed metabolic pools (compartments) of constant size, *i.e.*, constant quantity of ordinary S. Now when interconnected metabolic pools are of constant size, the system must be in a steady state such that the loss of S from any compartment is just equal to the gain of S by the compartment. Without the use of isotopic tracers it is always difficult, and often impossible, to study the dynamics of such a system because no observable changes in the concentrations of S occur unless we deliberately add an appreciable amount of S to one of the compartments. But the moment we add S to one of the compartments, we destroy the steady state which is the very thing we want to study! It is the prime virtue of isotopic tracers, and particularly radioactive tracers, that they allow us to avoid this difficulty.

To describe the transport and metabolic transformation of an isotopic tracer, the three assumptions previously discussed must again be made. They are listed, with the necessary modifications, in Table 9-1. But in addition we must assume that the isotope, though distinguishable ana-

TABLE 9-1

Analogous symbols, assumptions, and equations for the kinetics of transport and of metabolic transformation

	Transport of S by Fluid Flow	Transport of S by Clearance	Metabolic Transformation of S
Symbols	V_W = volume of Compartment W	V_W = volume of Compartment W	Q_W = quantity of S in Compartment W ("metabolic pool" W)
	$\dot{V}_{W \to X}$ = flow from W to X	$(\dot{V}\mathrm{cl})_{W \to X}$ = volume of W cleared of S by transfer to X per unit time	$\dot{Q}_{W \to X}$ = quantity of S in W transformed to X per unit time
	$Q_{W.t}$ = quantity of S in W at time t	$Q_{W.t}$ = quantity of S in W at time t	$Q^*_{W.t}$ = quantity of isotopic S in W at time t
	$C_{W.t} = Q_{W.t}/V_W$ = concentration of S in W at time t	$C_{W.t} = Q_{W.t}/V_W$ = concentration of S in W at time t	$(\mathrm{Sp.ac})_{W.t} = Q^*_{W.t}/Q_W$ = specific activity of S in W at time t
	$k_{W \to X} = \dot{V}_{W \to X}/V_W$ = proportion of S in W transferred to X per unit time	$k_{W \to X} = (\dot{V}\mathrm{cl})_{W \to X}/V_W$ = proportion of S in W transferred to X per unit time	$k_{W \to X} = \dot{Q}_{W \to X}/Q_W$ = proportion of S in W transformed to X per unit time
Assumptions	S is instantaneously distributed throughout W	S is instantaneously distributed throughout W	S^* is instantaneously distributed throughout W
	V_W remains constant	V_W remains constant	Q_W remains constant
	$\dot{V}_{W \to X}$ (and hence $k_{W \to X}$) remains constant	$(\dot{V}\mathrm{cl})_{W \to X}$ (and hence $k_{W \to X}$) remains constant	$\dot{Q}_{W \to X}$ (and hence $k_{W \to X}$) remains constant
Equations	$-dQ_W/dt = V_W(-dC_W/dt) = k_{W \to X}Q_{W.t}$	$-dQ_W/dt = V_W(-dC_W/dt) = k_{W \to X}Q_{W.t}$	$-dQ^*_W/dt = Q_W[-d(\mathrm{Sp.ac})/dt] = k_{W \to X}Q^*_{W.t}$
	$C_{W.t} = C_{W.0}\exp[-(\dot{V}_{W \to X}/V_W)t]$	$C_{W.t} = C_{W.0}\exp\{-[(\dot{V}\mathrm{cl})_{W \to X}/V_W]t\}$	$(\mathrm{Sp.ac})_{W.t} = (\mathrm{Sp.ac})_{W.0}\exp[-(\dot{Q}_{W \to X}/Q_W)t]$

lytically, is indistinguishable biochemically from the ordinary element. *We must be able to observe* the isotopic "label" in the laboratory, but cellular mechanisms must *not* be able to "observe" it. When (as is usual) this assumption is valid, the *observed* behavior of the isotope in a particular compound, given as a single dose at time zero into Compartment W, will faithfully indicate the *unobserved* behavior of the corresponding atoms of the normal unlabeled compound entering W at the same time. Furthermore, methods of detecting the isotope are often so sensitive that only a very minute amount of the labeled compound, too small to upset the steady state, need be given.

In Table 9-1 analogous symbols, definitions, assumptions, etc., are listed side by side. Symbols applying specifically to the isotopic tracer are distinguished by an asterisk. For example, the quantity of isotopic tracer in W at time t (usually expressed as a proportion of the administered dose) is symbolized by $Q^*_{W \cdot t}$. Notice that the quantity of ordinary S in compartment W, Q_W, rather than a volume, represents the "size" of a compartment when an isotopic tracer is used. It may at first seem confusing to use Q_W both for the size of a metabolic pool, W, and, as has been done hitherto, for the quantity of S in a compartment, W. Actually, however, the two quantities are fundamentally the same, and the same symbol *must* be used. For example, $Q_{I \cdot t}$ might mean "the quantity of iodide ion in the compartment of inorganic iodide," when the quantity is being treated as a function of time (perhaps after a single dose of iodide). On the other hand, Q_I might mean "the quantity of iodide ion in the compartment of inorganic iodide" regarded as the constant steady-state amount through which a dose of radioactive iodide would be distributed. The subscript "t" should, of course, be used only when Q_W is treated as a time-dependent variable.

Notice that in experiments with isotopic tracers "specific activity" takes the place of "concentration." Just as concentration is the quantity of S per unit volume of solution, so specific activity is the quantity of isotopic S^* per unit quantity of total S, *i.e.*, S plus S^*. However, in experiments with tracers, the contribution of S^* to the total is usually so minute that we can regard specific activity as the quantity of isotopic tracer S^* per unit quantity of ordinary S. In calculating specific activity—indeed in interpreting the results of tracer experiments in general—it is essential to define precisely what S is. For example, if only the carboxyl carbon of acetate is labeled with radioactive carbon, C_{14}, then S is not acetate, nor even the carbon in acetate, but rather the carboxyl carbon of acetate. It is only the fate of that particular carbon which is traced by the radioactive isotope. Furthermore, specific activity should be expressed as quantity of tracer per unit *mass* of ordinary S (*e.g.*, per cent of the dose of radioactive iodine *per microgram of thyroxine iodine*) and not as quantity per mole (*e.g.*, not per cent of dose per micromole). To express specific activity in terms of molar amounts of the compound containing S would invalidate the important relationships between the specific activity of a precursor and its product which are discussed below. *The truth is that with isotopes we can trace only atoms, not molecules!*

If these points are borne in mind, any of the previous equations can be applied to the behavior of an isotopic tracer in a steady-state system simply by substituting in the equations the analogous terms from Table 9-1. Two such equations are given at the bottom of Table 9-1.

9-8. Precursor-Product Relationships

Zilversmit *et al.* (124) have pointed out that the variation of specific activity with time in a series of compounds which become labeled with an isotope may indicate whether one compound is an immediate precursor of another compound in the series. Precisely the same relationships hold true for the concentrations of S when S is being transferred from compartment to compartment by actual flow. In the following analysis, we shall deal with actual transfer by flow merely because the symbols are a bit simpler. We shall consider a series of compartments of constant volume interlinked by pathways of constant flow. We shall restrict the discussion to systems in which each compartment has but one immediately preceding compartment from which it can obtain S. This restriction rules out loops of compartments which would allow S to pass from one compartment to another via two or more different routes. It does not rule out branched chains of compartments, for we shall allow any compartment to *donate* S to any number of other compartments. We stipulate only that it can *receive* S from but one preceding compartment.

Consider first the arrangement shown in Figure 9-5 in which only unidirectional flow between compartments is permitted. As usual, the letters stand for compartments and the arrows for pathways, but for simplicity

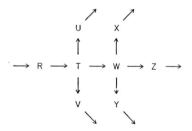

Fig. 9-5. A System of Compartments with Unidirectional Transfer between Them

the rectangles and the symbols for the rate constants, etc., have been omitted. Initially there is no S in any compartment. At time zero, Q_{tot} of S is placed into the first compartment, R, of the series, and is allowed to flow from compartment to compartment at whatever rates are determined by the rate constants. In such a system, any adjacent pair of compartments can be divided into a *donor* (or precursor) compartment, and a *recipient* (or product) compartment, the donor being the one nearer the compartment into which S was originally placed.

Let us see what will happen in the first donor-recipient pair, R and T in Figure 9-5. As long as the concentration of S in R exceeds the concentration of S in T, the concentration in T will rise. As long as the concentration of S in T continues to rise, so also must the concentration of S in U, V, W, and all other "downstream" compartments. But as the S in R continues to be diluted by S-free fluid flowing in from somewhere outside the system, there will come a time when the falling concentration in R exactly equals the concentration in T. *The concentration of S in T must then be maximal* because a moment later the concentration in R will have fallen *below* the concentration in T, and, then and thereafter, flow from R to T can only dilute the S in Compartment T. The same argument applies equally well to the relation between T and W. So long as C_T is greater than C_W, C_W will rise. When $C_T = C_W$, C_W will be at its maximum. When C_T falls below C_W, C_W must fall. Or, in general for *any* adjacent donor-recipient pair:

> In a series of compartments with no loops and with unidirectional flow, if C_{donor} and $C_{recipient}$ are both plotted against time on a single sheet of graph paper, the curve for $C_{recipient}$ will rise as long as it lies below the curve for C_{donor}. The two curves will intersect at precisely the point where $C_{recipient}$ is maximum. (At this point, the curve for C_{donor} must therefore already be falling.) Thereafter, both curves will fall, the curve for $C_{recipient}$ remaining higher than the curve for C_{donor}.

When several compartments are arranged in series (*e.g.*, R, T, W, and Z in Figure 9-5) the following features are worth noting (Fig. 9-6): 1) As S moves "downstream" from its original compartment, the nth compartment in the series always reaches its maximum sooner than the $(n + 1)$th or any subsequent compartment. 2) Each successive peak is lower than the last. 3) As stated above, the concentration of S in a compartment which is the *immediate* recipient of a donor compartment reaches its maximum when $C_{recipient} = C_{donor}$. But at that time, the concentration in more remote recipient compartments "downstream" will still be rising. It is thus perfectly possible for a *remote* recipient to achieve a maximum concentration higher than the concentration simultaneously present in a *remote* donor.

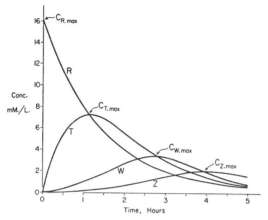

FIG. 9-6. Concentration of Solute in Four Successive Compartments of Figure 9-5 as a Function of Time

A single dose of solute was placed in Compartment R at time zero. Note the lower and lower peaks of concentration in successive "downstream" compartments at the point where the concentration curves for immediate precursor-product compartments cross. (The four curves of this figure were drawn freehand. Consequently, their shapes are not precisely correct.)

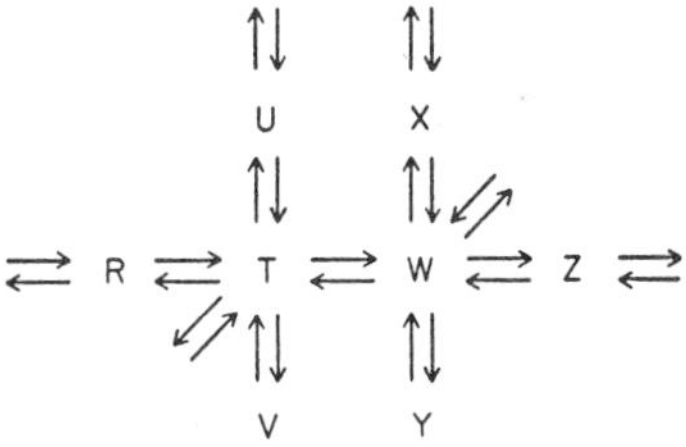

FIG. 9-7. A System of Compartments with Many Possible Routes of Transfer

Is it essential to retain the rather stringent restriction that all flows must be unidirectional, or will the conclusions reached above hold for compartments between which flow can occur in both directions? To answer this question, let us analyze the system of compartments shown in Figure 9-7. The general arrangement is the same as in Figure 9-5, but pathways between compartments in both directions are permitted. Indeed, every possible pathway for flow is illustrated, except that for variety Compartments V and Y communicate with the "outside world" only via other compartments in the system. Loops are still forbidden, so that every pair of adjacent compartments can again be divided unequivocally into donor and recipient according to which is nearer to the original source of S. As before, the concentration in any recipient compartment cannot rise if it exceeds the concentration in its immediately adjacent donor compartment. And as before, if the concentration in a given compartment is still rising, the concentration in its recipient compartments (and in any compartments still more remote in the series) must also still be rising.

Consider the recipient Compartment W in Figure 9-7. Let its donor compartment be T. C_W will be maximum when dC_W/dt, the slope of the curve of C_W plotted against time, is zero. We will therefore find an expression for dC_W/dt, set it equal to zero, and solve for the ratio $C_{W.\max}/C_{T.t'}$, in which t' is used to mean "time when C_W is maximum." We shall then be able to see under what general circumstances the concentrations in donor and recipient compartments are equal when $C_{W.t}$ is maximum.

By inspection of Figure 6-7, the differential equation for Compartment W is

$$\frac{dQ_W}{dt} = k_{T \to W}Q_{T.t} + k_{X \to W}Q_{X.t} + k_{Y \to W}Q_{Y.t} + k_{Z \to W}Q_{Z.t} - k_W Q_{W.t} \quad (9\text{-}22)$$

where, as before,

k_W = the sum of the rate constants for all pathways leading *out* of W
Replacing each Q in Equation 9-22 by the appropriate product of concentration and volume,

$$V_W \frac{dC_W}{dt} = k_{T \to W}V_T C_{T.t} + k_{X \to W}V_X C_{X.t}$$
$$+ k_{Y \to W}V_Y C_{Y.t} + k_{Z \to W}V_Z C_{Z.t} - k_W V_W C_{W.t} \quad (9\text{-}23)$$

Setting dC_W/dt equal to zero and rearranging terms,

$$k_{T \to W}V_T C_{T.t'} + k_{X \to W}V_X C_{X.t'} + k_{Y \to W}V_Y C_{Y.t'}$$
$$+ k_{Z \to W}V_Z C_{Z.t'} = k_W V_W C_{W.\max} \quad (9\text{-}24)$$

But $k_{T \to W}V_T = \dot{V}_{T \to W}$, $k_{X \to W}V_X = \dot{V}_{X \to W}$, etc. Substituting these values in Equation 9-24,

$$\dot{V}_{T \to W}C_{T.t'} + \dot{V}_{X \to W}C_{X.t'} + \dot{V}_{Y \to W}C_{Y.t'} + \dot{V}_{Z \to W}C_{Z.t'}$$
$$= (\dot{V}_{W \to T} + \dot{V}_{W \to X} + \dot{V}_{W \to Y} + \dot{V}_{W \to Z} + \dot{V}_{W \to \text{out}})C_{W.\max} \quad (9\text{-}25)$$

Since the volume of W is assumed to be constant, total flow out of W must equal total flow into W:

$$(\dot{V}_{W \to T} + \dot{V}_{W \to X} + \dot{V}_{W \to Y} + \dot{V}_{W \to Z} + \dot{V}_{W \to \text{out}})$$
$$= (\dot{V}_{T \to W} + \dot{V}_{X \to W} + \dot{V}_{Y \to W} + \dot{V}_{Z \to W} + \dot{V}_{\text{in} \to W}) \quad (9\text{-}26)$$

Substituting this sum of inflows into Equation 9-25 and rearranging terms,

$$\dot{V}_{T \to W} C_{T.t'} = (\dot{V}_{T \to W} + \dot{V}_{X \to W} + \dot{V}_{Y \to W} + \dot{V}_{Z \to W} + \dot{V}_{\text{in} \to W}) C_{W.\max}$$
$$- \dot{V}_{X \to W} C_{X.t'} - \dot{V}_{Y \to W} C_{Y.t'} - \dot{V}_{Z \to W} C_{Z.t'} \tag{9-27}$$

Dividing both sides of Equation 9-27 by $\dot{V}_{T \to W} C_{W.\max}$ and rearranging,

$$\frac{C_{T.t'}}{C_{W.\max}} = 1 + \frac{\dot{V}_{X \to W}}{\dot{V}_{T \to W}} \left(1 - \frac{C_{X.t'}}{C_{W.\max}}\right) + \frac{\dot{V}_{Y \to W}}{\dot{V}_{T \to W}} \left(1 - \frac{C_{Y.t'}}{C_{W.\max}}\right)$$
$$+ \frac{\dot{V}_{Z \to W}}{\dot{V}_{T \to W}} \left(1 - \frac{C_{Z.t'}}{C_{W.\max}}\right) + \frac{\dot{V}_{\text{in} \to W}}{\dot{V}_{T \to W}} \tag{9-28}$$

Equation 9-28 is the desired equation for the ratio of concentrations in the donor, T, and recipient, W, compartments when the concentration of S in the recipient compartment is maximal. Each of the terms on the right represents the influence of a particular pathway for flow *into* the recipient compartment. The first term, unity, represents the influence of flow from the donor compartment, T. Each of the next three terms represents the influence of flow back to W from one of the three compartments for which W is itself the donor. When C_W, previously rising, has just reached its maximum, the concentration in any such "downstream" compartment cannot be higher than the concentration in W. Therefore, each of the expressions in parentheses must be equal to or greater than zero. The parenthetical expression is greater than zero when the concentration in the "downstream" compartment is less than the concentration in W. The corresponding term then *increases* the ratio $C_{T.t'}/C_{W.\max}$. This makes sense because back-flow of fluid with a lower concentration can have only a diluent effect. The influence of an S-free fluid entering Compartment W from outside the system is shown by the last term, $\dot{V}_{\text{in} \to W}/\dot{V}_{T \to W}$, in Equation 9-28. Since the concentration of S in this stream is zero at all times, the last term contains no factor corresponding to the parenthetical expressions of the other terms.

From Equation 9-28 it is obvious that when C_W is maximal, the concentration in T will equal the concentration in W only when all terms on the right (except the first term, unity) are zero. This will be true under two circumstances: 1) $C_{T.t'}$ will equal $C_{W.\max}$ if the concentration of S is identical in T, W, X, Y, and Z. This is an almost trivial result, because after a single dose of S the concentrations can become equal only when the system is a *closed* system, with flow occurring only *between* compartments. There would then be no loss of S from the system, and a uniform concentration throughout would be approached asymptotically. Notice that even though the last term is independent of concentration, it, too, would be zero because in a closed system all inflows are zero. 2) $C_{T.t'}$ will equal

$C_{W.\max}$ if there are no pathways for flow into the recipient compartment, W, except from the donor compartment, T. This is the result we are seeking. It suggests the following rule which, being more general, can supersede the one given previously on page 234:

Consider any adjacent donor-recipient pair of compartments in an open steady-state system without loops. Let Q_{tot} of S be introduced at time zero into the donor compartment or some more remote compartment on the donor side of the system, and let S be carried from compartment to compartment by actual flow. If C_{donor} and $C_{recipient}$ are plotted against time on a single sheet of graph paper, $C_{recipient}$ will always pass through a maximum. *If the recipient compartment receives flow only from its donor*, the curve for $C_{recipient}$ will lie below the curve for C_{donor} until $C_{recipient}$ reaches its maximum. At the maximum, the curves will cross, the curve for $C_{recipient}$ thereafter lying above the curve for C_{donor} . *But if the recipient receives any flow in addition to the flow from its donor*, the curve for $C_{recipient}$ will remain below the curve for C_{donor} for at least some time after $C_{recipient}$ has passed its maximum, and, depending upon the parameters of the system, the two curves may or may not eventually cross.

Since these relationships are most commonly used in studying metabolic transformations with isotopic tracers, the preceding rule will now be restated in terms of specific activities.

Consider any precursor and its *immediate* product which form part of an open, steady-state, metabolic chain without loops. Let Q_{tot}^* of an isotope, A^*, be used at time zero to label the A atoms of the precursor compound or of some more remote compound on the precursor side of the chain. If $(Sp.ac)_{precursor}$ and $(Sp.ac)_{product}$ are plotted against time on a single sheet of graph paper, $(Sp.ac)_{product}$ will always pass through a maximum. *If the product receives its A atoms only from the labeled precursor*, the curve for $(Sp.ac)_{product}$ will lie below the curve for $(Sp.ac)_{precursor}$ until $(Sp.ac)_{product}$ reaches its maximum. At the maximum, the curves will cross, the curve for $(Sp.ac)_{product}$ thereafter lying above the curve for $(Sp.ac)_{precursor}$. *But if the product receives its A atoms from any source besides the precursor*, the curve for $(Sp.ac)_{product}$ will remain below the curve for the precursor for at least some time after $(Sp.ac)_{product}$ has passed its maximum, and, depending upon the parameters of the system, the two curves may or may not eventually cross.

The reader can hardly fail to notice how hedged about with restrictions is the phenomenon of the curves crossing at the maximum specific activity of the product. Because of these restrictions, tracer experiments designed to identify precursors and their products must be interpreted cautiously.

If the metabolic chain *has* loops, so that the suspected product might receive A^* from more than one source, the conclusions given above are wholly invalid. Even when the chain has no loops, so that the product is known to have but one *labeled* precursor, interpretation may not be easy. If the maximum specific activity of the suspected product is *higher* than the simultaneous specific activity of the suspected precursor, an *immediate* precursor-product relationship (but not a more remote one) is clearly ruled out. Crossing of the curves just at the maximum specific activity of the suspected product is strong evidence in favor of the suspected relationship. But if the maximum specific activity of the product is *lower* than the simultaneous specific activity of the suspected precursor, one must be quite sure that there are no other sources of the A atoms in the product before one can conclude that the compound under investigation is not an immediate precursor.

9-9. The Use of Analog Computers for the Analysis of Complex Systems of Compartments

Since the mathematical analysis of systems of three or more compartments is likely to be both complex and tedious, it is fortunate that such systems can be simulated by a suitable arrangement of electrical elements in an analog computer (Section 3-3 and Fig. 3-1). Just as concentrations and specific activities are usually the variables most easily studied as functions of time in a biological system, so the analogous voltages are the electrical variables most easily measured as functions of time in the electrical analog computer. The voltages across the several capacitances—each representing a different compartment—can be displayed together on an oscilloscope or a suitable recorder. This makes it easy to see whether a particular arrangement assumed for the compartments and pathways can adequately explain the observed behavior of concentrations or specific activities. If the voltage-time curves fail to match the concentration-time or specific activity-time curves, the electrical model can easily be modified, either by changing the magnitudes of the existing capacitances and resistances or by adding new ones, until good agreement between the electrical and the biological curves is obtained. It should be clearly recognized that this kind of electrical curve-fitting is subject to the very same mathematical and statistical uncertainties as were discussed previously for other methods of fitting empirical curves to biological data.

A detailed discussion of how analog computers can be used to analyze biological systems has been published by Brownell and his collaborators (19). Solomon (102) discusses many aspects of compartmental analysis which are too complex for inclusion in the present text.

EXERCISES. CHAPTER 9

Exercise 1

A substance, X, is being secreted by the liver into the bile.
   A. Neglecting lymph flow, write an otherwise general conservation
      equation ("Fick principle") for the quantity of X entering the
      liver.
   B. Also write a conservation equation for the volume of fluid entering
      the liver. (Assume that the volume of the liver remains constant.)

Exercise 2

Assume that the air in the lungs represents a single compartment with
a constant volume of 3.0 L. which is washed through by a constant flow
of 5.0 L. per minute of alveolar ventilation. What will be the half-time for
washout of nitrogen from the lungs when pure oxygen is substituted for
room air? (Neglect any transfer of nitrogen across the alveolar membrane.)

Exercise 3

The renal plasma clearance of Diodrast is often taken as a measure of
"effective renal plasma flow." How can the general equation for the Fick
principle (Equation 9-6) be simplified to cover this particular case? What
assumptions must be made?

Exercise 4

In discussing indicator-dilution curves for the estimation of cardiac
output, Fox and Wood (39) define the "mean or average concentration of
indicator during the first circulation" as the area under the curve (corrected
for recirculation) between the time at which the dye first appears and the
time at which it disappears, divided by the interval of time between
appearance and disappearance. What is the significance of this "mean
concentration of indicator"?

Exercise 5

In Section 9-5 it was pointed out that one can calculate the total area
beneath the exponential curve which is assumed to form the "infinitely
long tail" of a dye-dilution curve. What is the equation for this area?

Exercise 6

When given in adequate amounts, certain drugs, such as propylthiouracil, completely prevent the formation of organic iodine compounds in
the thyroid gland but do not interfere with the mechanism which maintains
a relatively high concentration of iodide ion in the thyroid. Under these

circumstances, especially when the gland is hyperplastic, it may be possible to distinguish between radioactive iodide ion in the thyroid gland, measured by external counting, and radioactive iodide present in the general iodide pool of the rest of the body. If a tracer dose of iodide Q_{tot}^* is given at time zero into the "body" iodide compartment, a portion of the dose rapidly enters the thyroid gland compartment of inorganic iodide, approaching a reasonably steady state of equilibrium with a half-time of a few minutes. From this half-time, $t_{1/2}$, from the fraction of the dose distributed into the gland, F_{thy}, and from the equilibrium concentration of radioactive iodide in plasma, $C_{body.eq}^*$, derive an equation for thyroid blood flow, $\dot{V}_{thy}$. Do you think that this equation could be put to practical use?

Exercise 7

Each of the following diagrams depicts a metabolic chain of compounds, A, B, C, etc., after the manner of Figures 9-5 and 9-7 in the text. We shall assume that the chain is in a steady state. An isotopic tracer, too small to disturb the steady state, is introduced *into pool* A at time zero. For which adjacent donor-recipient pairs will it be true that the specific activity of the donor will equal the specific activity of the recipient at the time when the specific activity of the recipient has reached its maximum value?

A.
$$\downarrow \quad \uparrow \quad \uparrow \quad \uparrow$$
$$\to A \rightleftarrows B \to C \to D \to$$
$$\downarrow \quad \downarrow \quad \downarrow$$

B.
$$\to A \to B \rightleftarrows C \to D \to$$

C.
$$C$$
$$\searrow \updownarrow$$
$$B \leftrightarrows A \to E \rightleftarrows F \to$$
$$\updownarrow$$
$$D$$
$$\downarrow$$

Exercise 8

Rall *et al.* (85) found that for sulfanilamide the equilibrium value of the ratio of concentration in plasma to concentration in cerebrospinal fluid from the cisterna magna was unity. Yet in Figure 2 of the paper cited, they show that after a single dose of sulfanilamide the concentration in cerebrospinal fluid continued to rise for some time after it had exceeded the plasma concentration. Can you offer an explanation for this observation?

Exercise 9

Skinner *et al.* (97) have published graphs showing the values they calculated* for the specific activity in the several compartments of various assumed models. Consider their Figure 3, *C*, and comment upon the curves which show the theoretical specific activity of Compartment 1, and of Compartment 2 as a function of time.

Exercise 10

Green and Lowther (51) incubated slices of carrageenin granuloma with C^{14}-labeled proline, and measured the specific activities of the proline and hydroxyproline in neutral-salt-soluble collagen after various periods of incubation. The specific activities of both amino acids increased during the whole time of observation, and were roughly equal, the specific activity of the hydroxyproline sometimes being less than, and sometimes greater than, the specific activity of the proline. In commenting upon these results, the authors say, "If all the collagen hydroxyproline had been derived directly from proline (free or bound) without accumulation of intermediates, a constant specific activity ratio of unity would have been expected, so that this variable ratio indicates some more complex process." Is this statement justified?

* Incorrectly called "analog computer solutions" in the first printing of this book.

10

THE LAW OF MASS ACTION

How far-reaching can be the consequences of a simple but fundamental concept is nowhere more convincingly demonstrated than by the law of mass action. In the present chapter, we will begin with a rather elementary derivation of the law, and thereafter we will examine a few of its applications which are important for physiological scientists.

10-1. The Kinetics of a Monomolecular Reaction

A monomolecular reaction is one in which *single* molecules of a reactant change "spontaneously" into one or more products. In Section 6-8, the kinetics of a monomolecular reaction (illustrated by the decay of a radioactive isotope) were deduced by considering the probability of disappearance of a single unit. The equation (6-25) thus derived,

$$-dN/dt = (P/dt)N_t \tag{10-1}$$

states that at any instant of time the rate of decrease in the number of units is proportional to the number of units remaining at that time, the proportionality constant being simply the probability that a single unit will disappear during the interval dt, divided by the interval. It is thus a probability per unit time. Let the units now be the individual molecules of a reactant, W, which change into one or more products. As a specific example, let the reaction be

$$W \xrightarrow{\ k_W\ } 2Y + Z \tag{10a}$$

where

k_W = the rate constant for the reaction. (We shall distinguish rate constants for different reactions by subscripts which designate the reactants.)

Let the experimental conditions, notably temperature,* remain constant so that P/dt will be constant. P/dt is the probability per unit time that a given molecule of W will decompose into Y and Z during the infinitesimal interval dt. Then the rate constant for Reaction 10a is

$$k_W = P/dt \tag{10-2}$$

Let $N_{W.t}$ be the number of molecules of W distributed throughout a constant volume, V at time t. Rewriting Equation 10-1 for W, combining it with Equation 10-2, and dividing both sides by $N_{Avog}V$, we obtain

$$- (dN_W/N_{Avog}V)/dt = k_W N_{W.t}/N_{Avog}V \tag{10-3}$$

or,

$$-dC_W/dt = k_W C_{W.t} \tag{10-4}$$

where

N_{Avog} = Avogadro's number (number of molecules in a gram molecular weight), a universal constant

$C_{W.t}$ = molar concentration of W at time t

Equation 10-4 is the usual kinetic expression for a monomolecular reaction. It states that *the rate of a monomolecular reaction is directly proportional to the concentration of the reactant.* In Equation 10-4, the rate of the reaction is expressed as the rate of decrease of the concentration of W. Reaction 10a shows that 2 molecules of Y and 1 of Z *appear* for each molecule of W which *disappears.* Hence the rate of reaction could equally well have been expressed as $+\frac{1}{2}dC_Y/dt$ or as $+dC_Z/dt$, both of which are equal to $-dC_W/dt$. Alternatively, the rate of reaction could be expressed in terms of the number of moles changing per unit time simply by multiplying both sides of Equation 10-4 by V.

10-2. The Kinetics of a Bimolecular Reaction

In deriving the equation for the kinetics of a monomolecular reaction, we first defined the probability per unit time that a single molecule would disappear. We then multiplied the number of molecules present at time t by this probability to obtain the number of molecules disappearing per unit time, *i.e.*, the rate of disappearance of the single reactant (Equation 10-1). We will now use the same kind of approach to derive an expression for the kinetics of a bimolecular reaction. A bimolecular reaction is a reaction in which 2 molecules, either of the same substance or of different

---

* The effect of a change in temperature upon the rate of a monomolecular reaction will be discussed in Section 10-7.

substances, combine with each other to form an *activated complex* which then decomposes into the products of the reaction.

Let us consider the bimolecular reaction:

$$W + X \xrightarrow{k_{W+X}} Y + Z \tag{10b}$$

where

k_{W+X} = the rate constant for the reaction

The meaning of k_{W+X} will become evident during the subsequent discussion.

Now it is obvious that a molecule of W and a molecule of X cannot react with each other unless they collide. We must therefore consider the probability per unit of time, P_{coll}/dt, that a single pair of molecules, 1 of W and 1 of X, will collide. This probability multiplied by the number of opportunities for collision will give the number of collisions per unit of time, *i.e.*, the frequency of collision. But collision alone is not enough to ensure reaction. For a reaction to occur, the two colliding molecules must also have between them at least the minimum *energy of activation* needed to form the active complex. Consequently, we must also consider the probability, P_{act}, that in any single collision the necessary energy of activation will be available. By multiplying the number of collisions per unit time by the probability that a single collision will form an active complex, we shall obtain the number of reactions per unit time. (According to this simplified argument, every activated complex is assumed to decompose into the products, Y and Z.)

10-3. The Probability of Collision

We shall begin by considering the probability of collision. Let us imagine a single molecule of W, W_1, and a single molecule of X, X_1, wandering about in a small volume of well-stirred solution. What determines the probability that W_1 and X_1 will bump into each other? The kinetic theory of *gases* suggests that the probability of collision will depend upon the weights and the diameters of the 2 molecules, upon the temperature, and upon the volume in which W_1 and X_1 are free to move about. But in a liquid solution, matters are very much more complex because of collisions and interactions between W_1 and the solvent molecules, and between X_1 and the solvent molecules. However, under constant conditions, that is to say with given reactants in a given solvent at constant temperature and volume, the probability of collision per unit of time between W_1 and X_1 ought at least to be constant. Let us therefore define P_{coll}/dt as the probability per unit time that W_1 and X_1 will collide when they are dissolved in a particular solvent at some specified temperature, T_0, (degrees Kelvin or absolute) and in a specified volume, V_0. Under these constant conditions, what is the *number of opportunities for collision*? Clearly, if W_1 and X_1 are the only reactant molecules present in V_0, there is but *one* opportunity for collision, namely, collision between W_1 and X_1. But if we introduce another molecule of W, W_2, there will be *two* opportunities for collision: W_1 may collide with X_1, and W_2 may collide with X_1. If we now add a second molecule of X, X_2, there will be *four* opportunities: W_1 with X_1, W_2 with X_1, W_1 with X_2, and W_2 with X_2. Similarly, with 3 molecules of W and 2 molecules of X there would be $(3)(2) = 6$ opportunities for collision. Thus each *different pair* of reactant molecules represents a separate opportunity for collision. The total number of different pairs, and hence of opportunities for collision, is simply the product of the number of molecules of W and the number of molecules of X, *i.e.*, $N_{W.t}N_{X.t}$. Accordingly, at time t the expected number of collisions per unit time will be

$$(N_{coll.}/dt)_t = (P_{coll}/dt)N_{W.t}N_{X.t} \quad \text{at} \quad T_0 \quad \text{and} \quad V_0 \quad (10\text{-}5)$$

where

N_{coll} = the number of collisions occurring between t and $t + dt$

By writing Equation 10-5 for an infinitesimal period of time, we imply that the numbers of molecules involved are so large that we can treat collision as a continuous process without introducing any appreciable error. This will usually be true.

10-4. The Effect of Changing the Volume

The probability of collision per unit of time, P_{coll}/dt, was defined for the particular volume V_0. But we can easily calculate the effect of having the same number of molecules in a different volume. Suppose that the volume containing $N_{W.t}$ and $N_{X.t}$ were increased by a factor of n to a new volume, V:

$$V = nV_0 \quad (10\text{-}6)$$

In this more dilute solution, a volume equal to V_0 will contain $(1/n)N_{W.t}$ molecules of W and $(1/n)N_{X.t}$ molecules of X. Substituting these numbers into Equation 10-5, we find that now the frequency of collision *in a volume of V_0* is $(P_{coll}/dt)(1/n^2)N_{W.t}N_{X.t}$. But the *total* rate of collision for the whole volume, V, is n times the rate for a volume of V_0 because V is n times as large as V_0. Hence the total rate of collision in the more dilute solution is

$$(N_{coll}/dt)_t = (P_{coll}/dt)(1/n)N_{W.t}N_{X.t} \quad \text{at} \quad nV_0 \quad \text{and} \quad T_0 \quad (10\text{-}7)$$

Combining Equations 10-6 and 10-7 with the elimination of n,

$$(N_{coll}/dt)_t = (P_{coll}/dt)(V_0/V)N_{W.t}N_{X.t} \quad \text{at} \quad T_0 \quad (10\text{-}8)$$

10-5. The Effect of Changing the Temperature

In gases, the average speed of a molecule, and hence the number of collisions per unit time, varies directly with the square root of the absolute temperature. If we assume that temperature has the same effect upon the frequency of collision in a liquid solution, we can find the rate of collision at any temperature, T, by multiplying the right side of Equation 10-8 by the factor $\sqrt{T}/\sqrt{T_0}$:

$$(N_{coll}/dt)_t = (P_{coll}/dt)(V_0/V)(\sqrt{T}/\sqrt{T_0})N_{w.t}N_{x.t} \qquad (10\text{-}9)$$

Notice that the effect of temperature upon the frequency of collision is quite small. For example, an increase from 273°K. (0°C.) to 313°K. (40°C.) would increase the frequency of collision by only about 7 per cent. As we shall see, by far the more important effect of temperature is upon the probability, P_{act}, that a given collision will form an activated complex. We must now turn our attention to this probability.

10-6. The Probability of Forming an Active Complex

At any given temperature, the *average* kinetic energy of any large number of molecules remains constant. But the kinetic energy of an *individual* molecule does not. Because energy is continually being exchanged between molecules when they collide, any particular molecule may at one moment have a very small kinetic energy, whereas a moment later, if it happens to gain kinetic energy in several successive collisions, it may have a very large kinetic energy. However, the probability that several random collisions in succession will endow a particular molecule with a high kinetic energy is small, so that only a small proportion of the molecules will have a kinetic energy much larger than the average. Yet it is just these few molecules which can form the activated complex. Fortunately, it is possible to derive from kinetic theory an equation for the fraction of the molecules whose kinetic energy is equal to, or greater than any given energy. Although the derivation is too complex to undertake here, the resulting equation is surprisingly simple (77):

$$N_{\geq E}/N_{tot} = e^{-E/RT} \qquad (10\text{-}10)$$

where

N_{tot} = the total number of molecules
$N_{\geq E}$ = the number of molecules whose energy is equal to or greater than E
E = a given energy
T = the absolute temperature (degrees Kelvin)
R = the ideal gas law constant (energy per degree-mole); a uni-

versal constant whose value for present purposes may be taken as 1.99 calories per degree-mole

Notice that the fraction $N_{\geq E}/N_{tot}$ is independent of molecular species and depends solely upon the absolute temperature and the chosen energy. It varies from 1 when the energy chosen is zero (*i.e.*, *all* molecules have an energy equal to or greater than zero) to zero when the energy chosen is infinity (*i.e.*, *no* molecules have an energy equal to or greater than infinity). Notice also that the fraction approaches zero as the temperature approaches absolute zero, and approaches unity as the temperature approaches infinity.

Now suppose we choose E to be E_{act}, the activation energy or the minimum energy which a molecule must have to form an activated complex by collision with another molecule. Then $N_{\geq E_{act}}/N_{tot}$ is the fraction of all molecules having the necessary energy, and hence also the fraction of collisions which will result in the formation of an active complex. But by our definition of probability, this fraction is also the probability, P_{act}, that any single collision will form an active complex. Accordingly we may rewrite Equation 10-10:

$$\left(\frac{N_{\geq E_{act}}}{N_{tot}}\right) = \left(\frac{N_{coll.act.t}}{N_{coll.t}}\right) = P_{act} = e^{-E_{act}/RT} \qquad (10\text{-}11)$$

where

E_{act} = the minimum energy required to form an activated complex

$N_{\geq E_{act}}$ = the number of molecules with at least the minimum activation energy

$N_{coll.act.t}$ = the number of collisions between time t and time $t + dt$ which result in the formation of an active complex

Solving Equation 10-11 for $N_{coll.act.t}$,

$$N_{coll.act.t} = P_{act}N_{coll.t} = N_{coll.t}e^{-E_{act}/RT} \qquad (10\text{-}11a)$$

or, dividing both sides by dt,

$$(N_{coll.act.t}/dt) = (N_{coll.t}/dt)e^{-E_{act}/RT} \qquad (10\text{-}11b)$$

Combining Equations 10-11b and 10-9 with the elimination of $(N_{coll}/dt)_t$,

$$\frac{N_{coll.act.t}}{dt} = (P_{coll}/dt)(V_0/V)(\sqrt{T}/\sqrt{T_0})(e^{-E_{act}/RT})N_{W.t}N_{X.t} \qquad (10\text{-}12)$$

Since we are assuming that every activated complex decomposes into products,* 1 molecule of W and 1 molecule of X must disappear, and 1

---

* This assumption is not strictly true. For most reactions, the probability that the complex will decompose into products rather than back into reactants lies between 0.5 and 1.0 ((77) p. 567).

molecule of Y and 1 molecule of Z must appear, for each activated complex which is formed. In other words, we assume that

$$\frac{N_{\text{coll.act.}t}}{dt} = \frac{-dN_W}{dt} = \frac{-dN_X}{dt} = \frac{dN_Y}{dt} = \frac{dN_Z}{dt} \quad (10\text{-}13)$$

Combining Equations 10-12 and 10-13,

$$\frac{-dN_W}{dt} = (P_{\text{coll}}/dt)(V_0/V)(\sqrt{T}/\sqrt{T_0})(e^{-E_{\text{act}}/RT})N_{W.t}N_{X.t} \quad (10\text{-}14)$$

Equation 10-14 is the mass action equation for a bimolecular reaction written in terms of the number of molecules reacting per unit of time. But it is usually much more convenient to write the equation in terms of concentration. To do this, we must divide both sides of Equation 10-14 by $N_{\text{Avog}}V$. We will also multiply the right side by $N_{\text{Avog}}/N_{\text{Avog}}$, and rearrange the terms to suit our convenience:

$$\frac{-dN_W/N_{\text{Avog}}}{dtV} = (P_{\text{coll}}/dt)(V_0/\sqrt{T_0})N_{\text{Avog}}\sqrt{T}(e^{-E_{\text{act}}/RT})$$
$$\cdot \left(\frac{N_{W.t}/N_{\text{Avog}}}{V}\right)\left(\frac{N_{X.t}/N_{\text{Avog}}}{V}\right) \quad (10\text{-}14a)$$

or, since $(N_t/N_{\text{Avog}})/V$ is the molar concentration at time t,

$$-dC_W/dt = (P_{\text{coll}}/dt)(V_0/\sqrt{T_0})N_{\text{Avog}}\sqrt{T}(e^{-E_{\text{act}}/RT})C_{W.t}C_{X.t} \quad (10\text{-}15)$$

Now let us *define*:

$$k_{W+X} = (P_{\text{coll}}/dt)(V_0/\sqrt{T_0})N_{\text{Avog}}\sqrt{T}(e^{-E_{\text{act}}/RT}) \quad (10\text{-}16)$$

Notice that for a given pair of reactants in a given solute, (P_{coll}/dt), V_0, T_0, and E_{act} are all constant so that the only variable on the right of Equation 10-16 is the temperature. If the temperature is also kept constant, k_{W+X} is constant, as indeed is implied by calling it the rate constant for the reaction. Combining Equations 10-15 and 10-16,

$$-dC_W/dt = k_{W+X}C_{W.t}C_{X.t} \quad (10\text{-}17)$$

Equation 10-17 is the usual expression of the law of mass action for a bimolecular reaction. Its derivation has given us some insight into the actual physical meaning of the rate constant. Furthermore, Equation 10-16 provides a theoretical explanation for the effect of a change in temperature upon the rate of reaction, a matter to which we shall return in Section 10-7.

Equation 10-17 states that *the rate of a bimolecular reaction is directly proportional to the product of the molar concentrations of the two reactants*. It is important to note that neither the argument nor the conclusion is substantially altered when both reactants are the same substance. Had the reaction been

$$W + W \xrightarrow{k_{W+W}} Y + Z \tag{10c}$$

we would have found that the rate of reaction was proportional to $C_{W.t}C_{W.t}$, in other words, to the square of the concentration of W (see Exercise 1).

The above derivation can be extended to any number of reactants although, in fact, one-step reactions requiring the interaction of more than 2 molecules are rare. The derivation leads to the following general (but somewhat erroneous) statement of the *law of mass action*:

> The rate of a one-step chemical reaction is directly proportional to the product of the molar concentrations of each of the molecular reactants, whether the reactants are the same substance or different substances.

This statement is in error because, properly speaking, it is the product of the thermodynamic *activities* of the components rather than their molar concentrations which determines the rate of reaction. However, in the ranges of concentration which are ordinarily of interest to physiological scientists, the formulation and use of mass action expressions in terms of molar concentration are usually not seriously misleading.* We shall continue to work with concentrations rather than with activities.

10-7. The Effect of Temperature upon the Rate of a Chemical Reaction

According to the simplified collision theory discussed above, the rate constant of a bimolecular reaction should vary with temperature in the manner shown by Equation 10-16. It has already been pointed out that the major effect of temperature is upon the proportion of molecules which have an energy high enough for reaction. Indeed, the influence of temperature upon the frequency of collision is so small that for most purposes it can be neglected, *i.e.*, the expression $\sqrt{T/T_0}$ may be treated as a constant, and Equation 10-16 can accordingly be rewritten in the form

$$k_{W+X} = Ae^{-E_{\text{act}}/RT} \tag{10-18}$$

where

$$A = (P_{\text{coll}}/dt)V_0 N_{\text{Avog}} \sqrt{T/T_0} \tag{10-19}$$

An expression equivalent to Equation 10-18 was first obtained by Arrhenius as an empirical description of the effect of temperature upon rate of reaction.

---

* But occasionally the use of concentrations in place of activities can be *disastrously* misleading. See, for example, Neuman and Neuman (78).

It is therefore known as the *Arrhenius equation*. The application of the Arrhenius equation to experimental data is facilitated by a linear transformation. Taking logarithms on both sides of Equation 10-18,

$$\ln k_{W+X} = \ln A - (E_{act}/R)(1/T) \qquad (10\text{-}20)$$

If k_{W+X} is measured at several different temperatures, and $\ln k_{W+X}$ is plotted against $(1/T)$, one should obtain a straight line whose slope is $-E_{act}/R$ and whose intercept is $\ln A$. From these parameters one can calculate the energy of activation and the probability that any particular pair of reactant molecules will collide when they are confined within the volume, V_0, at a temperature of T.

In the majority of cases, a straight line actually is obtained when $\ln k_{W+X}$ is plotted against $(1/T)$. This agreement between fact and theory enhances the plausibility of Equation 10-20 and strongly supports the concept of activation energy. But there seems little excuse for extending the notion of a critical activation energy to highly complex physiological processes merely because a plot of the logarithm of the rate of the process against the reciprocal of the absolute temperature yields a straight line. Over the very limited range of temperature tolerated by physiological systems, any process whose rate is *for any reason at all* increased by a rise in temperature is quite likely to give such a straight line. The slope of the line has the dimensions of temperature, not of energy. It requires an act of faith bordering upon mysticism to plot the logarithm of the heart rate of a frog against the reciprocal of the absolute temperature, to divide the slope of the resulting straight line by the gas constant, and to interpret the value so obtained as the energy of activation of a "master reaction" which determines the heart rate (20). The reader may find it amusing to list the assumptions which would have to be made to justify such an interpretation.

If a *monomolecular* chemical reaction were due to a truly spontaneous rearrangement of chemical bonds, its rate—like that of radioactive disintegration of atoms—should be independent of temperature. In fact, however, the Arrhenius equation is just as applicable to monomolecular reactions as it is to bimolecular reactions. Only the occasional atom which has acquired the requisite high energy of activation can decompose into products. The "frequency factor," A, in the Arrhenius equation can no longer be interpreted as depending upon the frequency of collision, but is instead a function of the frequencies of the several modes of vibration of the molecule (77).

10-8. The Temperature Coefficient

The influence of temperature upon the rate of a reaction is often expressed as the *temperature coefficient*, (Q_{10}). The (Q_{10}) is the factor by which the rate constant is increased by a rise in temperature of 10°C.:

$$(Q_{10}) = \frac{\text{rate constant at } T + 10°}{\text{rate constant at } T} \qquad (10\text{-}21)^*$$

Since the (Q_{10}) is itself a function of temperature, the temperature at which it was measured should be specified. Furthermore, since the relationship between temperature and rate of reaction is not linear, one cannot calculate the (Q_{10}) by doubling the factor by which a 5° rise in temperature increases the rate constant (see Exercise 2).

10-9. Inadequacies of the Simplified Collision Theory

It would be wrong not to emphasize the crudity of the simplified collision theory as here discussed. A number of difficulties have been deliberately glossed over. Almost nothing has been said about the factors which determine the probability of collision in liquid solutions, nor about the factors which determine how high the energy of activation must be. Nothing has been said about the complications which arise when one or more of the reactants is ionic. Nothing has been said about thermodynamic aspects of reaction rates. Unfortunately, we cannot pursue the important modern developments in the theory of reaction rates without reference to the quantum mechanics of intramolecular vibrations and like matters which are *terra incognita* for the author and presumably for at least some of his readers, also. But even we who are uninitiated can gain some insight into the intricacies of the field by reading the chapter on "Chemical Kinetics" in Moore's *Physical Chemistry* (77).

10-10. What Is Meant by the Order of a Reaction?

Suppose that the reactants, say W and X, in a particular reaction are known, but the precise mechanism of the reaction is not. It may not even be known whether the reaction proceeds in one step or in several steps. If the rate of the reaction is studied as a function of the concentration of the reactants, it may be possible to fit the observations by an empirical equation of the form

$$-dC_W/dt = k_{W+X} C_{W.t}^m C_{X.t}^n \qquad (10\text{-}22)$$

Then the order of the reaction is the sum of the exponents of the concentration terms on the right, in this example, $m + n$. Since m and n are deter-

---

* If the rates of reaction at the two temperatures are both measured at the *same* concentration of the reactants, the *observed* rate of reaction (expressed, for example, as rate of disappearance of one of the reactants) will be directly proportional to the rate constant. The (Q_{10}) may then be calculated as the ratio of the observed rates of reaction.

mined empirically from the parameters of the curve which best fits the data, it is perfectly possible to observe reactions whose order is fractional or zero. This stands in sharp contrast to the theoretical *molecularity* of a one-step reaction which can only be integral, and which in practice is one, two, or, very rarely, three. For example, suppose that W combines with X only in the presence of a catalyst, perhaps an enzyme, whose capacity for substrate is limited. If the rate of disappearance of W is observed when there is a great excess of W and X, so that the enzyme is fully occupied with substrate, the system will be capacity-limited (Section 8-3) and dC_W/dt will be constant. According to Equation 10-22 the observed rate of reaction can be constant only if $C_{W \cdot t}^m$ and $C_{X \cdot t}^n$ are both constant. But since a reaction which removes W and X is taking place, $C_{W \cdot t}$ and $C_{X \cdot t}$ *cannot* be constant. Hence m and n must both be zero, and the reaction is a reaction of zero order. As another example, suppose that C_W is enormously high compared with C_X. Then the amount of W used up as the reaction proceeds may be so small that for all practical purposes C_W remains constant. Regardless of its exponent, the effect of C_W would not be distinguishable experimentally and the disappearance of X would then be observed as an nth-order process:

$$-dC_X/dt = (k_{W+X}C_W^m)C_{X \cdot t}^n \qquad (10\text{-}23)$$

with an observed "rate constant" equal to the quantity in parentheses. This situation occurs very frequently when one of the reactants is the solute, usually water.

These examples should make it clear that the order of a reaction is an empirical description of the observed kinetic behavior of the reactants, and implies nothing at all about the molecular mechanism underlying the reaction nor about the number of individual steps which may actually be involved in the overall reaction.

Several of the techniques available for determining the order of a reaction are summarized by Moore (77).

10-11. Chemical Equilibria. Equilibrium Constants

Thus far we have considered only reactions proceeding in one direction. In a great many reactions of physiological importance the products of the reaction can themselves recombine to yield the original reactants. Such reversible reactions never proceed to completion in either direction unless one of the components is continually removed. If the original reactants and their products are allowed to remain in solution together, a steady state of *chemical equilibrium* is approached asymptotically. At equilibrium the concentration of each reactant and each product is determined by the mass-action equations for the two opposing reactions.

Consider the reversible reaction

$$W + X \xrightleftharpoons[k_{Y+Z}]{k_{W+X}} Y + Z \tag{10d}$$

taking place in a closed system under constant conditions. By Equation 10-17, the kinetic expressions for the two reactions involved are

$$(dC_W/dt)_{W+X} = -k_{W+X}C_{W.t}C_{X.t} \tag{10-24}$$

for the rate of decrease in the concentration of W which would be caused by the "forward" reaction, $W + X \to Y + Z$, acting alone, and

$$(dC_W/dt)_{Y+Z} = k_{Y+Z}C_{Y.t}C_{Z.t} \tag{10-25}$$

for the rate of increase in the concentration of W which would be caused by the "backward" reaction, $Y + Z \to W + X$, acting alone. The total or net rate of change in the concentration of W at any time is equal to the algebraic sum of these two opposing rates:

$$dC_W/dt = k_{Y+Z}C_{Y.t}C_{Z.t} - k_{W+X}C_{W.t}C_{X.t} \tag{10-26}$$

By definition, at equilibrium, the decrease of C_W due to the forward reaction is exactly balanced by the increase of C_W due to the backward reaction, so that the net rate of change is zero, and Equation 10-26 becomes

$$k_{Y+Z}C_{Y.eq}C_{Z.eq} = k_{W+X}C_{W.eq}C_{X.eq} \tag{10-27}$$

or,

$$\frac{C_{Y.eq}C_{Z.eq}}{C_{W.eq}C_{X.eq}} = \frac{k_{W+X}}{k_{Y+Z}} \tag{10-27a}$$

We shall now define the *equilibrium constant* for the forward reaction, $W + X \to Y + Z$:

$$(Keq)_{W+X} = k_{W+X}/k_{Y+Z} \tag{10-28}$$

Combining Equations 10-27a and 10-28,

$$\frac{C_{Y.eq}C_{Z.eq}}{C_{W.eq}C_{X.eq}} = (Keq)_{W+X} \tag{10-29}$$

Notice that by convention the concentrations of the *products* of the reaction for which (Keq) is defined always appear in the *numerator* of the ratio. Accordingly, in Equation 10-29 $(Keq)_{W+X}$ is the equilibrium constant for the reaction $W + X \to Y + Z$. But it would be the *reciprocal* of the equilibrium constant for the reverse reaction, $Y + Z \to W + X$. By this convention, *the larger the (Keq) for a given reaction, the more nearly complete is that reaction at equilibrium*. Notice also how ambiguous it would be to say,

"The constant for the equilibrium between W, X, Y, and Z is 75." Although there is but one equilibrium, there are two equilibrium constants, and we must specify whether 75 is $(Keq)_{W+X}$ or $(Keq)_{Y+Z}$.

10-12. Chemical Equilibria in Which the Hydrogen Ion Is One of the Reactants. The Henderson-Hasselbalch Equation

A large number of physiologically important chemical equilibria involve a proton donor, a proton acceptor, and a proton, *i.e.*, a hydrogen ion. A *proton donor* is any compound which in solution is able to give up ("donate") or *dissociate* a hydrogen ion. In the Brønsted terminology, it is an acid. A *proton acceptor* is any compound which in solution can accept, or associate a hydrogen ion. In the Brønsted terminology, it is a base. A *hydrogen ion* in aqueous solution is not a free proton but has been accepted by a water molecule (which thus serves as a proton acceptor or Brønsted base) so that it actually exists as the hydronium ion, $(H_3O)^+$. For simplicity, we will ordinarily neglect this fact and continue to speak of hydrogen ions, rather than of hydronium ions.

Note well that the definitions of proton donor and proton acceptor given above have nothing at all to do with whether the donor or acceptor is charged or uncharged. To preserve this highly desirable generality we shall continue whenever possible to use the unmistakable descriptive terms "proton donor" and "proton acceptor."

Let H = the hydrogen (properly, the hydronium) ion. It must, of course, carry a positive charge, but for the present we can take this for granted and omit it from the symbol.

AH = any proton donor, or Brønsted acid.

A = the proton acceptor or Brønsted base, which is formed when AH dissociates its hydrogen. In Brønsted's terminology, A is the *conjugate base* of the acid, AH.

Consider the reversible reaction

$$AH \xrightleftharpoons[k_{A+H}]{k_{AH}} A + H \qquad (10e)$$

By Equation 10-29, the equilibrium constant for this reaction is

$$(Keq)_{AH} = K_a = C_A C_H / C_{AH} \qquad (10\text{-}30)$$

where $(Keq)_{AH}$ is, of course, the equilibrium constant for the *dissociation* of the proton donor, AH. $(Keq)_{AH}$ is usually called the acid dissociation constant, and, as indicated in Equation 10-30, is usually symbolized K_a. Solving Equation 10-30 for the hydrogen ion concentration, C_H,

$$C_H = (Keq)_{AH} C_{AH} / C_A \qquad (10\text{-}30a)$$

Taking logarithms on both sides of Equation 10-30a,

$$\log_{10} C_H = \log_{10}(Keq)_{AH} + \log_{10}(C_{AH}/C_A) \tag{10-31}$$

or, changing signs,

$$
\begin{aligned}
-\log_{10} C_H &= -\log_{10}(Keq)_{AH} - \log_{10}(C_{AH}/C_A) \\
&= -\log_{10}(Keq)_{AH} + \log_{10}(C_A/C_{AH})
\end{aligned} \tag{10-31a}
$$

By definition

$$pH = -\log_{10} C_H \tag{10-32}$$

and by definition

$$pK_a = -\log_{10} K_a = -\log_{10}(Keq)_{AH} \tag{10-33}$$

Substituting these symbols into Equation 10-31a,

$$pH = pK_a + \log_{10}(C_A/C_{AH}) \tag{10-34}$$

Equation 10-34 is the well-known *Henderson-Hasselbalch equation.*

It cannot be too strongly emphasized that Equation 10-34 is applicable to *any* proton donor-proton acceptor pair, whether the proton donor is the ionized "salt" of a "base" such as epinephrine (adrenaline), or the un-ionized form of an "acid" such as acetic acid. Notice, however, that *substances which are ordinarily called acids become negatively charged by dissociating a hydrogen ion, whereas substances which are ordinarily called bases become positively charged by associating a hydrogen ion.* It is in this sense that we will henceforth use the terms "acid" and "base." However, the ions of "strong acids" and "strong bases," ions such as Cl^-, Na^+, K^+, the cations of quaternary amines, etc., are *always* ions, and they cannot act as proton donors or proton acceptors. For example, HCl in aqueous solution dissociates so completely that its pK_a cannot be measured. It is an acid by virtue of its hydronium ions which act as proton donors. Strong bases such as NaOH and choline, $[(CH_3)_3{\equiv}N{-}CH_2{-}CH_2{-}OH]OH$, dissociate so completely in aqueous solution that their pK_a's cannot be measured. They are bases by virtue of their dissociated hydroxyl ions which act as proton acceptors.

By a slight rearrangement, the Henderson-Hasselbalch equation may be written

$$\log_{10}(C_A/C_{AH}) = pH - pK_a \tag{10-34a}$$

Equation 10-34a makes it clear that *the ratio of the concentration of the proton acceptor to the concentration of the proton donor is determined solely by the difference between the pH and the pK_a.* The relationship is the same whatever the pH or the pK_a, whatever the chemical identity of the donor-

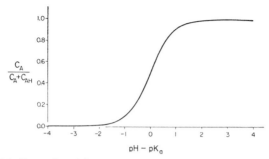

FIG. 10-1. Proportion of Total Proton Donor Plus Proton Acceptor in the Acceptor Form as a Function of the Difference between pH and pK_a

The relationship shown here is valid for any proton donor-proton acceptor pair.

acceptor group, and whatever the concentrations. In Figure 10-1 the ratio, $C_A/(C_{AH} + C_A)$, *i.e.*, the proportion of the total compound which is in the acceptor form, is plotted against pH $-$ pK_a. Notice that as the pH increases, by far the greater part of the change from proton donor to proton acceptor occurs within one pH unit on either side of the point where pH $=$ pK_a. Two pH units below the pK_a, only about 1 per cent of the total donor-plus-acceptor is in the acceptor form. One pH unit below the pK_a, about 9 per cent is in the acceptor form. *When the pH equals the pK_a the concentrations of donor and acceptor are equal.* One pH unit above the pK_a about 91 per cent of the total is in the acceptor form. Two pH units above the pK_a about 99 per cent is in the acceptor form. But it is well to remember that even when the pH is very far from the pK_a, so that the predominance of one form is overwhelming, there will still be a small concentration of the other form.

The great general utility of Equation 10-34a will now be illustrated by two examples.

Example 1

What per cent of acetylsalicylic acid (aspirin) will be in the un-ionized form in gastric juice at pH 1.5? The pK_a of acetylsalicylic acid is 3.5 (16). From Equation 10-34a, $\log_{10}C_A/C_{AH} = 1.5 - 3.5 = -2$; hence $C_A/C_{AH} =$ antilog of $-2 = 1/100$. With an acid, the proton donor is the un-ionized form. Therefore, slightly more than 99 per cent (actually 100/101) of the total is un-ionized.

Example 2

Consider a local anesthetic which is a tertiary amine with a pK_a of 8.2. If the total concentration is 0.04 mole per liter, what will be the concentration of the un-ionized form at a pH of 7.4? From Equation

10-34a $\log_{10}(C_A/C_{AH}) = 7.4 - 8.2 = -0.8$. Hence $C_A/C_{AH} =$ antilog of $-0.8 = 0.158$. The total concentration, $C_A + C_{AH} = 0.04$ mole per liter. Simultaneous solution of the two equations: $C_A/C_{AH} = 0.158$ and $C_A + C_{AH} = 0.04$ mole per liter gives $C_A = 0.0055$ mole per liter; $C_{AH} = 0.0345$ mole per liter. The concentration of the un-ionized form is therefore 0.0055 mole per liter, the un-ionized form of an amine being the proton acceptor.

10-13. "Salts"

The custom of writing the Henderson-Hasselbalch equation in terms of the concentration of the "salt" of an acid or base can lead only to confusion, and (like the equally confusing "base dissociation constant," K_b) should be abandoned. The transport mechanisms for ions which the physiologist studies commonly deal with single ionic species, and often manifest a quite extraordinary ability to distinguish between such apparently similar ions as sodium and potassium, or bromide and iodide. Aside from having to fulfill the electrostatic requirement that in volumes of any considerable size the sum of the positive charges must equal the sum of the negative charges, these mechanisms know nothing about "salts." The physiological scientist should be at least as discriminating as the mechanisms he studies!

10-14. Buffers

Any proton donor-proton acceptor pair (except $(H_3O)^+$—H_2O, and H_2O—OH^-) acts as a buffer in the range of pH near its pK_a. If H^+ is added to a solution containing a buffer (for example, by adding HCl), much of the added H^+ is removed by combination with the proton acceptor. If H^+ is subtracted from the solution (for example by adding NaOH) much of the H^+ is replaced by dissociation of the proton donor. Mole for mole, all substances with a single donor-acceptor group and a given pK_a are equally efficient as buffers at a given pH whether they are acids or bases. The efficiency of any buffer is maximal when the pH of the solution equals the pK_a of the buffer, and decreases fairly rapidly as the pH departs from the pK_a in either direction. This is really all that need be said about buffers, except that the behavior of molecules such as proteins with many proton donor-proton acceptor groups is naturally more complex.

Although water is both a proton donor and a proton acceptor, it cannot serve as a buffer. When any of the three monomeric components of water, OH^-, H_2O, and H_3O^+, participate in proton exchanges, there is necessarily an equivalent change in OH^- and H_3O^+. But changes in OH^-, and H_3O^+ are exactly what buffers are supposed to minimize! When a molecule of water "accepts" a proton to become a hydronium ion, it cannot oppose the change in pH because the hydronium ion *is* the hydrogen ion.

10-15. The Dissociation Constant of Water. The pK_w

It is possible, but not very useful, to calculate two pK_a's for water (see Exercise 4). It is much more important to recognize that the product $(C_{H^+})(C_{OH^-})$ is constant for a given temperature. (From now on, the subscripts for ions will bear the proper sign for charge.) The dissociation of water may be symbolized

$$2H_2O \rightleftarrows OH^- + H_3O^+ \tag{10f}$$

or, in less rigorous but more familiar form,

$$H_2O \rightleftarrows OH^- + H^+ \tag{10g}$$

The equilibrium equation for the dissociation of water according to 10g is

$$(Keq)_{H_2O} = (C_{OH^-})(C_{H^+})/C_{H_2O} \tag{10-35}$$

Since C_{H_2O} is virtually constant, it may be combined with $(Keq)_{H_2O}$, to define a new constant, K_w :

$$K_w = (Keq)_{H_2O}C_{H_2O} = (C_{OH^-})(C_{H^+}) \tag{10-36}$$

Taking logarithms on both sides of Equation 10-36 and changing signs,

$$-\log_{10}K_w = -\log_{10}C_{H^+} - \log_{10}C_{OH^-} \tag{10-37}$$

By defining $pK_w = -\log_{10}K_w$ we can write Equation 10-37 in the form

$$pK_w = pH - \log_{10}C_{OH^-} \tag{10-38}$$

The actual value of pK_w varies from nearly 15 at 0°C. to about 13 at 60°C. Since equal numbers of hydrogen ions and hydroxyl ions are produced when pure water ionizes, the pH of pure water is just one-half of the pK_w and therefore varies with temperature, also. However, for most purposes pK_w may be taken as 14, and the pH of pure water (which, by definition, is neutral) as 7, though these values are exact only at 24°C.

10-16. A Special Equation for the Bicarbonate-Carbonic Acid-CO_2 System

One buffer system of great biological importance—the HCO_3^-, H_2CO_3, CO_2 system—is complex because it involves two interrelated equilibria.*

---

* There is actually a third equilibrium,

$$H-O-C-O^- \rightleftarrows H^+ + {}^-O-C-O^-$$
$$\quad\quad \| \quad\quad\quad\quad\quad\quad\quad \|$$
$$\quad\quad O \quad\quad\quad\quad\quad\quad\quad O$$

but the bicarbonate ion is such a weak acid (pK_a = about 10.4) that we can usually neglect its dissociation to the carbonate ion.

When carbon dioxide dissolves in water, by far the greater part of it remains in simple physical solution as CO_2. However, a small proportion combines with water to form carbonic acid:

$$O=C=O + H-O-H \rightleftarrows H-O-\underset{\underset{O}{\parallel}}{C}-O-H \qquad (10h)$$

the equilibrium constant for the hydration of CO_2 being

$$(Keq)_{CO_2+H_2O} = C_{H_2CO_3}/C_{CO_2}C_{H_2O} \qquad (10\text{-}39)$$

We shall later need to use Equation 10-39 in the alternative form:

$$\frac{C_{H_2O}(Keq)_{CO_2+H_2O}}{1 + C_{H_2O}(Keq)_{CO_2+H_2O}} = \frac{C_{H_2CO_3}}{C_{H_2CO_3} + C_{CO_2}} \qquad (10\text{-}39a)$$

In ordinary aqueous media the equilibrium between CO_2, H_2O, and H_2CO_3 is approached so slowly that the body has developed a special enzyme, carbonic anhydrase, to speed the attainment of equilibrium. In contrast, the dissociation of carbonic acid,

$$H-O-\underset{\underset{O}{\parallel}}{C}-O-H \rightleftarrows H^+ + H-O-\underset{\underset{O}{\parallel}}{C}-O^- \qquad (10i)$$

is practically instantaneous. The *true* acid dissociation constant for carbonic acid is

$$K_{a.H_2CO_3} = C_{H^+}C_{HCO_3^-}/C_{H_2CO_3} \qquad (10\text{-}40)$$

with a pK_a of about 3.4. Solving Equation 10-40 for $C_{H_2CO_3}$,

$$C_{H_2CO_3} = C_{H^+}C_{HCO_3^-}/K_{a.H_2CO_3} \qquad (10\text{-}40a)$$

Substituting the value for $C_{H_2CO_3}$ from Equation 10-40a into the *numerator* only, on the right of Equation 10-39a, and rearranging terms so as to assemble all of the constant quantities on the left of the equation,

$$\left[\frac{C_{H_2O}(Keq)_{CO_2+H_2O}K_{a.H_2CO_3}}{1 + C_{H_2O}(Keq)_{CO_2+H_2O}}\right] = C_{H^+}\left[\frac{C_{HCO_3^-}}{C_{H_2CO_3} + C_{CO_2}}\right] \qquad (10\text{-}41)$$

Taking logarithms on both sides of Equation 10-41a, and changing signs,

$$-\log_{10}\left[\frac{C_{H_2O}(Keq)_{CO_2+H_2O}K_{a.H_2CO_3}}{1 + C_{H_2O}(Keq)_{CO_2+H_2O}}\right]$$
$$= -\log_{10} C_{H^+} - \log_{10}\left[\frac{C_{HCO_3^-}}{C_{H_2CO_3} + C_{CO_2}}\right] \qquad (10\text{-}42)$$

Now let us define

$$pK'_{a.H_2CO_3} = -\log_{10}\left[\frac{C_{H_2O}(Keq)_{CO_2+H_2O}K_{a.H_2CO_3}}{1 + C_{H_2O}(Keq)_{CO_2+H_2O}}\right] \quad (10\text{-}43)$$

Substituting this value, and the definition of pH into Equation 10-42, and rearranging terms,

$$pH = pK'_{a.H_2CO_3} + \log_{10}\left[\frac{C_{HCO_3^-}}{C_{H_2CO_3} + C_{CO_2}}\right] \quad (10\text{-}44)$$

where $pK'_{a.H_2CO_3}$ has a value of about 6.3 in dilute solutions or about 6.1 in plasma at 37°C. (12).

Notice that $pK'_{a.H_2CO_3}$ is a complex constant-of-convenience which, unlike the usual pK_a, is *not* the negative logarithm of an acid dissociation constant, but is the negative logarithm of a complicated expression containing an acid dissociation constant, another equilibrium constant, and the molar concentration of water. Moreover, the sum $C_{H_2CO_3} + C_{CO_2}$ has replaced the molar concentration of undissociated acid which appears in the usual Henderson-Hasselbalch equation. Now in dilute solutions at 37°C., C_{CO_2} is some *800 times* as great as $C_{H_2CO_3}$, so that the concentration of the true acid, carbonic acid, in Equation 10-44 is practically negligible! But even though the dissolved CO_2 is not a proton donor (and indeed contains no hydrogen at all) *it behaves as if it were a proton donor* with an effective pK_a equal to the $pK'_{a.H_2CO_3}$ defined above. Thus, Equation 10-44 not only shows how the respiratory regulation of CO_2 concentration can play an important role in the maintenance of a constant pH in body fluids, but also enables one to calculate either $C_{H_2CO_3} + C_{CO_2}$, or $C_{HCO_3^-}$, or pH from the $pK'_{a.H_2CO_3}$ and measurements of the other two quantities.

10-17. The Influence of Ionization upon the Distribution of Substances across Cell Membranes

The ease with which lipid-soluble substances cross cell membranes has already been pointed out in Section 7-6. In sharp contrast, the permeability of the cell membrane to most ions is severely restricted, partly because the lipid-water distribution ratio for charged molecules is low, the ions tending to remain in the polar solvent. Thus, in general, even the small inorganic ions cross cell membranes comparatively slowly and the larger organic ions, such as acetylcholine and other quaternary ammonium compounds, for all practical purposes are unable to cross at all.* Therefore, unless it can take

* Although this statement is justified as an overall summary, there are many important exceptions. Small anions such as chloride and bicarbonate enter and leave erythrocytes very rapidly. Special mechanisms for the active transport of sodium,

advantage of some special transport mechanism the *charged* member of a proton donor-proton acceptor pair will scarcely be able to enter cells. But the *uncharged* member is much more soluble in lipid. As a result, other things being equal, those proton donor-proton acceptor compounds which are least ionized at the prevailing pH are the ones which enter cells most rapidly. But the distribution of the compound across the membrane *at equilibrium* is not a function of time. Once the uncharged form has crossed the membrane and has re-entered an aqueous phase on the other side, it must again obey the Henderson-Hasselbalch equation. Hence, in theory at least, any compound with enough un-ionized form to cross at all should ultimately approach an equal concentration on both sides of the membrane *if the pH is the same on both sides*. Again we must say "other things being equal," for the equilibrium concentration of the substance on one side or the other may be subject to other conditions, notably binding to proteins (Section 10-20). Moreover, the substance may be destroyed or excreted so rapidly that a steady state of equilibrium is never established.

10-18. The Distribution of Proton Donor-Proton Acceptor Compounds between Aqueous Phases of Different pH

We will now consider in a more general way the equilibrium distribution of proton donor-proton acceptor pairs between two aqueous phases separated by a lipid membrane. *The following discussion applies only to acids*, but precisely the same kind of derivation can be used to obtain analogous equations for bases (Exercise 8B).

Let AH be an acid which can cross the lipid membrane only in the uncharged form. Suppose that the membrane separates two aqueous compartments, W and X, whose pH's, pH_W and pH_X, respectively, are kept constant by suitable buffer systems. Finally, assume for simplicity that the total concentration of the compound in Compartment W, *i.e.*, $C_{AH.W} + C_{A^-.W} = C_{tot.W}$, remains constant. If pH_W, pH_X, $C_{tot.W}$, and the pK_a are known, what will be the concentrations of AH and of A^- in the two compartments at equilibrium? To begin with, by definition

$$C_{tot.W} = C_{AH.W.eq} + C_{A^-.W.eq} \tag{10-45}$$

In Compartment W, the relative concentrations of A^- and AH will be de-

---

or potassium, or both are present in all cells that have been studied, and the kidney has a whole array of transport mechanisms for various organic and inorganic ions. Special mechanisms also exist in the thyroid gland (for iodide ion) in the stomach (for HCl), etc. Within the terminals of cholinergic nerves, acetylcholine seems to be "packaged" in minute vesicles, covered with a layer of (?) lipid, which presumably carry the quaternary ions across the cell membrane in spite of their charge. But most of these exceptions are due to the *active transport* of particular ions, and it is still fair to say that most cell membranes are practically impermeable to ions whose only means of entry is *passive diffusion*.

termined by the pK_a and by the pH_W according to the Henderson-Hassel-balch equation:

$$pH_W = pK_a + \log_{10}(C_{A^-.W.eq}/C_{AH.W.eq}) \qquad (10\text{-}46)$$

or,

$$\text{antilog } (pH_W - pK_a) = C_{A^-.W.eq}/C_{AH.W.eq} \qquad (10\text{-}47)$$

Now let us assume that the uncharged form, AH, diffuses across the membrane from W to X until it has attained the same concentration on both sides. Then,

$$C_{AH.X.eq} = C_{AH.W.eq} \qquad (10\text{-}48)$$

Finally, the relative concentrations of AH and of A^- in Compartment X at equilibrium will be determined by the pK_a and by the pH_X according to the Henderson-Hasselbalch equation:

$$pH_X = pK_a + \log_{10}(C_{A^-.X.eq}/C_{AH.X.eq}) \qquad (10\text{-}49)$$

or,

$$\text{antilog } (pH_X - pK_a) = C_{A^-.X.eq}/C_{AH.X.eq} \qquad (10\text{-}50)$$

We now have four simultaneous equations, 10-45, 10-47, 10-48, and 10-50, for the four unknown concentrations, $C_{A^-.W.eq}$, $C_{AH.W.eq}$, $C_{A^-.X.eq}$, and $C_{AH.X.eq}$. Solving Equations 10-45 and 10-47 for $C_{AH.W.eq}$,

$$C_{AH.W.eq} = \frac{C_{tot.W}[\text{antilog}\,(pK_a - pH_W)]}{1 + \text{antilog}\,(pK_a - pH_W)} \qquad (10\text{-}51)$$

But by Equation 10-48 this is also $C_{AH.X.eq}$:

$$C_{AH.X.eq} = \frac{C_{tot.W}[\text{antilog}\,(pK_a - pH_W)]}{1 + \text{antilog}\,(pK_a - pH_W)} \qquad (10\text{-}52)$$

Since all of the terms on the right of Equations 10-51 and 10-52 are known, $C_{AH.W.eq}$ and $C_{AH.X.eq}$ can be calculated. $C_{A^-.W.eq}$ may then also be calculated by subtracting $C_{AH.W.eq}$ from $C_{tot.W}$ according to Equation 10-45. Lastly, we can solve Equation 10-50 for $C_{A^-.X.eq}$:

$$C_{A^-.X.eq} = C_{AH.X.eq}[\text{antilog}(pH_X - pK_a)] \qquad (10\text{-}50a)$$

and thus calculate $C_{A^-.X.eq}$ from the known values on the right of Equation 10-50a.

Example

Brodie and Hogben (16) have obtained evidence that many weak acids and bases are distributed between acid gastric juice and plasma in accordance with the assumption that only the un-ionized form of the compound can diffuse across the gastric mucosa between the bloodstream and the lumen of the stomach. Consider the equilibrium distribution of barbital (5,5 diethylbarbituric acid) which was one of the compounds they studied. Let us assume that barbital is not appreciably bound to plasma protein. If the pH of the stomach contents is 1.0, the pH of the plasma 7.4, and the total concentration of barbital in the blood plasma 12 mg. per liter, what will be the equilibrium concentrations of the ionized and of the un-ionized forms of barbital in plasma and in gastric juice? The pK_a of barbital is 7.8.

Let the blood plasma be Compartment W and the gastric juice be Compartment X. We then have the following known values:

$$C_{tot.W} = 12 \text{ mg./L.}$$

$$pH_W = 7.4$$

$$pH_X = 1.0$$

$$pK_a = 7.8$$

Substituting these values into Equation 10-51 (or 10-52),

$$C_{AH.W.eq} = C_{AH.X.eq} = \frac{12[\text{antilog}(7.8 - 7.4)]}{1 + [\text{antilog}(7.8 - 7.4)]} = \frac{30.144}{3.512} = 8.58 \text{ mg./L.}$$

By Equation 10-45:

$$C_{A^-.W.eq} = 12.00 - 8.58 = 3.42 \text{ mg./L.}$$

By Equation 10-50a:

$$C_{A^-.X.eq} = 8.58[\text{antilog}(1.0 - 7.8)] = 8.58(1.585)(10^{-7})$$

$$= 1.36(1 0^{-6})\text{mg./L}$$

It is by no means always necessary to calculate all four concentrations separately as was done in the above example. Often one is interested only in the ratio of the equilibrium concentrations of *total* compound present in the two compartments, *i.e.*, $C_{tot.X.eq}/C_{tot.W.eq}$. We can easily derive an equation for this ratio by combining Equations 10-47 and 10-50 with definitions of $C_{tot.X.eq}$ and $C_{tot.W.eq}$:

$$\frac{C_{tot.X.eq}}{C_{tot.W.eq}} = \frac{1 + \text{antilog}(pH_X - pK_a)}{1 + \text{antilog}(pH_W - pK_a)} \tag{10-53}$$

The influence of pH on the distribution of proton donors and acceptors is likely to be important wherever two body fluids of different pH's are separated from each other by a membrane or by a thin layer of cells. Of particular interest are the differences in pH between urine and plasma. In

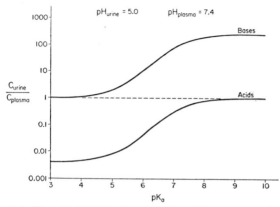

Fig. 10-2. Theoretical Distribution of Acids and Bases between Plasma and an Acid Urine

The curves show the *equilibrium* ratio of total concentration of acid or base in urine of pH 5.0 to total concentration in plasma of pH 7.4 as a function of pK_a. It is assumed that the renal tubular epithelium is permeable only to the un-ionized forms.

Figure 10-2, the ratio $C_{tot.urine.eq}/C_{tot.plasma.eq}$ calculated for weak acids from Equation 10-53 has been plotted on a logarithmic scale against pK_a, assuming a plasma pH of 7.4 and a urine pH of 5.0. Figure 10-3 presents a similar plot for a plasma pH of 7.4 and a urine pH of 8.0. In each figure the corresponding curves for weak bases are also given. According to these theoretical curves, changes in urine pH ought to cause very considerable changes in the equilibrium distribution of compounds whose pK_a's lie within certain ranges. But it must be remembered that these curves were calculated by assuming that 1) the tubules of the kidney are completely impermeable to the ionic form, 2) no processes of active transport are involved, and 3) after the pH of the urine has been adjusted by the kidney, there is still time for complete equilibration between the plasma and the flowing tubular fluid. It is unlikely that all of these assumptions can be justified for any particular compound, and it is therefore hardly surprising that the ratios predicted from these assumptions are not always the ratios actually observed. Nevertheless, the actual effect of changes in urinary pH upon the excretion of weak acids and bases may be very large, and are in the direction predicted by the theoretical curves. The theory thus helps to explain not only the influence of urinary pH upon the rate of excretion of certain drugs (75) but also the *immediate* changes in the excretion of the ammonium ion which always occur when the pH of the urine is abruptly altered (80).

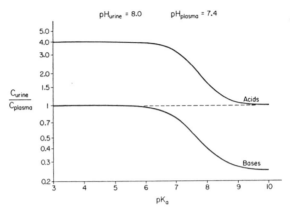

FIG. 10-3. Theoretical Distribution of Acids and Bases between Plasma and an Alkaline Urine

The curves show the *equilibrium* ratio of total concentration of acid or base in urine of pH 8.0 to total concentration in plasma of pH 7.4 as a function of pK_a. It is assumed that the renal tubular epithelium is permeable only to the un-ionized forms.

10-19. Ionization and the Potency of Drugs

We have just seen how the extent of ionization may influence the distribution of weak acids and bases across biological membranes. Indeed, it is the complete, or practically complete, ionization of drugs such as streptomycin and quaternary ammonium compounds which prevents their absorption from the gastrointestinal tract and hence makes them inactive, or at best unreliable, for systemic therapy when administered orally. But in addition, the potency of a drug is often strongly influenced by the extent to which it is ionized at its site of action. For example, the local anesthetics, most of which are tertiary amines with pK_a's of about 8.0 or 9.0, are believed to act chiefly in the un-ionized form (for a contrary view, see Reference 91). Although the *total* concentration of a given local anesthetic which is just sufficient to block nerve conduction increases as pH decreases, the concentration of the un-ionized portion, calculated from the Henderson-Hasselbalch equation, remains about the same. Certain other drugs, for example, the bacteriostatic acridines, are most potent when in the ionized form (3). Presumably their site of action is on the surface of cells, rather than within the cell itself. Still other drugs, notably the sulfonamides are not so easily classified, for when the antibacterial potency of a large number of sulfonamides was tested at pH 7, the most effective compounds were those whose pK_a was about 7, *i.e.*, those which were about half ionized and half un-ionized. Sulfonamides which were almost fully ionized, or hardly ionized at all, were much less potent (8). For more information about the relationships between ionization and the biological activity of drugs the reader should consult the excellent monographs by Albert (3, 4).

10-20. The Reversible Binding of Solutes to Plasma Proteins

It is very common to find that when a solute, which may be a drug, or a hormone, or even an inorganic ion, enters the bloodstream, it becomes more or less extensively bound to plasma proteins. (Presumably the binding of solute molecules to tissue proteins is also common, but it is much more difficult to study.) The extent of binding to protein varies from almost 100 per cent (for example, thyroxine) to practically zero (for example, urea) and must be determined experimentally for each particular compound. Binding to protein is important because protein-bound molecules are unable to leave the bloodstream to exert physiological or pharmacological effects. Consequently *the effective concentration of the substance is the concentration of the unbound form.** But since the binding to protein is freely reversible, the bound substance is only temporarily *hors de combat*, and, in fact, represents a kind of circulating reserve from which the substance will be dissociated whenever the concentration of the unbound form decreases.

If the substance is a drug, protein binding may prolong its duration of action. For example, drug which is protein-bound is not available for filtration through the glomeruli of the kidney. Filtration removes unbound drug and water simultaneously from the plasma, leaving the *concentration* of the unbound drug essentially unchanged. The equilibrium between bound and unbound drug is therefore not upset, and there is no reason for the drug-protein complex to dissociate. In contrast, removal of unbound molecules without a corresponding volume of water, for example by active tubular transport in the kidney, or by metabolism of the drug in the liver, will upset the equilibrium and cause the dissociation of bound drug. Thus, in theory at least, such processes might be able to remove *all* of the drug brought to them by the bloodstream, provided both the dissociation of bound drug and the process itself are rapid in comparison with the rate of passage of blood through the capillaries of the region concerned. Sometimes this is true. For example, even though a considerable fraction of penicillin is bound to plasma protein, the renal tubules are able to remove penicillin practically completely from the blood flowing past them. But with certain other drugs, protein binding is so extensive as to hinder complete removal. For example, the mercurial diuretics are about 95 per cent bound

* With the exception of substances, such as the anticoagulant, heparin, whose sites of binding to plasma proteins may also represent sites of action.

to plasma protein, and only a small fraction of the total drug is removed from the blood as it flows past the renal tubules. If a substance is actually 100 per cent bound (*i.e.*, irreversibly bound) to plasma protein, there is no way in which the kidney can "get hold" of it at all, and it cannot be excreted in the urine. This seems to be true of iron which is bound to the special plasma protein, transferrin.

For a complete review of the kinetics of protein binding, together with an authoritative discussion of its significance, the reader should consult the paper by Goldstein (48). We will here be concerned only with deriving a simple equation to show that the proportion of the total drug which is bound is not a constant but varies with the concentration of unoccupied binding sites on the protein (see also Exercise 9).

Let S be a solute which is bound to a particular plasma protein, Pr. Though a single protein molecule may well have more than one binding site for the solute, we will assume that the affinity of all binding sites is the same, *i.e.*, that they are all governed by the same equilibrium dissociation constant. We also assume that if two or more different proteins are involved, the same equilibrium constant holds for all. These assumptions enable us to treat the phenomenon of protein binding as a simple mass action equilibrium where C_{Pr} will *not* be the molar concentration of protein molecules but rather the "molar" concentration of unoccupied binding sites:

C_{Pr} = (molar concentration of protein)(number of unoccupied binding sites per molecule)   (10-54)

Accordingly, we may write the reaction for the dissociation of bound substance, SPr, as

$$SPr \rightleftarrows S + Pr \qquad (10j)$$

the corresponding mass action expression for equilibrium between bound and unbound substance being

$$(Keq)_{SPr} = C_S C_{Pr}/C_{SPr} \qquad (10\text{-}55)$$

where

C_S = concentration of unbound substance
C_{Pr} = concentration of unoccupied binding sites as defined above
C_{SPr} = concentration of bound substance = concentration of occupied binding sites
$(Keq)_{SPr}$ = equilibrium constant for the dissociation of SPr

We now note that for a given *total* concentration of S, $C_{S.tot}$,

$$C_{S.tot} = C_S + C_{SPr} \qquad (10\text{-}56)$$

Combining Equations 10-55 and 10-56 with the elimination of C_S,

$$(Keq)_{SPr} = C_{Pr}\left(\frac{C_{S.tot}}{C_{SPr}} - 1\right) \tag{10-57}$$

The fraction of the total drug which is bound, F_{bound}, is by definition

$$F_{bound} = C_{SPr}/C_{S.tot} \tag{10-58}$$

Combining Equations 10-57 and 10-58 by replacing $C_{S.tot}/C_{SPr}$ by $1/F_{bound}$,

$$(Keq)_{SPr} = C_{Pr}\left(\frac{1}{F_{bound}} - 1\right) \tag{10-59}$$

Solving Equation 10-59 for F_{bound},

$$F_{bound} = \frac{C_{Pr}}{(Keq)_{SPr} + C_{Pr}} \tag{10-59a}$$

According to Equation 10-59a, if $(Keq)_{SPr}$ is small compared with C_{Pr} (*i.e.*, if there is comparatively little tendency for the substance to dissociate from the protein), the fraction bound approaches 1.0. If $(Keq)_{SPr}$ is large compared with C_{Pr}, the fraction bound approaches zero. But the fraction bound also depends upon the concentration of unoccupied binding sites, increasing from zero when C_{Pr} is zero, toward 1.0 as C_{Pr} approaches infinity. Now for a given concentration of a given solute, C_{Pr} is itself determined by the total concentration of the binding protein, and hence the fraction of S which is bound will also be influenced by the concentration of protein. This is why, for example, the *total* concentration of calcium in the blood plasma varies with the concentration of plasma protein, even though the concentration of *unbound* calcium is kept constant by the homeostatic mechanism of the parathyroid glands.

The derivation of the much more important *general* equation for the dependence of the fraction bound upon the *total concentration of S*, the *total concentration of binding sites*, and the *equilibrium dissociation constant* is not difficult, and is assigned as Exercise 9. Figure 10-4 illustrates the influence of each of these three variables upon the fraction bound.

EXERCISES. CHAPTER 10

Exercise 1

In deriving the law of mass action for a bimolecular reaction between W and X, we found that the number of opportunities for collision between molecules of W and X was $N_W N_X$. Now suppose that the reaction is between 2 molecules of the *same* substance, W. What will be the number of opportunities for collision?

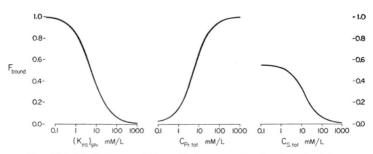

FIG. 10-4. The Influence of Three Factors upon the Fraction of a Solute
Which Is Bound to Plasma Protein

Left. Fraction bound as a function of the dissociation constant of the solute-protein
complex when the total concentration of binding sites is 6 mM. per liter and the total
concentration of solute is 1.0 mM. per liter. *Middle.* Fraction bound as a function of
total concentration of binding sites on the protein when the dissociation constant
is 5 mM. per liter and the total solute concentration is 1.0 mM. per liter. *Right.* Fraction
bound as a function of total concentration of solute when the dissociation constant
is 5 mM. per liter and the total concentration of binding sites is 6 mM. per liter.

Exercise 2

Derive an equation for calculating the (Q_{10}) from E_{act}, the energy of activation, and from the observed rates of reaction at *any* two temperatures,
T and $T + \Delta T$.

Exercise 3

The rate constant (*not* the equilibrium constant) for a certain reaction
was reported to be 1.5 L. per mole-second. What was the order of the reaction?

Exercise 4

The pH of pure water at 24°C. is 7.00. What is the pK_a for water acting
as a proton donor (*i.e.*, as an acid) at this temperature? What is the pK_a
for water acting as a proton acceptor or "base"?

Exercise 5

The pK_a for the tertiary nitrogen in morphine is 8.2. The pK_a for the
phenolic hydroxyl in morphine is 9.9. Assuming that ionization of these
two groups is independent (*i.e.*, that the pK_a values are unchanged by
changes in pH), what proportion of the total number of molecules of morphine will carry no charge at all at pH 7.4? At pH 9.05?

Exercise 6

The Handbook of Chemistry and Physics ((58), p. 1721) gives the pH of
"saturated" carbonic acid at 25°C. as 3.8. Assuming that this means the

pH of a solution which was originally pure water but which has come to equilibrium with pure CO_2 at a partial pressure of 760 mm. Hg minus the vapor pressure of water, estimate the pK_a' of carbonic acid.

Exercise 7

Richards (87) states that "about 5 % of the total carbon dioxide in cells and plasma is in the form of carbonic acid, H_2CO_3 , this being maintained in equilibrium with the partial pressure of CO_2 in blood, tissues, and alveolar air." What is meant by this statement?

Exercise 8

A and B are two well-stirred chambers, each containing distilled water and a suitable buffer. A and B are separated by a very thin lipid membrane which is completely impermeable to ions (and to all components of the buffer systems). A contains 1 L. at pH 4.8; B contains 3 L. at pH 2.8. The membrane is equally permeable in both directions to lipid-soluble substances.

    A. At time zero, 10 mg. of an acid, Y, with a pK_a of 7.3 is introduced into Compartment A. The initial rate of change of concentration of Y in A is (0.01 mg. per L.) per minute. About how long will it take for the concentration in B to rise to half its equilibrium value?

    B. If an organic base with a pK_a of 4.0 were present in this system, what would be the concentration ratio, $C_{tot.A.eq}/C_{tot.B.eq}$, for the base at equilibrium?

Exercise 9

Accepting the simplifying assumptions outlined in the text which allowed us to write Equation 10-55, derive an equation for F_{bound} , the fraction of substance S which is bound to protein, as a function of $C_{S.tot}$, the total concentration of S (bound plus unbound), $C_{Pr.tot}$, the total concentration of binding sites (occupied plus unoccupied), and $(Keq)_{SPr}$, the dissociation constant for the drug-protein complex.

11

SUBSTRATE-ENZYME AND DRUG-RECEPTOR INTERACTIONS

In the present chapter we shall see how, by making suitable assumptions, we can apply the law of mass action to the analysis of enzyme kinetics, and even (though less plausibly) to the combination of drugs with their cellular receptors. Traditionally, the mathematical approach to enzyme-substrate interactions has differed somewhat from the approach to drug-receptor interactions. However, to the limited extent that the action of drugs can be related directly to simple mass-action equilibria, the two types of interactions are fundamentally so similar that the equations derived for the one will also be applicable, with little modification, to the other. The similarities which will be emphasized in this chapter are summarized in Table 11-1.

11-1. Enzymes as Reactants

An enzyme is a complex organic catalyst which increases the rate of some often highly specific chemical reaction by decreasing the energy of activation which a molecule of substrate must acquire before it can participate in the reaction. As a catalyst, the enzyme is not "used up" in the reaction. After combining as a reactant with its substrate to form an enzyme-substrate complex, the enzyme is regenerated intact during one or more subsequent reactions. Even the simplest analysis of enzyme kinetics must therefore take two steps into consideration. Moreover, the whole reaction is very often reversible, and the enzyme hastens the "backward" combination of products to form substrate just as much as it hastens the "forward" decomposition of substrate to form products. As a result, the equilibrium constant is not altered at all. With any such reversible reaction, the choice of which reactants to regard as substrate and which as product is obviously a matter of sheer convenience.

The two consecutive steps in a reversible reaction catalyzed by an en-

TABLE 11-1

Analogous symbols, assumptions, equations, etc., for enzyme-substrate and for drug-receptor relationships

	Substrate—Initial Reaction Rate	Concentration—Effect
Symbols	S = substrate E = enzyme SE = enzyme-substrate complex X, Y = products v = *initial* velocity of reaction	D = drug R = receptor DR = drug-receptor complex (Eff) = effect of drug, "response"
Reaction scheme assumed	$S + E \rightleftarrows SE \rightarrow X + Y + E$	$D + R \rightleftarrows DR$
Assumptions	1. Initially, no back reaction, $X + Y + E \rightarrow SE$; confine attention to this initial period 2. C_{SE} very much smaller than C_S 3. v proportional to C_{SE} 4. $v_{\max}$ is reached only when $C_{SE} = C_{E.\text{tot}}$	No analogous assumption needed C_{DR} very much smaller than C_D (Eff) proportional to C_{DR} (Eff)$_{\max}$ is achieved only when $C_{DR} = C_{R.\text{tot}}$
Independent variable	C_S	C_D
Dependent variable	v or $v/v_{\max}$ or $100(v/v_{\max})$ (per cent of maximum velocity)	(Eff) or (Eff)/(Eff)$_{\max}$ or $100[(\text{Eff})/(\text{Eff})_{\max}]$ (per cent response)
Constants	$v_{\max}$ Total enzyme concentration: $C_{E.\text{tot}} = C_E + C_{SE}$ Michaelis constant, k_M	(Eff)$_{\max}$ Total receptor concentration: $C_{R.\text{tot}} = C_R + C_{DR}$ Dissociation constant, (Keq)$_{DR}$
Fundamental equation	$v/v_{\max} = C_S/(k_M + C_S)$ (Equation 11-9b) $v/v_{\max} = \frac{1}{2}$ when $C_S = k_M$	(Eff)/(Eff)$_{\max} = C_D/[(\text{Keq})_{DR} + C_D)]$ (Equation 11-25) (Eff)/(Eff)$_{\max} = \frac{1}{2}$ when $C_D = (\text{Keq})_{DR}$

zyme can be represented as follows:

$$S + E \; \underset{k_{SE.\text{back}}}{\overset{k_{S+E}}{\rightleftarrows}} \; SE \; \underset{k_{X+Y+E}}{\overset{k_{SE.\text{fore}}}{\rightleftarrows}} \; X + Y + E \tag{11a}$$

where

S = the substrate
E = the enzyme not combined with substrate
SE = the substrate-enzyme complex
X and Y = the products of the reaction
the k's = the rate constants for the designated pathways

Suppose we want to study the effect of the concentration of the substrate upon the rate at which the substrate is converted into products. For simplicity, let us arrange the experiment so that at time zero only S and E are present in the reaction mixture. As the reaction proceeds, we shall ordinarily not be able to measure the concentration of either E or SE; but let us suppose that S is measurable so that we can study the rate of reaction by observing how rapidly C_S decreases. Now it is mathematically impossible to derive a general equation for C_S as a function of time ((100) Chap. 25, p. 9). But the problem becomes much more tractable if we confine our attention to the *initial* rate of reaction, for we can then assume that the concentrations of X and of Y are so small that their combination to form SE can be neglected. (Let us call this Assumption 1.) The utility of this simplifying assumption was first brought to general notice by Michaelis and Menten, and it is to their analysis of the problem that we must now turn.

11-2. The Michaelis-Menten Formulation

According to Assumption 1, by considering only the *initial* rate of disappearance of S, Reaction 11a can be simplified to

$$S + E \; \underset{k_{SE.\text{back}}}{\overset{k_{S+E}}{\rightleftarrows}} \; SE \; \overset{k_{SE.\text{fore}}}{\longrightarrow} \; X + Y + E \tag{11b}$$

But if we are really confining our attention to the very first moment of reaction, when X and Y are zero, are we not logically compelled to assume that SE is also zero, and should we not accordingly write the initial reaction as

$$S + E \; \overset{k_{S+E}}{\longrightarrow} \; SE \tag{11c}$$

The logic of this argument is unassailable. But Reaction 11b can still be justified if we make the additional assumption (Assumption 2) that the actual concentration of enzyme is minute in comparison with the concentration of S and with the equilibrium concentrations of X and Y. When this is true, a negligible quantity of S will be involved in the initial combination with the enzyme, so that the quantity of substrate initially added, which is accurately known, divided by the volume of the solution, also accurately known, may be taken as the initial concentration of substrate when the initial rate of reaction is measured. Furthermore, the turnover of enzyme-substrate complex itself will be extremely rapid in comparison with the rate at which S decreases and X and Y increase. As a result C_{SE} will reach a virtual *steady state* almost instantaneously. The situation is closely analogous to the diffusion of solute across a thin membrane which was discussed in Section 7-6. Because the membrane was too small to *contain* an appreciable amount of solute, it could at all times be assumed that the rate of entry of solute into the membrane at one face was exactly equaled by the rate of exit of solute from the membrane at the other face. Similarly here, when the concentration of enzyme is sufficiently small we may assume that by the time we are able to measure the initial rate of reaction, C_{SE} will already have reached a steady state, its rate of synthesis from S and E equaling its rate of breakdown. However, as Straus and Goldstein (110) have clearly shown, with relatively large concentrations of enzyme and relatively small concentrations of substrate this assumption is not justified; the simple Michaelis-Menten formulation is then no longer valid, and a more elaborate analysis must be undertaken.

According to Reaction 11b, by the law of mass action the rate of change of C_{SE} will be

$$dC_{SE}/dt = k_{S+E}C_S C_E - (k_{SE.\text{fore}} + k_{SE.\text{back}})C_{SE} \tag{11-1}$$

But by Assumption 2, $dC_{SE}/dt = 0$. Hence, Equation 11-1 becomes

$$k_{S+E}C_S C_E = (k_{SE.\text{fore}} + k_{SE.\text{back}})C_{SE} \tag{11-2}$$

or,

$$C_S C_E/C_{SE} = (k_{SE.\text{fore}} + k_{SE.\text{back}})/k_{S+E} \tag{11-2a}$$

Now let us define the Michaelis constant, k_M, as

$$k_M = (k_{SE.\text{fore}} + k_{SE.\text{back}})/k_{S+E} \tag{11-3}$$

Notice that the Michaelis constant is not a rate constant, nor an affinity constant, nor a dissociation constant, but is merely a constant of convenience. Substituting k_M into Equation 11-2a and dividing both sides by C_S,

$$k_M/C_S = C_E/C_{SE} \tag{11-4}$$

By definition, the total concentration of enzyme, $C_{\text{tot}.E}$, is

$$C_{tot.E} = C_E + C_{SE} \tag{11-5}$$

Combining Equations 11-4 and 11-5 with the elimination of C_E,

$$k_M/C_S = (C_{tot.E}/C_{SE}) - 1 \tag{11-6}$$

Now since C_{SE} is assumed to be constant, the initial* rate (or velocity) of reaction, v, will be given by the rate at which the enzyme-substrate complex decomposes into the products X and Y:

$$(-dC_S/dt)_0 = v = k_{SE.fore}C_{SE} \tag{11-7}$$

Furthermore, the initial rate of reaction is assumed to be maximal when all of the enzyme is occupied by substrate, *i.e.*, when $C_{SE} = C_{tot.E}$. Applying Equation 11-7 to this special case,

$$v_{max} = k_{SE.fore}C_{tot.E} \tag{11-8}$$

Combining Equations 11-6, 11-7, and 11-8 with the elimination of C_{SE} and $C_{tot.E}$,

$$k_M/C_S = (v_{max}/v) - 1 \tag{11-9}$$

Solving Equation 11-9 for v,

$$v = v_{max}/[(k_M/C_S) + 1] \tag{11-9a}$$

Equation 11-9a is the desired expression for the rate of reaction as a function of concentration of substrate. By dividing both sides of Equation 11-9a by v_{max} we obtain a very simple expression for v/v_{max} which is the initial rate of reaction expressed as a proportion of the maximum initial rate:

$$v/v_{max} = C_S/(k_M + C_S) \tag{11-9b}$$

When the rate of reaction expressed as a fraction of the maximum rate is plotted according to Equation 11-9b against the concentration of substrate, a rectangular hyperbola is obtained (Fig. 11-1a). The fraction rises from zero when C_S is zero toward unity as C_S approaches infinity. Notice that when the rate is half-maximal, the Michaelis constant is numerically equal to the concentration of substrate.

11-3. Linear Transformations of the Michaelis-Menten Relationship

Though the form of Equation 11-9b is pleasingly simple, it is not well suited to the calculation of v_{max} and k_M from experimental observations of

---

* To emphasize that v and v_{max} as used here *always* mean the *initial* rates of reaction, it might be better to call them v_0 and $v_{max.0}$. Custom, however, sanctions the simpler notation.

the initial rate of reaction at various substrate concentrations. For this purpose it is usual to employ some linear transformation in which the quantities to be plotted against each other can be calculated entirely from the observed values of C_S and v. The parameters v_{max} and k_M can then be obtained from the slope and intercept of the straight line fitted to the points. There are three nonlogarithmic linear transformations of Equation 11-9.

1. Solving Equation 11-9 for $1/v$ we obtain

$$1/v = 1/v_{max} + (k_M/v_{max})(1/C_S) \qquad (11\text{-}9c)$$

according to which a plot of $1/v$ against $1/C_S$ will give a straight line with a slope of k_M/v_{max} and an intercept of $1/v_{max}$. This is the *Lineweaver-Burk method* of plotting the data.

2. Solving Equation 11-9 for C_S/v we obtain

$$C_S/v = (k_M/v_{max}) + (1/v_{max})C_S \qquad (11\text{-}9d)$$

Notice that the intercept in Equation 11-9c is the slope of Equation 11-9d, and the slope of 11-9c is the intercept of 11-9d.

3. Solving Equation 11-9 for v in terms of (v/C_S) we obtain

$$v = v_{max} - k_M(v/C_S) \qquad (11\text{-}9e)$$

according to which a plot of v against v/C_S would give a straight line with a slope of $-k_M$ and an intercept of v_{max}.

Figure 11-1 illustrates the results of plotting the same data according to Equation 11-9b and according to the three rectilinear methods just outlined. The "observed" values of v for various substrate concentrations (Table 11-2) were calculated from Equation 11-9a by assuming $v_{max} = (0.34 \text{ mM./L.})/\text{min.}$ and $k_M = 12 \text{ mM./L.}$ These values of v, evenly spaced at intervals of 10 per cent of v_{max}, are represented by the *solid dots* in Figure 11-1. Since they were calculated from an equation, they are errorless. If the actual experimental observations of the initial reaction velocity were equally free from error, it would make little difference which method of plotting the data was used. In fact, however, though it may be possible to control the initial concentration of substrate precisely, the measurement of initial velocity will always be subject to more or less experimental error. To illustrate this, broken lines have been drawn in each graph showing the deviations from the trend of the errorless points which would be caused by an error of $\pm(0.005 \text{ mM./L.})/\text{min.}$ (about ± 1.5 per cent of v_{max}) in the measurement of v. In the present example it is assumed that this error is constant regardless of the absolute magnitude of v.

Each method of plotting has its special advantages and disadvantages which, though not easy to describe quantitatively, are important to keep in mind when analyzing actual data. The Lineweaver-Burk method (Fig.

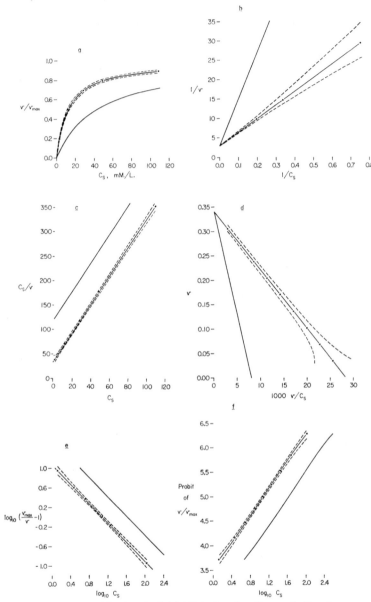

FIG. 11-1. Ways of Plotting the Initial Velocity of an Enzymatic Reaction as a Function of Substrate Concentration

The data of Table 11-2, calculated by assuming a maximum velocity of 0.34 mM. per liter per minute and a Michaelis constant of 12 mM. per liter, are here plotted as

TABLE 11-2

Initial velocity of reaction as a function of substrate concentration

The data tabulated here and plotted in Figure 11-1 were calculated from Equation 11-9a by assuming that $v_{max} = (0.34 \text{ mM./L.})/\text{min.}$ and that $k_M = 12 \text{ mM./L.}$

Substrate Concentration, C_S	Initial Velocity, v	Initial Velocity as Proportion of Maximum
mM./L.	*mM./L.-min.*	v/v_{max}
1.333	0.034	0.1
3.000	0.068	0.2
5.143	0.102	0.3
8.000	0.136	0.4
12.00	0.170	0.5
18.00	0.204	0.6
28.00	0.238	0.7
48.00	0.272	0.8
108.00	0.306	0.9

11-1*b*) has the advantage that the independent and dependent variables are completely separated so that there is no need to worry about any inevitable correlation. But the method suffers from the very serious defect that even a small experimental error has a relatively enormous influence upon the position of the points which represent the slower rates of reaction. The more reliable points, which should be given far more weight, are bunched together near the intercept. The dangers of ignoring the effects of this distortion of error were made abundantly clear in Section 3-15. *If the Lineweaver-Burk method of plotting is used, a least-squares line should never be fitted to unweighted data.* A line fitted by eye which is required to pass as nearly as possible through the more reliable points may give a good estimate of the intercept, $1/v_{max}$, but the slope, k_M/v_{max}, is likely to be subject to considerable uncertainty.

When C_S/v is plotted against C_S (Fig. 11-1*c*) some correlation between the plotted variables is inevitable so that the scattering of points due to

solid dots in the several ways described in the text. In each portion of the figure, the influence of a constant error of 0.005 mM. per liter per minute above and below the true value of the velocity is indicated by the *broken lines*. (The concentration of substrate is supposed to be measured without error.) The *solid line* which is unaccompanied by *broken lines* shows how the relationship between substrate concentration and velocity of reaction would be altered in the presence of 64 mM. per liter of a reversible competitive inhibitor with a dissociation constant for the enzyme-inhibitor complex of 26.2 mM. per liter.

errors in v tends to be unduly suppressed. Furthermore, the points corresponding to low initial rates of reaction are "bunched" toward the lower left of the curve. Although it looks from the graph as if the intercept could be established quite accurately, in fact the intercept lies so close to the origin that an apparently small uncertainty of position corresponds to a relatively large uncertainty percentagewise in estimating k_M/v_{max}. However, in comparison with the Lineweaver-Burk plot, the error in C_S/v produced by a given error in v is relatively independent of C_S. Consequently, the slope of the line, and hence v_{max}, can be calculated with considerable precision. With this method of plotting, an unweighted least-squares line might be fitted to the points without courting disaster.

When v is plotted against v/C_S, some correlation is again inevitable. Worse still, v, the variable subject to error, influences both coördinates. The method has the trivial advantages that v_{max} can be read directly from the intercept of the line on the v axis, and that k_M is simply the negative of the slope. Finally, this is the only one of the three methods of plotting in which the points (representing equal increments of velocity) are equally spaced. But weighting of the points would be needed if a line were to be fitted by the method of least squares.

The reader may be interested in a recent discussion of these methods (59). So far as the author is aware, no critical statistical analysis of their relative merits is available.

There are yet other methods by which these data can be plotted. Reiner (86) has shown that for certain purposes a plot of $(v/v_{max})/[1 - (v/v_{max})]$ as a function of C_S is advantageous (Exercise 5). Proof that

$$\log\left[\frac{v_{max}}{v} - 1\right]$$

is a linear function of $\log C_S$ (Fig. 11-1e) has been assigned as an exercise (Exercise 6). Moreover, a plot of the probit of v/v_{max} against $\log C_S$ will yield a line which is practically straight except at its extremes (Fig. 11-1f). Further discussion of such probit plots will be found in Section 11-12.

11-4. Enzyme Inhibitors

v_{max} and k_M are often estimated as the first steps in studying the mode of action of enzyme inhibitors. A complete account of the several mechanisms whereby inhibitory substances can reduce or abolish the activity of an enzyme must be sought elsewhere (86). Only two contrasting types of enzyme inhibition will be discussed here.

11-5. Reversible Competitive Inhibition

Many important inhibitors act by combining reversibly with the enzyme at the same site as the substrate itself. When the active group on the enzyme is thus occupied by a molecule of the inhibitor (which may or may not itself be a substrate), the usual substrate is excluded and that particular molecule of enzyme is inhibited. But since both the combination between enzyme and inhibitor and the combination between enzyme and substrate are reversible, the proportion of the enzyme molecules occupied by inhibitor and the proportion occupied by substrate will depend upon the concentrations of inhibitor and substrate, and upon the "affinities" of the enzyme for inhibitor and substrate. In a sense, the substrate and the inhibitor "compete" with each other for attachment to the same active site on the enzyme molecule—hence the term "reversible competitive inhibition" for this type of inhibition.

Let I be the inhibitor, and IE the enzyme-inhibitor complex. Then reversible competitive inhibition may be diagramed as follows:

$$E + S \; \underset{k_{SE.back}}{\overset{k_{E+S}}{\rightleftarrows}} \; SE \; \underset{k_{E+X+Y}}{\overset{k_{SE.fore}}{\rightleftarrows}} \; E + X + Y \qquad (11d)$$
$$+$$
$$I$$
$$k_{E+I} \; \updownarrow \; k_{IE}$$
$$IE$$

By applying our previous assumptions to this system, it may be simplified to

$$E + S \; \underset{k_{SE.back}}{\overset{k_{E+S}}{\rightleftarrows}} \; SE \; \overset{k_{SE.fore}}{\longrightarrow} \; E + X + Y \qquad (11e)$$
$$+$$
$$I$$
$$k_{E+I} \; \updownarrow \; k_{IE}$$
$$IE$$

Equations 11-4, 11-7, and 11-8 are still valid:

$$k_M/C_{S_I} = C_E/C_{SE} \qquad (11\text{-}4)$$

$$v = k_{SE.fore}C_{SE} \qquad (11\text{-}7)$$

$$v_{\max} = k_{SE.fore}C_{tot.E} \qquad (11\text{-}8)$$

where

C_{S_I} = the concentration of substrate which will produce an initial reaction velocity of v when the inhibitor, I, is present in a concentration of C_I.

However, Equation 11-5 now becomes

$$C_{E.tot} = C_E + C_{SE} + C_{IE} \qquad (11\text{-}10)$$

In addition, we must consider the equilibrium between enzyme, inhibitor, and enzyme-inhibitor complex:

$$C_E C_I / C_{IE} = (Keq)_{IE} \tag{11-11}$$

where

$(Keq)_{IE}$ = the equilibrium constant for the dissociation of the enzyme-inhibitor complex.

Now let us combine Equations 11-4, 11-7, 11-8, 11-10, and 11-11 to obtain an expression for $1/v$ in terms of the known variables C_I and C_{S_I} and various constants. (The individual steps are not given here, since the derivation of Equation 11-12 is assigned as Exercise 2A in Chapter 13):

$$\frac{1}{v} = \frac{1}{v_{max}} + \frac{k_M}{v_{max}} \left(1 + \frac{C_I}{(Keq)_{IE}}\right) \frac{1}{C_{S_I}} \tag{11-12}$$

Equation 11-12 is the counterpart of Equation 11-9c, and, in fact, reduces to equation 11-9c when $C_I = 0$. Equations for reversible competitive inhibition which are analogous to Equations 11-9d and 11-9e may also be derived (Exercise 8). Figure 11-1 illustrates how a reversible competitive inhibitor with a $(Keq)_{IE}$ of 26.2 mM. per liter at a constant concentration of 64 mM per liter would influence the velocities of reaction in the previous example, and how the various transformations would be altered. Notice that a reversible competetive inhibitor does not change the maximum rate of reaction which is approached as the concentration of substrate is increased toward infinity, but merely increases the concentration of substrate needed to achieve a given proportion of the maximum rate. This is indeed just what one would expect when substrate and inhibitor compete with each other for attachment to the same active group on the enzyme molecule. Even with a high concentration of an inhibitor which has a high affinity for the enzyme, i.e., a small dissociation constant, $(Keq)_{IE}$, the inhibition can be practically abolished if we can increase substrate concentration until the molecules of enzyme are much more likely to combine with substrate than with inhibitor.

We have just seen (Fig. 11-1) how a *constant* concentration of a reversible competitive inhibitor changes the concentration of substrate needed to achieve various reaction velocities. It is also instructive to calculate the concentrations of substrate needed to achieve a constant fraction of the maximum velocity in the presence of *various* concentrations of inhibitor Solving Equation 11-12 for C_{S_I},

$$C_{S_I} = k_M \left(\frac{v/v_{max}}{1 - (v/v_{max})}\right) \left(\frac{C_I}{(Keq)_{IE}} + 1\right) \tag{11-12a}$$

For convenience, define $F_{v_{max}}$ as the velocity expressed as a fraction of the maximum velocity:

$$F_{v_{max}} = v/v_{max} \qquad (11\text{-}13)$$

Then Equation 11-12a may be written

$$C_{S_I} = k_M \left(\frac{F_{v_{max}}}{1 - F_{v_{max}}} \right) \left(\frac{C_I}{(Keq)_{IE}} + 1 \right) \qquad (11\text{-}14)$$

But when $C_I = 0$, Equation 11-14 reduces to

$$C_{S_{I=0}} = k_M \frac{F_{v_{max}}}{1 - F_{v_{max}}} \qquad (11\text{-}15)$$

where

$C_{S_{I=0}}$ = the concentration of substrate required to achieve $F_{v_{max}}$ when no inhibitor is present

Combining Equations 11-14 and 11-15,

$$C_{S_I} - C_{S_{I=0}} = \frac{C_{S_{I=0}}}{(Keq)_{IE}} (C_I) \qquad (F_{v_{max}} \text{ constant}) \qquad (11\text{-}16)$$

Equation 11-16 shows that when a reversible competitive inhibitor is present, the *additional* concentration of substrate, $C_{S_I} - C_{S_{I=0}}$, which is needed to achieve a given proportion of the maximum rate of reaction is a constant fraction, $C_{S_{I=0}}/(Keq)_{IE}$, of the concentration of inhibitor. For instance, in the example discussed above, a substrate concentration of 12 mM. per liter would be needed for a reaction velocity which is half of the maximum velocity ($F_{v_{max}} = 0.5$). Equation 11-16 shows that in the presence of an inhibitor whose $(Keq)_{IE} = 26.2$ mM. per liter, an *additional* concentration of substrate equal to 12/26.2 of the concentration of inhibitor would be needed for the same half-maximal velocity. This example is further illustrated in Figure 11-2, where C_{S_I} (*solid line*), and $C_{S_I} - C_{S_{I=0}}$ (*broken line*) have been plotted against C_I. For convenience, both scales in Figure 11-2 are logarithmic. Notice that when the concentration of inhibitor is large, so that $C_{S_{I=0}}$ is negligible in comparison with C_{S_I}, the two lines practically coincide. Under these circumstances, the ratio of substrate concentration to inhibitor concentration for a given rate of reaction is practically constant. If the concentration of inhibitor is doubled, the concentration of substrate must also be doubled to restore the original rate of reaction. But this simple relationship does not hold for small concentrations of the inhibitor. In contrast, the ratio of the *increment* of substrate concentration, $C_{S_I} - C_{S_{I=0}}$, to the concentration of inhibitor remains constant throughout the entire range of inhibitor concentrations. In passing, notice that when two variables whose ratio is constant are plotted against each other on log-log graph paper, the slope of the resulting straight line is unity (see, for example, the *broken line* in Fig. 11-2). The reader should have no difficulty in explaining why this is true.

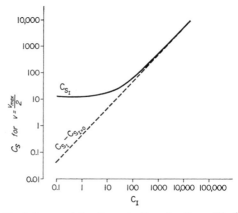

FIG. 11-2. The Influence of the Concentration of a Reversible Competitive
Inhibitor upon the Concentration of Substrate Needed for a
Half-Maximal Reaction Velocity

As in Figure 11-1, v_{max} is 0.34 mM. per liter per minute, the Michaelis constant is 12 mM. per liter, and the dissociation constant of the enzyme-inhibitor complex is 26.2 mM. per liter. Notice that both scales are logarithmic. $C_{S_I} - C_{S_{I=0}}$ is the *additional* substrate concentration needed for a half-maximal velocity in the presence of the inhibitor. It is this increment of substrate concentration (*broken line*), not the substrate concentration itself (*solid curve*), which gives a straight line when plotted on log-log paper against concentration of inhibitor.

Two of the exercises at the end of this chapter deal further with the general problem of reversible competitive inhibition.

11-6. Irreversible Competitive Inhibition

As an example of a quite different type of enzyme inhibition, let us consider an inhibitor which, as before, competes with substrate for attachment to the active group of the enzyme. However, let us now suppose that instead of combining reversibly with the enzyme (as the substrate does) the inhibitor combines *irreversibly*:

$$E + S \underset{k_{SE.back}}{\overset{k_{E+S}}{\rightleftarrows}} SE \underset{k_{E+X+Y}}{\overset{k_{SE.fore}}{\longrightarrow}} E + X + Y \qquad (11f)$$
$$+$$
$$I$$
$$\downarrow k_{E+1}$$
$$IE$$

Under these circumstances, competition between enzyme and substrate for occupation of the active group of the enzyme molecule can occur initially, but no mass-action equilibrium between substrate, enzyme, and inhibitor is possible. Once a molecule of inhibitor becomes irreversibly attached to the enzyme, no concentration of substrate, however high, can displace it, and that particular enzyme molecule is permanently *hors de combat.* Under these conditions, a high concentration of substrate can delay, but it cannot prevent, the total inactivation of enzyme which will eventually occur if there is at least one molecule of inhibitor for each active enzyme group in the reaction mixture.* However, the development of this type of irreversible enzyme inhibition is often slow enough so that the inhibitor can be removed or destroyed before all of the enzyme has been inactivated. If this be done, some of the enzyme molecules will retain their original affinity for substrate, just as if they had never been exposed to the inhibitor. The system will thus behave exactly as if the total concentration of enzyme, and consequently v_{max}, had been reduced. The original Equations 11-9 through 11-9e will therefore be perfectly valid, except that v_{max} must be replaced by v_{max_I}, the maximum velocity after the inhibitor has irreversibly inactivated some of the enzyme. This maximum velocity will be equal to the maximum velocity without inhibitor, $v_{max_{I=0}}$, multiplied by the fraction of the enzyme which remains active after inhibition:

$$v_{max_I} = \left(\frac{C_{E.tot} - C_{IE}}{C_{E.tot}} \right) v_{max_{I=0}} \tag{11-17}$$

For example, by substituting v_{max_I} from Equation 11-17 for v_{max} in Equation 11-9a, we obtain

$$v_I = \left(\frac{C_{E.tot} - C_{IE}}{C_{E.tot}} \right) (v_{max_{I=0}}) \bigg/ \left(\frac{k_M}{C_S} + 1 \right) \tag{11-18}$$

But by Equation 11-9a, the expression

$$(v_{max_{I=0}}) \bigg/ \left(\frac{k_M}{C_S} + 1 \right)$$

is $v_{I=0}$. Hence, Equation 11-18 may be written

$$v_I = \left(\frac{C_{E.tot} - C_{IE}}{C_{E.tot}} \right) v_{I=0} \tag{11-19}$$

Equation 11-19 shows that with irreversible inhibition, the initial velocity of reaction produced by *any* concentration of substrate is simply equal to the velocity without inhibitor multiplied by the fraction of enzyme remaining active.

---

* We are here assuming what is not, in fact, very likely; namely, that the inhibitor undergoes no reaction except with the active groups on the enzyme molecules.

11-7. The Combination between a Drug and Its Receptor

With rare exceptions, a drug can produce an effect only by combining with some more-or-less specific cellular constituent at its site of action. For a still select, but gradually increasing number of drugs, the cellular constituent has been clearly identified. For example, several drugs act (at least in part) by inhibiting the enzyme acetylcholinesterase; the drug acetazolamide acts by inhibiting carbonic anhydrase; and carbon monoxide acts by competing with oxygen for attachment to the ferrous iron atoms in hemoglobin. But for the great majority of drugs the precise chemical nature of the cellular constituent with which they combine is still unknown. We must then employ the perfectly general term "drug receptor" or simply "receptor" for the molecular site of action of the drug.

A small minority of drugs form a firm, covalent bond with their receptor, so that the drug-receptor combination is essentially irreversible. But the great majority of drug-receptor combinations are freely reversible, the proportion of the receptors occupied by drug at any moment being determined in accordance with the law of mass action by the concentration of drug to which the receptors are exposed. It is hardly surprising that the action of such drugs can be terminated by washing the drug out of a piece of isolated tissue suspended in a bath, or by the more gradual removal of drug from an intact organism by metabolism, or excretion, or both. Furthermore, as one would expect, the intensity of the response to such drugs increases as the concentration of drug increases, at least over a very wide range of concentrations. Responses of this kind are called *graded* responses to distinguish them from the *all-or-none* responses described later.

11-8. Concentration-Effect Relationships

If an isolated tissue—let us say a bit of smooth muscle—is exposed to various concentrations of a drug, D, and the intensity of the effect (Eff) of the drug is plotted against the concentration of the drug, C_D, one usually finds empirically that the points can be fitted by a hyperbolic curve (Fig. 11-3a) which is strongly reminiscent of the curve shown in Fig. 11-1a. To convert such a curve into a symmetrical form, pharmacologists have long been accustomed to plot (Eff) as a function of $\log_{10} C_D$ rather than as a function of C_D itself (Fig. 11-3b), thus obtaining a symmetrical sigmoid curve whose mid-portion approximates a straight line. We shall now see how one particular curve of this sort can be derived by applying the law of mass action to the combination between drug and receptor. The derivation is closely analogous to the one used for Equation 11-9, the corresponding symbols, assumptions, etc., being listed in Table 11-1.

Assume that the reaction between drug D and its receptor, R, to form

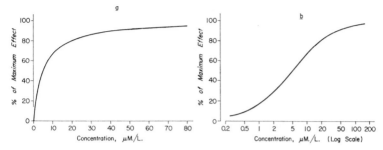

FIG. 11-3. The Relationship between the Concentration of a Drug and Its Effect

In most simple biological systems, the intensity of the effect of a drug varies with the concentration of drug in the fluid bathing the system. When the intensity of such a *graded* effect, expressed as per cent of the greatest possible effect, is plotted against the concentration of drug, one commonly obtains a rectangular hyperbola such as the one at the *left* of the figure. Because the first part of the curve rises very steeply, the gradation of response according to the concentration of drug is displayed to better advantage by plotting per cent of maximum effect against the logarithm of the concentration as is done at the *right*.

the drug-receptor complex, DR, can be represented as

$$D + R \xrightleftharpoons[k_{DR}]{k_{D+R}} DR \tag{11g}$$

(According to this assumption, one molecule of drug combines with one receptor, and all receptors are identical.) By the law of mass action, it follows directly from 11g that at equilibrium

$$\frac{C_D C_R}{C_{DR}} = \frac{k_{DR}}{k_{D+R}} = (Keq)_{DR} \tag{11-20}$$

where $(Keq)_{DR}$ is the dissociation constant for the drug-receptor complex. Equation 11-20 is analogous to Equation 11-4. Now since

$$C_{R.tot} = C_R + C_{DR} \tag{11-21}$$

Equation 11-20 may be rewritten as

$$(Keq)_{DR}/C_D = \frac{C_{R.tot}}{C_{DR}} - 1 \tag{11-22}$$

which is analogous to Equation 11-6.

In the previous derivation, the next equation, 11-7, stated that the velocity of the enzyme-catalyzed reaction is proportional to the concentration of the enzyme-substrate complex. In the present analysis of the relation of drug concentration to effect, we are not concerned with reaction velocity. Nevertheless, we must make a somewhat analogous assumption; namely, that the effect of the drug is directly proportional to the concentration of the drug-receptor complex:

$$(\text{Eff}) = kC_{DR} \tag{11-23}$$

where k is a constant of proportionality with appropriate dimensions. We shall also assume, in analogy with Equation 11-8, that the maximum effect, $(\text{Eff})_{max}$, occurs only when all receptors are occupied by drug, i.e., when $C_{DR} = C_{R.tot}$:

$$(\text{Eff})_{max} = kC_{R.tot} \tag{11-24}$$

Combining Equations 11-22, 11-23, and 11-24, with the elimination of C_{DR} and $C_{R.tot}$, and solving for (Eff),

$$(\text{Eff}) = \frac{(\text{Eff})_{max}C_D}{(K\text{eq})_{DR} + C_D} \tag{11-25}$$

Now Equation 11-25 is precisely analogous to Equation 11-9a and could accordingly be written in any one of the ways suggested by Equations 11-9b, c, d, and e. What is more, any concentration-effect relationship which can be described by Equation 11-25, could be depicted graphically in any one of the ways shown in Figure 11-1 merely by substituting C_D for C_S, (Eff) for v, and $(\text{Eff})_{max}$ for v_{max}.

11-9. Real Concentration-Effect Curves

Although a number of real concentration-effect relationships can be described very well by Equation 11-25 (40) the assumptions used in its derivation are more noteworthy for their simplicity than for their credibility. For example, there is substantial reason to suspect that, with certain receptors, a maximum effect can be attained when only a small proportion of the available receptors are actually occupied by drug (109). Furthermore, the notion that the intensity of the effect is directly proportional to the concentration of drug-receptor complex can most charitably be described as naïve. Finally, many a real concentration-effect relationship is *not* well fitted by Equation 11-25. If we insist upon using the law of mass action to account for such curves, we are forced to abandon the one drug-one receptor scheme of 11g and to postulate that n molecules of drug combine with each receptor. Now it is true that a wide variety of concentration-effect curves can be fitted by the corresponding general equation:

$$(\text{Eff}) = \frac{(\text{Eff})_{max}C_D{}^n}{K + C_D{}^n} \tag{11-26}$$

However, n does not necessarily turn out to be a simple fraction or a small integer as one might expect on the basis of reasonable molecular relationships. In our present state of ignorance about drug-receptor combinations, it would seem best to regard Equation 11-26 as a useful device for describing certain concentration-effect curves empirically rather than as a theoretical equation with a well-established fundamental meaning.

It would be entirely wrong to derogate the importance of trying to explain concentration-effect curves on the basis of mass-action equilibria. The concept that drug effects are in some way quantitatively related to the occupation of receptors by molecules of drug has proved enormously fruitful, particularly in accounting for the action of drug antagonists whose similarities to enzyme inhibitors are remarkable. Commendable ingenuity has been—and continues to be—shown in modifying the simple theory presented above so that it can be applied to a wider range of data. Details of this fascinating, at times bewildering, field must be sought elsewhere (6, 42, 109). We must now divert our attention to a quite different explanation of certain concentration-effect, or, more properly, dose-response relationships.*

11-10. Quantal ("All-or-None") Responses. Log-Normal Distributions of Sensitivities

Suppose that we want to measure the toxicity of a drug by finding out experimentally how much of the drug is needed for a lethal effect in mice. Now death is an all-or-none affair; a particular mouse is either dead or it is not dead. In this experiment, therefore, we shall be studying a *quantal* or *all-or-none* response to a drug. Suppose further that we could find, for each individual mouse, the smallest dose of the drug which would just suffice to kill that particular mouse. Finally, suppose that we could study a very large number, say 400, individual mice in this way. Let us now count up the number of mice for which the individual lethal dose lies between 3 and 4 mg. per kilogram, between 4 and 5 mg. per kilogram, between 5 and 6 mg. per kilogram, etc., and plot the number of mice in each such dose in-

---

* The *dose* of a drug is the total quantity given at one time to a whole animal. The receptors upon which the drug acts are never "aware of" this total dose but respond only to the *concentration* of the drug to which they are exposed. However, with a particular drug and a particular animal we may usually assume that the concentration of drug at the receptors will be directly proportional to the dose administered. When dealing with intact animals (rather than with isolated tissues bathed in a medium of known drug concentration) it is therefore both common and correct to speak of a *dose-effect* or *dose-response* relationship rather than a concentration-effect relationship. But if the assumption that there is a direct proportionality between dose and concentration is valid, the fundamental significance of a dose-response curve is exactly the same as that of a concentration-effect curve.

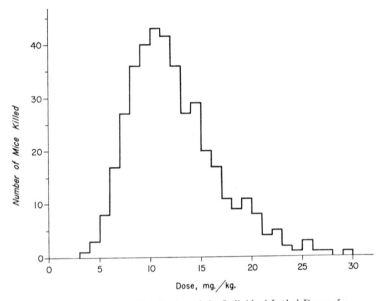

Fig. 11-4. Frequency Distribution of the Individual Lethal Doses of a
Drug Tested in a Group of 400 Mice

Notice that the distribution is not symmetrical but is skewed to the *right*.

terval against the dose. We would thus obtain a curve (or actually, with a
finite number of mice, the tops of many narrow bars, in this example, 1
mg. per kilogram wide) showing the frequency distribution of individual
lethal doses among the mice studied (Fig. 11-4). Now like most biological
variables, these lethal doses are not normally distributed unless we plot
the frequency against the *logarithm* of the lethal dose (Fig. 11-5). We may
therefore conclude that the sensitivity* of the mice to the lethal effect of
the drug is log-normally distributed, rather than normally distributed.

It is rarely practicable to determine the individual lethal dose for each
animal as was suggested above. Much more commonly, the available mice
are divided into several groups. Each mouse in a particular group is then
given the same dose of drug per kilogram of body weight, but different
doses, spaced equally apart on a logarithmic scale, are given to the various

---

* The *sensitivity* of a single animal to the lethal effect is inversely proportional to
the dose needed to kill that animal. We might equally well have said that the *tolerance*
of the mice for the lethal effect was log-normally distributed, tolerance being directly
proportional to the individual lethal dose.

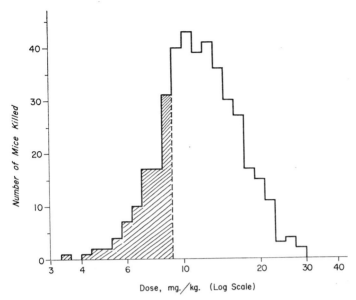

Fɪɢ. 11-5. Frequency Distribution of the *Logarithms* of the Individual Lethal Doses Shown in Figure 11-4

This distribution is much more symmetrical than the one in Figure 11-4, and conforms reasonably well to a curve of normal distribution. Studies of this kind show that the individual dose of a drug required to produce an "all-or-none" response (such as death) is usually log-normally distributed in a population of animals, and that the tolerance for (or sensitivity to) the drug is therefore log-normally distributed. There is nothing unusual about this, for in fact most biological characteristics are log-normally rather than normally distributed (41). The *shaded area* in this figure comprises 23 per cent of the total area.

groups. We would then plot the *per cent mortality* within a group as a function of the log dose. Let us use the data from the much more elaborate experiment described above to predict what the relationship between the *logarithm of the dose* and the *per cent mortality* would be. Suppose, for example, that all 400 mice in the previous experiment had received a dose of exactly 9.1 mg. of drug per kilogram of body weight. We can logically assume that this dose would have killed all of the 92 mice which were in fact killed by this dose, or by some smaller dose. These 92 mice, falling in the *shaded area* of Figure 11-5 represent 23 per cent of the total group of 400 mice. The *shaded area* is the integral of the (approximately) normal distribution curve in Figure 11-5 between the limits of dose = 0 (log dose =

$-\infty$), and the dose of 9.1 mg. per kilogram (log dose $= 0.959$). It represents 23 per cent of the total area under the curve between the limits of log dose $-\infty$ and log dose $+\infty$. By cumulating the results for the mice tested individually, we thus reach the conclusion that a dose of 9.1 mg. per kilogram kills about 23 per cent of mice. Presumably we would have reached the same conclusion had the mice actually been tested as a single group.

By using the method just described, let us find what per cent of the 400 mice which would have been killed by various single doses. We can then plot the per cent killed (the "per cent response") against the logarithm of the dose so as to construct a log dose-per cent response curve (Fig. 11-6). In its general shape, this curve is remarkably similar to the log concentration-effect curve of Figure 11-3.

11-11. A Linear Transformation of Log Dose-Per Cent Response Curves. Probit Plots

A log-normal distribution of tolerances to the lethal effect of a drug upon mice was discussed in detail above. This kind of log-normal distribution of tolerance to *any* all-or-none effect of a drug is so very common that we can usually *assume* it to be present even when the number of animals studied is too small to provide any real evidence for a log-normal distribution. The assumption makes it possible for us to use a simple linear transformation of log dose-per cent response curves.

In any normal distribution, the area beneath the curve which is bounded by $-\infty$ at the extreme left and by any particular value of the variable on the abscissa (here log dose) can be found as a proportion of the total area by consulting a table of areas of the curve of normal distribution. But to use the table, we must express the value of the variable on the abscissa not in the original units of log dose but rather as its deviation (measured in units of standard deviation) from the mean of the distribution. (Using this "standard measure" allows a single table to serve for all normal distributions.) Such a table shows, for example, that 15.87 per cent of the total area lies to the left of the abscissa which is -1.00 standard deviation away from the mean, 93.82 per cent of the total area lies to the left of the abscissa which is $+1.54$ standard deviations away from the mean, etc. Now if our frequency distribution is log-normally distributed, we should be able to *work the table backwards*, so to speak, and look up for each *observed* per cent response its *normal equivalent deviate* which is simply the corresponding tabulated deviation from the mean, expressed in units of standard deviation. If these normal equivalent deviates are then plotted, instead of the actual percentages, against the log dose, we should obtain a straight line whose slope will be the reciprocal of the standard deviation, and whose intercept on the x axis, where the standard deviation is zero,

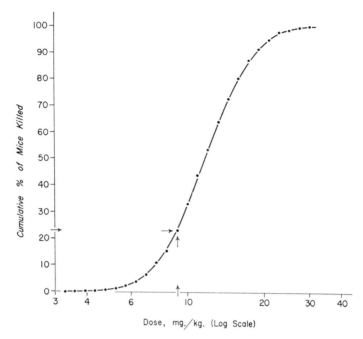

Fig. 11-6. Cumulative Per Cent Mortality as a Function of Logarithm of the Dose

Each point in this graph represents the percentage of the entire group of 400 mice (Figs. 11-4 and 11-5) which were killed by the corresponding dose *or by a smaller dose*. For example, as was shown by the *shaded area* in Figure 11-5, 92 mice, or 23 per cent of the whole group, were killed by a dose equal to or smaller than 9.1 mg. per kilogram. In the present figure, therefore, the point indicated by the arrows has been plotted at a cumulative mortality of 23 per cent and at a dose of 9.1 mg. per kilogram. Notice that the log dose-per cent response curve so generated is a symmetrical sigmoid curve quite similar to the log concentration-graded effect curve of Figure 11-3.

will correspond to the mid-point of the distribution and will therefore estimate the log of the LD_{50} , *i.e.*, the dose which would kill half the popula-lation of mice. Merely to avoid working with the negative standard devia-tions which occur to the left of the mean, it is usual to add an arbitrary value of 5.00 to each normal equivalent deviate, the resulting sum then being known as a *probit* ("*probability unit*"). Thus, in our example, a mortality of 23 per cent corresponds to a normal equivalent deviate of -0.74, or to a probit of $-0.74 + 5.00 = 4.26$. This value has been plotted against the log of the corresponding dose, 9.1 mg. per kilogram, in Figure 11-7 together with similarly calculated probits for the other per cent re-

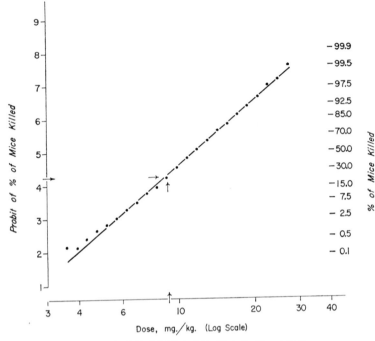

Fig. 11-7. A Probit Plot of the Data in Figure 11-6

Provided the sensitivity of individual mice to the lethal effect of the drug is log normally distributed, the probit of the cumulative per cent mortality shown in Figure 11-6 will give a straight line when plotted against the logarithm of the dose. The point indicated by the *arrows* at a probit of 4.26, corresponding to a mortality of 23 per cent, is the same point as was indicated by *arrows* in the previous figure.

sponses deduced from the frequency distribution of Figure 11-5. Except for the points corresponding to very small or very large percentages (which carry little weight), the data are well fitted by a straight line. The standard deviation for the normal distribution of log doses in Figure 11-5 can now be estimated as 0.155 log unit, the reciprocal of the slope of the straight line. Furthermore, the value of the log dose where the line crosses the ordinate for probit = 5.00 is 1.06, whose antilog, 11.5 mg. per kilogram, is the LD$_{50}$ estimated from the probit plot.

Statistical methods for dealing with probits, *including the essential weighting factors*, have been worked out in great detail. They not only enable one to estimate the LD$_{50}$ or the ED$_{50}$ ("effective dose for 50 per cent of animals") and the standard deviation as we have just done, but they also provide estimates of the reliability with which these parameters have been approximated (37). Probit paper is commercially available and saves a good deal of time if much use is to be made of probit analysis.

11-12. Probit Plots of Graded Responses

We began our discussion of concentration-effect and dose-response relationships by considering the graded response of an isolated tissue to varying concentrations of a drug. It is a curious fact that plotting probits for such a graded response, expressed as per cent of the maximum response, against the logarithm of the concentration usually gives as good a straight line as is obtained from all-or-none responses. Now it is true that any macroscopic bit of tissue contains a great many cells; one might therefore with justice argue that the gradation of response with dose can be ascribed, at least in part, to a log-normal distribution of sensitivities among the individual cells. But this is by no means an inevitable conclusion. *Even if the gradation of effect were due entirely to simple mass-action equilibria* of the sort discussed above, *a plot of the probit of per cent response against log concentration would still give a practically straight line* (Fig. 11-1*f*). Indeed the curvature would be so small as scarcely to be noticed in plotting real data unless the data were exceptionally accurate and extensive. So, although the distinction between gradations of response due to mass-action equilibria and gradations of response due to differences in sensitivity of the biological units being tested is of profound theoretical importance, in practice, either of these mechanisms, or both together, will produce much the same kind of symmetrical, sigmoid, log dose-response curve.

11-13. A Logarithmic View of the Universe

People who are awed by the vastness of space and by the minuteness of the atom may well be inclined to feel that biological systems do not cut much of a figure in the general scheme of things. This is less than fair. The prevalence in nature of exponential ("logarithmic") growth and decay, and the overwhelming preponderance of log-normal distributions (41) suggest the propriety of adopting a logarithmic view of the universe. If we accordingly place lengths upon a logarithmic scale (Fig. 11-8), we see that biological objects actually cover about one-quarter of the total range of lengths at present comprehended by science. Even a single human being grows from a zygote some 2×10^{-4} meter "tall" to an adult slightly less than 2×10^{0} meters tall, a span of 4 log units or nearly one-tenth of the total. Besides offering a tongue-in-cheek defense of the importance of biology, Figure 11-8 shows that man, whose unquenchable curiosity bids him look as far as possible in both directions, is rather favorably situated within 5 or 6 log units of the mid-point of the scale. But we shall have to

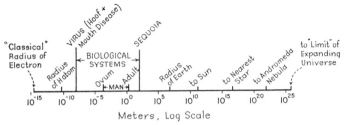

FIG. 11-8. A Logarithmic View of the Universe

See Section 11-13.

leave to the philosophers the question of whether man really sits so close to the center of things, or whether his apparently strategic position is the result of his being able to see only just so far up and down a scale which, for all we know, may actually extend an infinite distance in both directions.

EXERCISES. CHAPTER 11

Exercise 1

A biochemist working with a new enzyme preparation studied the initial rate at which the enzyme hydrolyzed its substrate, as a function of the substrate concentration. His measurement of the initial velocity of reaction, v (μmoles of substrate hydrolyzed per minute) was, of course, subject to experimental error, but the initial concentration of substrate (C_s mM. of substrate per liter) was adjusted without appreciable error. The following data were observed:

Experiment No.	C_s	v
1	1.000	1.67
2	0.500	1.33
3	0.250	0.95
4	0.125	0.66
5	0.080	0.41
6	0.050	0.33

In order to estimate the maximum initial velocity and the Michaelis constant, the biochemist plotted $1/v$ against $1/C_s$ according to the method of Lineweaver and Burk, so that the data, thus transformed, could be fitted by a straight line. He decided to use the method of least squares to find the best straight line through the points.

A. Assuming that the value of v in each of the experiments is subject to an experimental error of ± 0.03 μmole per minute (constant absolute error), what weight should the biochemist give to the data of experiment 6 relative to the weight assigned to the data of experiment 1?

B. Assuming that the value of v in each of the experiments is subject to an experimental error of ± 6.0 per cent (constant percentage error) what weight should the biochemist give to the data of experiment 6 relative to the weight assigned to the data of experiment 1?

Exercise 2

Consider the following hypothetical data:

Using an isotonic lever marking upon a smoked drum, a pharmacologist recorded the ability of acetylcholine with and without tubocurarine to cause contraction of the isolated rectus abdominis muscle of the frog. The response to acetylcholine was measured as the millimeters of excursion of the lever upon the drum. For convenience, the concentration of acetylcholine is expressed in micromoles per liter.

Concentration of Acetylcholine	Height of Contraction in Millimeters	
$\mu moles/L.$	Without tubocurarine	With 10 μM./$L.$ of tubocurarine
0.03	0.0	
0.1	0.5	
0.3	1.2	
1.0	6.2	0.0
3.0	12.5	0.9
10	26.2	3.6
30	33.8	6.3
100	36.3	19.5
300	40.0	28.9
1,000	40.0	34.6
3,000		39.3
10,000		40.0

Assume that the data can be explained by the law of mass action with one molecule of acetylcholine combining with one receptor.

A. What is the dissociation constant for the drug-receptor complex?
B. What kind of inhibitor is tubocurarine?
C. What is the dissociation constant for the tubocurarine-receptor complex?

D. Assuming that the data can be explained by a log-normal distribution of tolerances of the individual muscle fibers, what is the standard deviation of the tolerance as measured by the $\log_{10}$ dose?

Exercise 3

If the line obtained from a Lineweaver-Burk plot of $1/v$ against $1/C_s$ is extrapolated until it intersects the line for $1/v = 0$, what will the intercept at that point be?

Exercise 4

Prove that the dissociation constant, $(Keq)_{IE}$, for a reversible competitive inhibitor is numerically equal to the concentration of inhibitor, C_I, which doubles the concentration of substrate needed for a given initial velocity of reaction.

Exercise 5

Reiner (86) points out that if v_{max} is known, one can calculate the function $(v/v_{max})/[1 - (v/v_{max})]$, which, when plotted against C_s will give a straight line with a slope of $1/k_M$. Show that this is true.

Exercise 6

Show that $\log[(v_{max}/v) - 1]$ is a linear function of $\log C_s$.

Exercise 7

Suppose that you wanted to fit an equation of the general form of Equation 11-26 to a set of experimental observations of per cent response as a function of concentration or of dose. What method of plotting the data would allow you to evaluate the exponent n?

Exercise 8

What are the variants of Equation 11-12 corresponding to Equations 11-9d and 11-9e?

Exercise 9

Prove that when C_s/v is plotted against C_s (Fig. 11-1c) the error of C_s/v for a given error in v is smallest when $C_s = k_M$. (Assume that C_s is measured without error.)

12

THE DERIVATION OF EQUATIONS.
A GENERAL METHOD

The equations used in biological work are of two broad types: empirical and theoretical.

12-1. Empirical Equations

An *empirical equation* is one which has been fitted to experimental data so as to describe a relationship which has actually been observed between two or more variables. It implies nothing about the underlying reason for the relationship. This does not mean that empirical equations have nothing to contribute to the development of biological theory. As a matter of fact, an equation which summarizes an observed relationship may well serve as the point of departure for the elaboration of a general theory. Examples of this are commoner in the physical than in the biological sciences, probably because the systems dealt with in physics and chemistry are far less complicated and more easily controlled than are biological systems. As was pointed out in Section 3-9, a particular set of observations can be adequately described by any one of an infinite number of empirical equations. When no guidance is available from theory, the law of parsimony requires that the simplest equation which satisfactorily fits the data should be chosen.

12-2. Theoretical Equations

A *theoretical equation* is derived not from a particular group of experimental observations but from some theory or hypothesis about the fundamental nature of a biological system. While an empirical equation may describe a *particular* set of data faithfully enough, a well-founded theoretical equation may be able to describe *all* such sets of data and, what is more, may explain why the observed relationship exists. Though an equation derived from theory may also fail to fit a wide range of experimental values, this very failure can be turned to advantage if it suggests how the underlying theory can be refined and extended. Theoretical equations play an essential part in the formulation and solution of a host of fundamental biological problems which are too complex to solve non-mathematically. Because of their generality and because of their power to aid the logical development of fundamental knowledge, theoretical equations have an elegance and an intellectual appeal quite lacking in empirical equations.

When undertaking a theoretical analysis of a biological system, it is best to follow an orderly sequence of steps; otherwise, much time is likely to be wasted in fruitless hacking at the edges of the problem.

12-3. The First Step

The *first step* is to state the object of the theoretical analysis as precisely as possible. Is it merely desired to find an equation for calculating an unknown quantity from other quantities which are directly measurable? Will the analysis be used to formulate a new hypothesis, or to deduce from an existing hypothesis certain logical consequences which can then be tested experimentally? Or is an attempt being made to find a formal connection between two or more variables which are thought to be inter-related? The purpose of the derivation will often determine the mode of approach. An equation for calculating an unknown quantity *may* be derivable so easily from well-known relationships that the mode of derivation is practically self-evident. On the other hand, deducing the conse-quences of a hypothesis or finding a theoretical connection between a number of variables may require an elaborate theoretical attack upon a surprising variety of interlinked problems.

12-4. The Second Step

The *second step* in the mathematical analysis of a biological system is the selection of a suitably simplified *model* of the system. It is this model, not the biological system itself, which will be dealt with mathematically. Start with a *preliminary model* embodying only those features of the actual living system which you believe may be important for the objective you have in mind. All other features must be disregarded, for they would only confuse the picture. The preliminary model should be sketched, however crudely, and carefully labeled. While the preliminary model will usually be too complicated for mathematical description, it serves both as a summary of the complex interrelationships which actually exist among the various factors, and as a guide to further simplification.

Next, the preliminary model should be examined critically. What factors are likely to have so small an influence that they may safely be dis-regarded? What features, though anatomically and physiologically distinct, may be "lumped together" for ease of mathematical description? Are there certain factors—perhaps even important ones—which influence the system in such a complicated way that they must be drastically simplified, or eliminated altogether? As these questions are answered, the preliminary model is gradually pruned of its more forbidding complexities until, at last, it becomes mathematically manageable. *Every simplifying assumption made in reducing the preliminary model to the final model should be stated explicitly, and a list of these simplifying assumptions should be compiled.* Again, *the final model should be sketched and labeled.* The sketch should be large enough so that the symbols to be used in the subsequent analysis can be written in their proper places directly upon the sketch.

The process of arriving at a suitably simplified model is, in practice, rarely as straightforward as might appear from the description just given. It is often by no means easy to see *a priori* just what features of the living system may be, or must be, neglected. One is naturally reluctant to discard a biological factor which is known to be important simply because it may complicate the mathematical manipulations. Much time and effort may then be lost in attempting to analyze a model which is faithful to the facts but intractably complex. At the other extreme, an oversimplified model which yields easily to mathematical analysis may be so unrealistic as to be misleading or useless when applied to actual biological events. Fortunately some of the thinking, if not of the mathematical work, devoted to an unprofitable model may be useful later during the analysis of a more appropriate one.

The importance of the steps just outlined cannot be exaggerated. If the subsequent mathematical analysis is carried out correctly, it will provide an *exact* description of the simplified model system. To the extent that the simplifying assumptions are reasonable, it will also provide a *reasonable* description of the original biological system. But it can never describe the living system exactly. In applying mathematical deductions to actual experimental data one must therefore keep clearly in mind just what simplifying assumptions have been made, and how they may distort the true picture.

12-5. The Third Step

The *third step* is to write down all of the simple relationships which characterize the theoretical model. Each relationship can be expressed as a brief equation, often so patently true that it hardly appears to be worth recording. But even the most complicated final equation is nothing more than a judicious combination of elementary equalities many of which, by themselves, seem trivial. It is these building blocks which you must now seek.

Scrutinize the model and try to visualize how it would work. Is it in a steady state? If so, there is likely to be at least one equation which expresses this "state of steadiness." If not, what is changing and why? For example, if factor z is decreasing with time, t, you should look for an explicit statement of the form:

$$\frac{-dz}{dt} = f(a, b, \text{etc.})$$

where a, b, etc., are the factors upon which the rate of change of z depends. Even when you are dealing with a transient state, there may be two or more adjacent parts of the system which are in immediate equilibrium with each other. Each such equilibrium should contribute an equation. Some general kinds of relationships worth looking for are listed in Table 12-1. While a checklist of this sort may help you to find the elementary equations needed as building blocks, it is no substitute for a thoughtful study of the model system itself.

Before attempting to go any further, make sure that all of your elementary equations are simultaneously true.* If they are not, combining them will give a result which is trivial or absurd.

Example

Consider separately the effects of pressure, P, and of absolute temperature, T, upon the volume, V, of 1 mole of an ideal gas. When T is constant, V is inversely proportional to P. Hence,

$$V = k_1(1/P) \tag{12-1}$$

where k_1 is a constant of proportionality.

Likewise, when P is constant, V is directly proportional to T. Hence,

$$V = k_2T \tag{12-2}$$

where k_2 is another constant of proportionality.

---

* Since all of the equations—both elementary and derived—which are based upon a particular model system must be *simultaneously* true, it is tempting to refer to them as a set of simultaneous equations. This would be misleading because the term "simultaneous equations" commonly means equations which contain the same unknowns and which are simultaneously true only for a strictly limited number of values of the unknowns, *i.e.*, the simultaneous solutions of the equations. In deriving theoretical equations, we do not want to obtain particular numerical solutions; we are interested in general relationships. Each of our equations must be valid for *any* values of its variables which are permitted by the model system. They constitute a set of noncontradictory *general* equations, not a set of simultaneous equations in the usual sense. See also Rule 6.

Combine Equations 12-1 and 12-2 by eliminating V:

$$k_1(1/P) = k_2 T \qquad (12\text{-}3)$$

or,

$$PT = K \qquad (12\text{-}4)$$

where K is a new constant defined as equal to k_1/k_2 .

Equation 12-4 states that for 1 mole of gas, the product of pressure and of absolute temperature is a constant. Now as a general statement this is false, although it was derived from two equations each of which, by itself, is true. The trouble is that Equations 12-1 and 12-2 are *simultaneously* true only when *both* T and P are constant. But when both T and P are constant, Equation 12-4 becomes a self-evident and totally uninteresting identity.

12-6. The Fourth Step

The *fourth step* is to combine the elementary equations so as to attain the desired objective. If there are only two or three simple equations with few terms in common, the way in which they should be combined may be clear from the outset. But with a larger number of equations, especially if they have several terms in common, it is all too easy (as the author knows from bitter personal experience) to flounder about for hours in a frustrating maze of true but irrelevant derivations before stumbling upon the right combination. Much of this floundering can be eliminated by displaying all of the equations and their interconnections on a single sheet of paper. Begin by writing each of the elementary equations, one beneath the other, in a single column at the left. The order in which they are listed is immaterial. Next, rule off an additional column for each different symbol used in the equations, and write each symbol as a heading for its particular column. Again the order is immaterial. Finally, in the row for each equation place a check mark in each column headed by a symbol used in that equation. Such a tabular array of equations and symbols shows at a glance *which* equations can be combined and *how* they can be combined. It is wise to leave plenty of room at the bottom of the table for new equations derived by combining the original ones. The symbols in these new equations should then be checked off in the appropriate columns. This often makes clear what the next step should be. It is astonishing how much confusion can be avoided by using this simple tabular arrangement!

12-7. Rules for Combining Equations

It is helpful to be familiar with certain rules for combining equations:

Rule 1. If two equations have no symbol in common, they cannot

TABLE 12-1

Some of the kinds of simple equations which may be useful as building blocks in the derivation of more complex equations

---

I. *Equations of conservation*

   A. The whole equals the sum of its parts.

   Example: Equation 10-56:

$$C_{s.tot} = C_s + C_{sPr}$$

   B. What goes in must either come out, or be stored, or be used up.

   Example: The Fick principle. Equation 9-6a:

$$dQ_w/dt = V_w(dC_w/dt) = \dot{V}(C_{in} - C_{out})_t - \dot{Q}_{other.t}$$

II. *Equations of equilibrium*

   Example: Equilibrium between "forward" and "backward" reactions. Equation 10-27:

$$k_{Y+Z}C_{Y.eq}C_{Z.eq} = k_{W+X}C_{W.eq}C_{X.eq}$$

III. *Definitions*

   Example: The renal plasma clearance of substance x. Equation 1-22:

$$(\dot{V}_P cl)_x = C_{x.U}\dot{V}_U/C_{x.P}$$

IV. *Statements of proportionality*

   A. The rate of change is proportional to the intensity of the factor causing the change.

   Example: Fick's law of diffusion. Equation 7-1:

$$dQ_S/dt = k(-dC_S/dx)$$

   where $k = D_S(dy)(dz)$

   B. The law of mass action: The rate of a reaction is proportional to the "active masses" (here assumed to be the concentrations) of the reactants.

   Example: Equation 10-17:

   If $W + X \xrightarrow{\quad k_{W+X} \quad} Y + Z$

   Then

$$-dC_W/dt = k_{W+X}C_{W.t}C_{X.t}$$

TABLE 12-1—*Concluded*

C. The rate of change is proportional to the thing changing (exponential growth or decay).

Example: N bacteria in the logarithmic phase of growth. Equation 6-27a:

$$dN/dt = kN_t$$

D. Steady-state proportionalities.

Example: The superficial area of a solid of a given shape is proportional to the $\frac{2}{3}$ power of its volume. Equation 2-40:

$$A = k_2 V^{2/3}$$

N.B.: Equations which include a constant of proportionality, k, must be used with great caution because serious restrictions may hide in the innocent-looking constant! (See the example in Section 12-5.) One must not be satisfied to think of k as just an arbitrary constant inserted for the sole purpose of turning a statement of proportionality into an equation. On the contrary, every effort should be made to find the separate factors of which k may be composed, or at least to state specifically how the general validity of the equation is limited. For example, the equation given above (Equation 2-40) for the relation of superficial area to the volume of a solid is valid *only* for a series of solids of the same shape.

V. *Assumptions.* Very frequently, progress in a mathematical analysis depends upon *assuming* that an equation is true even when it is not actually *known* to be true.

Example: (Section 4-10) If diffusion equilibrium for CO_2 between alveolar air and pulmonary capillary blood is *assumed:*

$$P_{CO_2 \cdot \text{art}} = P_{CO_2 \cdot A}$$

(Actually, this equation also involves other assumptions as well.)

VI. *Equations previously derived by others*

Example: Equation for the proportion of molecules having an energy equal to, or greater than E. Equation 10-10:

$$N_{\geq E}/N_{\text{tot}} = e^{-E/RT}$$

N.B.: As is true of any equation, an equation previously derived must be used with full knowledge of the limitations which restrict its general validity.

be combined except by using one or more additional equations as a link.

Example

Let

$$a + b - c = d \qquad (12\text{-}5)$$

and

$$e = f^2 \qquad (12\text{-}6)$$

Having no symbol in common, Equations 12-5 and 12-6 cannot be combined and are, by themselves, useless for further derivation. But if it is also known that

$$f = 1 - b \qquad (12\text{-}7)$$

then Equations 12-6 and 12-7 may be combined to give

$$e = (1 - b)^2 \qquad (12\text{-}8)$$

and Equations 12-5 and 12-8 may be combined to give

$$e = (1 + a - c - d)^2 \qquad (12\text{-}9)$$

Rule 2. If two equations have but one symbol in common, they may be combined only by the elimination of that symbol. Solve either equation for the common symbol and substitute the value so obtained into the other equation.

Example

In the preceding example, Equation 12-5 was solved for b:

$$b = d + c - a \qquad (12\text{-}5\text{a})$$

This value for b was then substituted into Equation 12-8, thus obtaining Equation 12-9 with the elimination of b.

Rule 3. If two equations have two different symbols in common, and these two symbols *can* be wholly replaced by an appropriately defined single symbol, the equations can be combined only by eliminating *both* common symbols, thus yielding but one derived equation.

Example

Let

$$b = (1/a) + c^2 - 4 \qquad (12\text{-}10)$$

and

$$d = \frac{\sqrt{(1/a) + c^2}}{4} \tag{12-11}$$

By defining

$$e = (1/a) + c^2 \tag{12-12}$$

a and c may be wholly replaced by e, both in Equation 12-10,

$$b = e - 4 \tag{12-13}$$

and in Equation 12-11,

$$d = (\sqrt{e}/4) \tag{12-14}$$

Now Equations 12-13 and 12-14 may be combined by eliminating e:

$$b = 16d^2 - 4 \tag{12-15}$$

If one tries to obtain one equation by eliminating only a, and another by eliminating only c, it will be found that both of the resulting equations reduce to Equation 12-15.

Rule 4. If two equations have two different symbols in common, and these two symbols *cannot* be wholly replaced by an appropriately defined third symbol, the equations may be combined by eliminating *either* common symbol, thus yielding two different derived equations. But the two equations cannot be combined by eliminating *both* symbols.

Example

Let

$$a^2 = (1/c) - b \tag{12-16}$$

and

$$d = b - c \tag{12-17}$$

Equations 12-16 and 12-17 may be combined by eliminating b:

$$a^2 = (1/c) - c - d \tag{12-18}$$

or they may be combined by eliminating c:

$$a^2 = \left(\frac{1}{b - d}\right) - b \tag{12-19}$$

but b and c cannot *both* be eliminated.

(Rules 3 and 4 may be generalized to cover equations with three or more symbols in common.)

Rule 5. Let c be a term common to n equations. If c is of no interest in the subsequent analysis, it can be eliminated by solving any one of the equations for c, and substituting the value so obtained in *all* of the other $n - 1$ equations. *The resulting $n - 1$ new equations contain all of the relevant information from the original n equations* which may therefore be discarded.

Example

Let it be desired to find an equation for k in terms of a, d, e, and f, when

$$a = b - 3c \qquad (12\text{-}20)$$

$$d = (b/c) + e \qquad (12\text{-}21)$$

$$k/f = c/5 \qquad (12\text{-}22)$$

and b and c are unwanted terms.

Solve Equation 12-22 for c:

$$c = 5k/f \qquad (12\text{-}22\text{a})$$

and substitute this value of c both in Equation 12-20,

$$b = a + (15k/f) \qquad (12\text{-}23)$$

and in Equation 12-21,

$$b = 5k(d - e)/f \qquad (12\text{-}24)$$

Now we can discard Equations 12-20, 12-21, and 12-22, for all of the relevant information which they contained has been incorporated into Equations 12-23 and 12-24. Combine Equations 12-23 and 12-24 by eliminating b and solving for k:

$$k = fa/5(d - e - 3) \qquad (12\text{-}25)$$

which is the equation desired.

(Notice that if we had begun by combining Equations 12-20 and 12-22 by eliminating c, we would have been left with two equations, 12-21 and 12-23, from which we could eliminate b:

$$k = f[c(d - e) - a]/15 \qquad (12\text{-}26)$$

but not c, except by going back and using Equation 12-22 again. Equations 12-21 and 12-23 clearly do *not* contain all of the relevant information from the original three equations. When a term is not wanted, it should be eliminated from *all* of the equations in which it appears.)

Rule 6. Consider any specific selection of symbols (variables, or parameters, or both), say w, b, d, and z, which appear in a set of general equations describing a particular model system. It may or may not be possible to derive a general equation containing w, b, d, and z, and no other symbols. However, if such an equation can be found, it is a *unique equation*. No other such equation exists, and it is therefore a waste of time to look for another.

Example

Suppose that

$$w = z - b - d \qquad (12\text{-}27)$$

and

$$w = bd/z \qquad (12\text{-}28)$$

were both valid *general* equations describing a particular model system. We should then be able to assign any values we please to b, d, and z, and calculate the same value of w from either equation. For example, let $b = 2$, $d = 3$, and $z = 4$. Then by Equation 12-27, $w = -1$, while by Equation 12-28, $w = 1.5$. Hence for these values (and for most other arbitrary values) the equations are contradictory. Only for certain *particular* values (*e.g.*, $b = 2$, $d = 3$, and $z = 6$) will *both* equations be true. Although Equations 12-27 and 12-28 are simultaneous equations in the ordinary mathematical sense, they cannot be *general* equations describing the same model system. (See also the footnote on page 302.)

In considering Rule 6, it is important to understand that two general equations, relating the same set of variables and parameters, and derived from the same model, may at first glance appear to be two quite different equations. But if both are correct, it will always be possible by appropriate rearrangement to show that they are, in fact, one and the same equation. As an illustration, let us examine the ideal gas law, the *unique* equation which, for the model system known as an "ideal gas," expresses the relationship between temperature, pressure, volume, and the number of moles:

$$PV = nRT \qquad (12\text{-}29)$$

where

P = pressure of the gas

V = volume of the gas

n = number of moles of gas

T = temperature of the gas in degrees Kelvin

R = a constant of proportionality, the "gas law constant"

Now by rearranging terms, Equation 12-29 could be written

$$R = PV/nT \qquad (12\text{-}29a)$$

or

$$n/V = P/RT \qquad (12\text{-}29b)$$

or

$$1/P = V/nRT \qquad (12\text{-}29c)$$

or

$$PV/nRT = 1 \qquad (12\text{-}29d)$$

or in many other equally simple ways. Clearly these are all variants of the *same* equation, not different equations. (To emphasize this, it is well to designate such variants by the number of the original equation followed by a letter, as is done throughout this book.) But the possible disguises under which this same equation might masquerade are, in fact, infinite, as the following *reductio ad absurdum* will demonstrate:

A graduate student in psychology was writing his doctoral thesis on the relationship between the ambient *temperature* and the psychological *pressure* to drink a large *volume* of beer. One particularly warm evening, while making some personal observations in a bar, he suddenly realized that the relationship was very much like the ideal gas law and might profitably be incorporated with it. Two hours and several bottles of beer later, he succeeded in formulating the *beer-gas law*:

$$P = - \left[\frac{(1 - \Delta \bar{V}_{\text{beer}})R}{V \Delta \bar{V}_{\text{beer}}} \right] \left[n - \left(\frac{n}{1 - \Delta \bar{V}_{\text{beer}}} \right) \right] T \quad (12\text{-}29e)$$

where $\Delta \bar{V}_{\text{beer}}$ = the mean daily volume of beer drunk in New York City during July, minus the corresponding mean for January, and the other symbols have the same meanings as before.

Now as it happens, this equation is just as valid as the gas law itself. But even cold sober, one may need a bit of time to show that

$$- \left[\frac{1 - \Delta \bar{V}_{\text{beer}}}{\Delta \bar{V}_{\text{beer}}} \right] \left[1 - \left(\frac{1}{1 - \Delta \bar{V}_{\text{beer}}} \right) \right] \equiv 1 \qquad (12\text{-}30)$$

and that the "beer-gas law" is therefore nothing but a preposterously bloated version of the gas law. There is, in fact, no "beer-gas law." Equation 12-29e is the *same* equation as Equation 12-29.

12-8. The Fifth and Last Step

The *fifth and last step* in the analysis is to check every final equation for correctness. This step is so important that it will be discussed in a separate chapter (Chap. 14). But first the steps thus far outlined will be illustrated in detail by a concrete example.

Note. Exercises bearing upon the material of this chapter will be found at the end of Chapter 13.

13

THE DERIVATION OF EQUATIONS.
A DETAILED EXAMPLE

The general procedure described in Chapter 12 will now be illustrated by a specific example.

13-1. Anesthetic Gases

Inhalation of a general anesthetic agent such as diethyl ether, cyclopropane, or nitrous oxide will cause more or less depression of the central nervous system. The degree of depression (the "depth" of anesthesia) will depend upon the concentration of the general anesthetic in the brain. In turn, the concentration in the brain will depend upon the concentration in the blood, and the concentration in the blood upon the concentration in the inspired air. Thus by varying the concentration of anesthetic in the inspired air, the anesthetist can control the depth of anesthesia. Because the inhalation anesthetics are not to any appreciable extent changed chemically in the body, they behave as "inert gases" whose distribution *at equilibrium* will be determined only by their solubility in various tissues. However, the rate at which equilibrium is approached when a constant concentration of anesthetic is inhaled will be determined by many factors. If the rate of approach to equilibrium is slow, induction of anesthesia and recovery from anesthesia will tend to be slow.* An analysis of how various factors influence the rate of approach to equilibrium is therefore of practical as well as theoretical interest. Furthermore, in most current textbooks of pharmacology, discussion of this problem is both inadequate and misleading

---

* During induction the anesthetist may be able to hasten the onset of anesthesia by administering a high concentration of anesthetic which would, if long continued, produce a dangerous or even fatal concentration in the brain. When the patient has become anesthetized deeply enough, and long before equilibrium is approached, the anesthetist reduces the concentration of anesthetic in the inspired air to or toward the "safe anesthetic concentration," *i.e.*, the concentration which, *at equilibrium*, would just suffice to maintain surgical anesthesia. It should be obvious that no such manipulation of concentrations to hasten recovery is possible.

(see Exercise 1). Let us, then, undertake a theoretical analysis of the approach to equilibrium when an inert gas is inhaled at a constant concentration.

13-2. Step 1. Precise Formulation of the Objective of the Analysis

We have already stated our objective in general terms. Now we must be more precise. What exactly is meant by "equilibrium," and how should "the rate of approach to equilibrium" be expressed? These questions are best answered by considering, in a very simple way, how an inert gas accumulates in the body.

When administration of the gas is begun, its concentration in the mixed venous blood entering the lungs is at first zero. Therefore, initially there is a considerable gradient of partial pressure between alveolar air and pulmonary capillary blood, and a correspondingly large net diffusion of the gas into the blood stream for distribution to the tissues. As the gas accumulates in the tissues (in accordance with the partial pressure gradients from blood to tissue), the concentration of the gas in the venous blood returning to the lungs will increase. The gradient of partial pressure from alveolar air to pulmonary capillary blood, and the net transfer of gas by diffusion along this gradient, will be proportionately reduced. For example, when the gradient has been reduced to one-third of the original gradient, the amount of gas absorbed per unit time will be one-third of the original amount. Finally (theoretically only after an infinite period of time), the gradient of partial pressure will disappear because the partial pressure of the gas in all of the tissues will have risen to the partial pressure of the gas in arterial blood. Hence, no more gas can be removed from arterial blood, and the partial pressure will be identical in alveolar air, arterial blood, tissues, and venous blood. This is the final state of equilibrium which is being approached.

At equilibrium, a certain total quantity of the gas, $Q_{tot.eq}$, will be distributed throughout the body. Let $Q_{tot.t}$ be the quantity present at any time, t, prior to the attainment of equilibrium. Then

$$F_{eq} = Q_{tot.t}/Q_{tot.eq} \qquad (13\text{-}1)$$

is the fraction of the equilibrium amount which has been attained at time t. The "rate of approach to equilibrium" may now be defined *precisely* as dF_{eq}/dt. And the objective of our analysis is *to find a general equation for F_{eq} as a function of time and other important variables.*

13-3. Step 2. Selection of a Simplified Model Which Can Be Described Mathematically

Figure 13-1 represents a *preliminary model* of the system which now concerns us. It illustrates most of the factors which may influence the rate

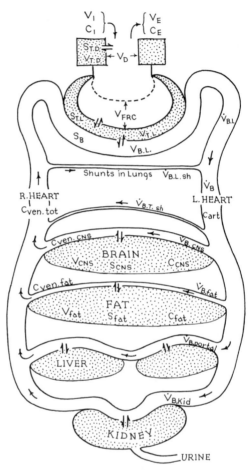

FIG. 13-1. A Preliminary Model of the Absorption and Distribution of Inert Gas

of approach to equilibrium when an inert gas is inhaled.* In the following description, the parenthetical numbers serve as a running tally of factors

---

* Although it sounds childish, construction of such a diagram is often aided by imagining that you yourself are passing through the system in an orderly fashion. For example, it is all too easy to think of the respiratory dead space as a lifeless series of branching tubes which are incapable of affecting gas exchange except by withholding from the alveoli a certain portion of the inspired air. But if, as a molecule of a water-soluble gas, you allow yourself to be drawn through these tubes, you will see at once that they are warm, moist, and supplied with blood.

which we may need to consider. Most of these factors are listed by number in Table 13-2, pages 325 to 327.

When administration of an inert gas is begun, a volume of air, V_I (1), containing the gas at a concentration, C_I (2), is inhaled. (For convenience, let all gases be at body temperature and pressure, and saturated with water vapor, i.e., BTPS.) The gas has a certain solubility in the tissues lining the dead space, $S_{T.D}$ (3), and accordingly some gas dissolves in that volume of tissue, $V_{T.D}$ (4). The remainder either stays in the volume of air in the dead space, V_D (5), or is drawn into the alveoli as they expand (dotted lines) by a volume equal to V_I. Mixing (6) of the inspired air with the air already in the alveoli, the "functional residual capacity," V_{FRC} (7), dilutes the gas to a concentration, C_A (8), and brings it into contact with the volume of lung tissue, $V_{T.L}$ (9), and the volume of capillary blood, $V_{B.L}$ (10), surrounding the alveoli. Into these the gas diffuses (11) in accordance with its solubility in lung tissue, $S_{T.L}$ (12), and blood, S_B (13). As gas is absorbed from the lungs, both C_A and the volume of air in the alveoli will change, so that the volume expired, V_E (14), will not equal V_I even when the respiratory exchange ratio is unity. The concentration of gas in the expired air is C_E (15). The respiratory rate, $\dot{N}$ (16), is the number of breaths per minute. When the inert gas is first administered, the residual air in the lungs must be "washed out" by the new gas mixture containing the inert gas. Many respiratory cycles may be needed before this process of lung washout (17) is substantially complete. The volume of blood flowing out of the lung capillaries, $\dot{V}_{B.L}$ (18), may be augmented by a volume of shunted blood, $\dot{V}_{B.L.sh}$ (19), which has not undergone gas exchange in the alveoli. The total volume of blood returned to the left heart, $\dot{V}_B$ (20), becomes the arterial blood containing gas, at a concentration C_{art} (21), for distribution to the rest of the body. Before leaving the lungs, we should note that the individual alveoli will differ one from another in the ratio of air flow (ventilation) to blood flow (perfusion) (22).

Now consider what happens to the gas in the capillary beds of the extrapulmonary organs and tissues. Of these, only a few representative examples need be illustrated (Fig. 13-1). Each has its particular volume, V_T (23), blood flow, $\dot{V}_{B.T}$ (24), and gas concentration, C_T (25). Furthermore, the solubility of a gas in a tissue, S_T (26), may not be the same as in blood. Tissues may also differ in the fraction of complete diffusion equilibrium, (27), which the inert gas achieves when diffusing into the tissue from the capillary blood. Because of these differences between individual tissues, the venous blood leaving each organ or tissue has its own particular gas concentration $C_{ven.T}$ (28), to contribute to the concentration of gas in the mixed ("total") venous blood, $C_{ven.tot}$ (29), returning to the lungs. For completeness, we should note that the portal circulation has two capillary beds in series (30), that some gas may be lost by extrapulmonary routes

(31), and that, as in the lungs, there may be arteriovenous shunts in the tissues through which blood may flow, $\dot{V}_{B.T.\text{sh}}$ *(32)*, without participating in gas exchange.

Not all of the factors mentioned above are independent variables. For example, the concentration of inert gas anywhere in the body is completely determined by the time, t *(33)*, during which the gas has been administered, by the concentration of gas in the inspired air, by the solubility of the gas in blood and other tissues, and by the physiological variables listed above. Hence, of the terms for concentration only C_I is not subservient to other factors. Even so, there remain far too many factors which can vary independently of each other, especially when we remember that *every* separate capillary bed (not just those illustrated) has its particular volume, blood flow, etc. The preliminary model is obviously much too complex for mathematical analysis. We must set about simplifying it.

Consider first the extrapulmonary factors. One of these, loss of gas in the urine or by other extrapulmonary routes, is known to be trivial and may be dismissed forthwith. Arteriovenous shunts could be treated as if they were the blood supply for a tissue of zero volume. But the remaining factors cannot be so easily dealt with. Differences in blood flow per gram of tissue, and differences in gas solubility among various tissues may be very large. These differences introduce troublesome complications which may best be indicated by an actual example.

13-4. The Effect of Differences between Tissues

Example

Let us pretend that the body is entirely made up of kidney, fat, and a uniform third tissue, "other tissue," which is a composite of all of the nonrenal, nonadipose tissues of the body. Let us further suppose that the inert gas in question is five times as soluble in fat as in blood, kidney, and other tissue. The important features of this imaginary body are given in Table 13-1. The values chosen for kidney, fat, and whole body are reasonable for an adult human. The values for "other tissue" have been obtained by difference.

We will now proceed to find an equation with which we can calculate, from the tabulated values, the concentration of gas in the mixed venous blood at any time, t.

In Chapter 9 Equation 9-11 was derived for the tissue concentration of a substance accumulating in a tissue by passive diffusion from a *constant* concentration in the arterial blood. (We assumed for simplicity that diffusion equilibrium is attained between blood and tissue by the time the blood reaches the venous end of the capillary.)

TABLE 13-1

Salient features of the imaginary tripartite human body discussed in the text

From time zero onward, an inert gas is present in arterial blood at a *constant* concentration of C_{art} μmoles per liter. The solubility of the gas in blood at body temperature is 0.01. At time zero, there is no inert gas in any tissue.

Variables	Symbol	Units	Kidney	Other Tissue	Fat	Total Body
Volume of tissue	V_T	L.	0.300	57.7	12.0	70.0
Blood flow through tissue	$\dot{V}_{B.T}$	L./min.	1.30	3.46	0.240	5.00
Solubility of gas in tissue	S_T		0.0100	0.0100	0.0500	
Quantity of gas at equilibrium	$Q_{T.eq}$	μmole	$0.300\,C_{art}$	$57.7\,C_{art}$	$60.0\,C_{art}$	$118.0\,C_{art}$
Fraction of total blood flow	$\dfrac{\dot{V}_{B.T}}{\dot{V}_{B.tot}}$		0.260	0.692	0.0480	1.00
Fraction of total gas in body at equilibrium	$\dfrac{Q_{T.eq}}{Q_{tot.eq.}}$		0.00254	0.489	0.508	1.00
Blood flow per liter of tissue	$\dfrac{\dot{V}_{B.T}}{V_T}$	(L./min.)/L.	4.33	0.0600	0.0200	0.0714
Distribution ratio: tissue/blood	$\dfrac{S_T}{S_B}$		1.00	1.00	5.00	
Exponent for Equations 13-10 and 13-12	$\dfrac{\dot{V}_{B.T} S_B t}{V_T S_T}$		$4.33\,t$	$0.0600\,t$	$0.00400\,t$	
Half-time ($t_{1/2}$)	$\dfrac{0.693\,V_T S_T}{\dot{V}_{B.T} S_B}$	min.	0.160	11.5	173.	

$$C_{T.t} = C_{\text{art}} R_{T/B} \left[1 - \exp\left(-\frac{\dot{V}_{B.T} t}{V_T R_{T/B}} \right) \right] \qquad (13\text{-}2)$$

For gases, the distribution ratio ("partition coefficient"), $R_{T/B}$, is equal to the ratio of the solubilities:

$$R_{T/B} = S_T / S_B \qquad (13\text{-}3)$$

By the assumption of diffusion equilibrium between venous blood and tissue,

$$C_{T.t} = C_{\text{ven}.T.t} (S_T / S_B) \qquad (13\text{-}4)$$

Combining Equations 13-2, 13-3, and 13-4,

$$C_{\text{ven}.T.t} = C_{\text{art}} \left[1 - \exp\left(-\frac{\dot{V}_{B.T} S_B t}{V_T S_T} \right) \right] \qquad (13\text{-}5)$$

The *quantity* of gas per unit time contributed to the mixed venous blood by the venous effluent from a given tissue is

$$\dot{Q}_{\text{ven}.T.t} = C_{\text{ven}.T.t} \dot{V}_{B.T} \qquad (13\text{-}6)$$

The *volume* of blood contributed per unit time by that tissue is obviously $\dot{V}_{B.T}$. The concentration of gas in the mixed venous blood, $C_{\text{ven.tot}}$, is the *sum of the quantities* contributed by the individual tissues divided by the *sum of the volumes* contributed by the individual tissues:

$$C_{\text{ven.tot}.t} = (\Sigma \dot{Q}_{\text{ven}.T.t}) / (\Sigma \dot{V}_{\text{ven}.T}) = (\Sigma \dot{Q}_{\text{ven}.T.t}) / \dot{V}_{B.\text{tot}} \qquad (13\text{-}7)$$

At equilibrium, *i.e.*, when $t = \infty$,

$$C_{\text{ven.tot.eq}} = C_{\text{art}} \qquad (13\text{-}8)$$

Now let us define

$$F_{(\text{eq})\text{ven.tot}} = C_{\text{ven.tot}.t} / C_{\text{ven.tot.eq}}. \qquad (13\text{-}9)$$

as the fraction of the equilibrium concentration which has been attained by the mixed venous blood at any time, t. Combining Equations 13-6, 13-7, 13-8, and 13-9,

$$F_{(\text{eq}.)\text{ven.tot}} = \frac{\Sigma (C_{\text{ven}.T.t} \dot{V}_{B.T})}{C_{\text{art}} \dot{V}_{B.\text{tot}}} \qquad (13\text{-}10)$$

Now let us apply Equation 13-5 specifically to each of the three tissues under consideration, and combine the resulting three equations with Equation 13-10:

$$F_{(eq)ven.tot} = \left(\frac{\dot{V}_{B.kid}}{\dot{V}_{B.tot}}\right)\left[1 - \exp\left(-\frac{\dot{V}_{B.kid}S_{Bl}}{V_{kid}S_{kid}}\right)\right]$$
$$+ \left(\frac{\dot{V}_{B.other}}{\dot{V}_{B.tot}}\right)\left[1 - \exp\left(-\frac{\dot{V}_{B.other}S_{Bl}}{V_{other}S_{other}}\right)\right] \quad (13\text{-}11)$$
$$+ \left(\frac{\dot{V}_{B.fat}}{\dot{V}_{B.tot}}\right)\left[1 - \exp\left(-\frac{\dot{V}_{B.fat}S_{Bl}}{V_{fat}S_{fat}}\right)\right]$$

All of the quantities on the right side of Equation 13-11 are given in Table 13-1. We can therefore calculate the concentration of inert gas present at any time in mixed venous blood as a fraction of its equilibrium concentration. The results are shown graphically in Figure 13-2 (*Curve A*). The initial very sharp rise in the gas concentration in mixed venous blood to about one-quarter of its equilibrium value is due almost entirely to the contribution made by the kidney (*Curve B*), whose small volume and large blood supply allow it to approach

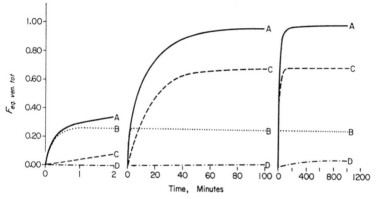

FIG. 13-2. The Fraction of the Equilibrium Concentration of Inert Gas in Mixed Venous Blood as a Function of Time

It is assumed that the concentration of inert gas in arterial blood achieves its equilibrium concentration instantaneously at time zero, and that the body consists entirely of the three tissues whose characteristics are listed in Table 13-1. In order to illustrate properly the influence of each tissue, the same curves have been plotted on three very different time scales. *Curve A* for the mixed venous blood itself has been calculated from the data of Table 13-1 by means of Equation 13-11. The contributions to Curve A by kidney (*Curve B*), by "other tissue" (*Curve C*), and by fat (*Curve D*) have been calculated from the corresponding individual terms of Equation 13-11. Compare these curves for *concentration in venous blood* with the corresponding curves in Figure 13-3 for the *quantity* of inert gas in the various tissues.

its own final equilibrium concentration with a half-time of about 10 sec. Indeed, after the first minute or so, the kidney behaves just like an arteriovenous shunt. The middle portion of the curve is largely controlled by the "other tissue" (*Curve C*) with an average sort of blood flow per volume of tissue. The very slow final approach to equilibrium is largely due to the fat (*Curve D*) which, with a sluggish blood flow and a large capacity for the inert gas approaches equilibrium with a half-time of 173 minutes—about 1000 times as slowly as the kidney! (Note that the high solubility of the gas in fat makes the *effective* volume of the adipose tissue five times as large as the actual volume.)

A similar argument leads to a similar equation for the *total quantity* of inert gas in the body at any time, t, as a fraction of the total quantity at equilibrium:

$$
\begin{aligned}
F_{\mathrm{eq}} = {} & \left(\frac{Q_{\mathrm{kid.eq}}}{Q_{\mathrm{tot.eq}}}\right)\left[1 - \exp\left(-\frac{\dot{V}_{B.\mathrm{kid}}S_B t}{V_{\mathrm{kid}}S_{\mathrm{kid}}}\right)\right] \\
& + \left(\frac{Q_{\mathrm{other.eq}}}{Q_{\mathrm{tot.eq}}}\right)\left[1 - \exp\left(-\frac{\dot{V}_{B.\mathrm{other}}S_B t}{V_{\mathrm{other}}S_{\mathrm{other}}}\right)\right] \qquad (13\text{-}12) \\
& + \left(\frac{Q_{\mathrm{fat.eq}}}{Q_{\mathrm{tot.eq}}}\right)\left[1 - \exp\left(-\frac{\dot{V}_{B.\mathrm{fat}}S_B t}{V_{\mathrm{fat}}S_{\mathrm{fat}}}\right)\right]
\end{aligned}
$$

Equation 13-12, be it noted, is an explicit equation for F_{eq} as a function of time, and is therefore the *kind* of equation we are seeking. But Equation 13-12 falls far short of our desire for a *general* equation, because it is valid only when the concentration of inert gas in the arterial blood remains constant throughout the entire period of equilibration. This is a very severe restriction indeed.

Using Equation 13-12 and the values of Table 13-1, we can calculate the total quantity of inert gas in the body at any time as a fraction of the equilibrium quantity (Fig. 13-3). Because the exponential terms are exactly the same in Equations 13-11 and 13-12, the proportional rate at which each individual tissue approaches its particular equilibrium is the same whether we are considering concentration in the venous effluent (Fig. 13-2) or quantity in the tissue (Fig. 13-3). But the influence which each tissue has upon the curve for the whole body is not at all the same for concentration in mixed venous blood and for total quantity in the body. To the *concentration* in mixed venous blood each tissue contributes in proportion to the blood flow it receives. (The coefficients of the exponential terms in Equation 13-11 are $\dot{V}_{B.T}/\dot{V}_{B.\mathrm{tot}}$.) To the *total quantity* in the body each tissue contributes in proportion to its equilibrium store of the inert gas.

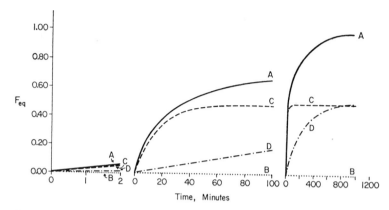

Fɪɢ. 13-3. The Fraction of the Equilibrium Quantity of Inert Gas in the Body as a Function of Time

The conditions assumed are exactly the same as those for Figure 13-2, and the same three time scales have been used. *Curve A* for the entire body has been calculated from the data of Table 13-1 by means of Equation 13-12. The contributions to Curve A by kidney (*Curve B*), by "other tissue" (*Curve C*), and by fat (*Curve D*) have been calculated from the corresponding individual terms of Equation 13-12. Compare these curves with the corresponding ones in Figure 13-2.

(The coefficients of the exponential terms in Equation 13-12 are $Q_{T.eq}/Q_{tot.eq.}$) For example, the slow accumulation of gas in fat depots affects the curve for concentration in mixed venous blood only slightly, because the blood flow through fat is only about 5 per cent of the total blood flow. But fat has a marked influence upon the curve for the total quantity of gas in the body, because at equilibrium the fat will contain about half of the total. Therefore, one cannot use the concentration of inert gas in mixed venous blood as an index of the approach to equilibrium.

When we abandon the highly restrictive assumption that the concentration of inert gas in arterial blood remains constant, the differences between tissues become even more difficult to deal with. The concentration of inert gas in arterial blood will then be a complex function not only of many pulmonary factors, but also of the concentration of gas in the mixed venous blood returning to the lungs. But the concentration in mixed venous blood is, as we have just seen, a function of how rapidly gas accumulates in the individual organs and tissues. And gas accumulation in each organ or tissue is, in turn, a function of the characteristics of the tissue, and of the concentration of gas in arterial blood. All of these functions are thus interdependent.

Even if we could find an explicit equation for F_{eq} which included terms for the individual characteristics of different tissues, it would be too unwieldy to be very useful.

13-5. Further Simplifying Assumptions

The only alternative to taking account of the differences between tissues is to assume that the body consists of a single homogeneous tissue, with a volume equal to the volume of distribution of the inert gas at equilibrium, and a blood flow equal to the cardiac output. We cannot be very much pleased with so unrealistic an assumption, but at least it allows us to get on with the analysis. Let us also assume for simplicity that diffusion equilibrium is attained between the peripheral tissue and venous blood.

Having reluctantly reduced most of the body to a featureless volume, let us now turn to the factors influencing gas exchange in the lungs. Fortunately, several of these can be eliminated rather easily. Solution of inert gas in the tissue lining the dead space is not likely to be important except for gases which are extremely soluble in water. Let us accordingly neglect Factors *3* and *4*. There is ample evidence (38) that mixing of inspired gas with air already in the alveoli (Factor *6*) is so very rapid that we can assume it to be instantaneous. There is also evidence that for *inert* gases diffusion equilibrium between alveolar air and pulmonary capillary blood is practically achieved by the time the blood leaves the capillaries. We may therefore assume Factor *11* to be instantaneous. The solubility of various gases in lung tissue (Factor *12*) is known to be similar to their solubility in blood (21). Let us assume that $S_{T.L} = S_B$. If diffusion into, and across, alveolar walls is assumed to be instantaneous and $S_{T.L} = S_B$, we can combine $V_{T.L}$ and $V_{B.L}$ into a single volume, V_L, representing the total volume in immediate diffusion equilibrium with alveolar air. Changes in alveolar volume due to absorption of inert gas may be neglected, even for highly soluble gases, if we stipulate that the fraction of inert gas in the inspired air be small. Then we may assume that $V_I = V_E = V_{tid}$,* where V_{tid} is the tidal volume. Arteriovenous shunts (Factor *19*) bypassing the alveoli are normally insignificant and may be neglected. This means that $\dot{V}_{B.L}$ is assumed to be equal to cardiac output and to total peripheral blood flow. We may therefore designate blood flow everywhere as $\dot{V}_B$. Finally, although alveoli undoubtedly differ in their ventilation-perfusion ratios (Factor *22*), these differences will probably not greatly affect inert gas exchange, at least in normal lungs. We may therefore assume that the alveoli are uniform.

---

* This assumption also implies that the respiratory exchange ratio, $\dot{Q}_{CO_2,\,B \to A} / \dot{Q}_{O_2,\,A \to B}$, is unity. Otherwise the tidal volume, V_{tid}, is strictly synonymous only with V_E, and should not be used for V_I.

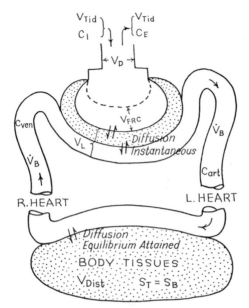

Fig. 13-4. First Simplification of the Model for Inert Gas Absorption

The simplifying assumptions which are necessary to achieve this degree of simplification are listed in Table 13-2.

13-6. A Manageable Model

Now let us redraw the model so as to incorporate the simplifying assumptions discussed previously (Fig. 13-4). As was mentioned before, except for C_I, the terms for concentration are wholly dependent upon the other factors. And if we confine our attention only to the time needed to attain a given *proportion* of equilibrium, even C_I is no longer a determining factor. As independent variables, we are therefore left with V_{tid}, V_D, V_{FRC}, V_L, $\dot{V}_B$, (Vdist), S_B, $\dot{N}$, and t, instead of the indefinitely large number we started with. An equation for F_{eq} as a function of these nine variables has, in fact, been derived (89) so that the model system illustrated in Figure 13-4 is no longer hopelessly complex. However, the derivation·is not completely rigorous, and the resulting equation is cumbersome. Its greatest value is to show that further simplification of the model system can be achieved with no appreciable loss of accuracy.

The most troublesome aspect of the model illustrated in Figure 13-4 is that its respiration is still cyclic. To describe the approach to equilibrium in this model, one must first describe gas transfer within a single respiratory

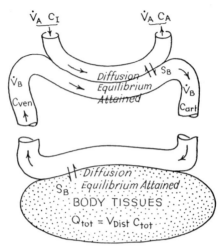

FIG. 13-5. Final Simplification of the Model for Inert Gas Absorption

The additional simplifying assumptions needed for this model are listed in Table 13-2.

cycle and then find some way to add up the contributions which the successive individual breaths make to the total quantity of inert gas in the body. It is much more convenient to assume that ventilation of the alveoli is continuous, not cyclic, and that in the lungs the inert gas is simply distributed, according to its solubility, between parallel streams of air and blood. If we make this assumption, the stationary volumes in the lung, V_L and V_{FRC} can be disregarded. The resulting *final simplified model* is illustrated in Figure 13-5. The sac which represented the alveoli in the two previous diagrams has been replaced by a tube through which air flows continuously at a rate equal to the alveolar ventilation per minute, $\dot{V}_A$, defined as

$$\dot{V}_A = \dot{N}(V_{tid} - V_D) \tag{13-13}$$

Table 13-2 lists the assumptions which were made as the preliminary model was progressively simplified and shows how few of the original factors survive in the final model.

TABLE 13-2

A summary of the assumptions made in reducing the preliminary model to the final simplified model

Preliminary Model		First Simplification			Final Simplification		
Factor	Symbols	Assumptions	Symbols		Assumptions	Symbols	
1. Volume of air inspired	V_I	$V_I = V_E = V_{tid}$	V_{tid}			V_{tid}	
2. Concentration of inert gas in inspired air	C_I		C_I			C_I	
3. Solubility gas, dead space tissue	$S_{T.D}$	Negligible					
4. Volume of dead space tissue	$V_{T.D}$	Negligible					
5. Volume of air in dead space	V_D		V_D			V_D	
6. Mixing of air in alveoli		Instantaneous					
7. Functional residual capacity	V_{FRC}		V_{FRC}				
9. Volume of tissue lining alveoli 10. Volume capillary blood in lungs	$V_{T.L}$ $V_{B.L}$	Pooled	V_L	Unimportant			
11. Diffusion of gas in lungs		Instantaneous					
12. Solubility of gas in lung tissue	$S_{T.L}$	$S_{T.L} = S_B$					
13. Solubility of gas in blood	S_B		S_B			S_B	

TABLE 13-2—*Continued*

Preliminary Model		First Simplification		Final Simplification	
Factor	Symbols	Assumptions	Symbols	Assumptions	Symbols
14. Volume of air expired	V_E	$V_E = V_I = V_{\text{tid}}$			
16. Rate of respiration	$\dot{N}$		$\dot{N}$		$\dot{N}$
17. Initial washout of air in lung		Accounted for		Neglected	
18. Capillary blood flow in lung	$\dot{V}_{B.L}$	$\dot{V}_{B.L} = \dot{V}_B$			
19. Flow of nonaerated blood (shunt)	$\dot{V}_{B.L.\text{sh}}$	Negligible			
20. Total blood flow (cardiac output)	$\dot{V}_B$		$\dot{V}_B$		$\dot{V}_B$
22. Inhomogeneity of alveoli		Negligible			

TABLE 13-2—*Concluded*

Preliminary Model		First Simplification		Final Simplification	
Factor	Symbols	Assumptions	Symbols	Assumptions	Symbols
23. Volume of tissue T	V_T		(Vdist)		(Vdist)
24. Blood flow through tissue T	$V_{B.T}$	Homogeneous one-tissue body			
26. Solubility of gas in tissue T	S_T				
27. Diffusion gas, blood to tissue		Instantaneous			
30. Portal circulation					
31. Extrapulmonary loss of gas		Negligible			
32. Flow through arteriovenous shunts	$\dot V_{B.T.sh}$	Negligible			
33. Elapsed time	t		t		t
Lung ventilation treated as		Cyclic		Continuous	

N.B.: Factors 8, 15, 21, 25, 28, and 29 are various terms for concentration and have been omitted from this tabulation because they are wholly dependent upon other factors.

13-7. Step 3. Finding the Elementary Equations Which Describe the Model*

In the model of Figure 13-5 there are but two places where gas exchange occurs: the lungs and the body tissues. For each we may write an equation of conservation and an equation of equilibrium.

Since the model admits of no gas storage in the lungs, *conservation* requires that the quantity of inert gas entering the lungs in an interval of time, dt, be equal to the quantity leaving the lungs in the same interval of time:

$$\dot{V}_A C_I dt + \dot{V}_B C_{\text{ven}} dt = \dot{V}_A C_A dt + \dot{V}_B C_{\text{art}} dt \qquad (13\text{-}14)$$

or,

$$\dot{V}_A(C_I - C_A) = \dot{V}_B(C_{\text{art}} - C_{\text{ven}}) \qquad (13\text{-}14a)$$

Diffusion *equilibrium* is assumed to be attained between air leaving the lungs and blood leaving the lungs:

$$C_{\text{art}} = S_B C_A \qquad (13\text{-}15)$$

In the body tissues, *conservation* requires that the quantity of gas entering in the arterial blood in an interval of time, dt, be equal to the quantity gained by the tissue, dQ_{tot}, plus the quantity leaving in the venous blood in the same interval of time:

$$\dot{V}_B C_{\text{art}} dt = dQ_{\text{tot}} + \dot{V}_B C_{\text{ven}} dt \qquad (13\text{-}16)$$

or,

$$dQ_{\text{tot}}/dt = \dot{V}_B(C_{\text{art}} - C_{\text{ven}}) \qquad (13\text{-}16a)$$

The *equilibrium* assumed is

$$C_{\text{tot}} = C_{\text{ven}} \qquad (13\text{-}17)$$

These four equations describe how gas accumulates in the tissues of the body. In addition, we will probably want to know the total quantity of gas which will be present in the body at equilibrium:

$$Q_{\text{tot.eq}} = (V\text{dist}) C_I S_B \qquad (13\text{-}18)$$

We might also need to use other expressions of the "$Q = CV$" variety. Indeed we have already done so in writing the conservation equations, 13-14 and 13-16, whose $C\dot{V}$ terms are, of course, quantities per unit of time. But we have not yet made use of the general relationship,

$$Q_{\text{tot}} = C_{\text{tot}}(V\text{dist}) \qquad (13\text{-}19)$$

of which Equation 13-18 is a special case.

* To conserve space in Table 13-3, the usual subscript, t, designating time-dependent variables is not used in the following derivation. But the nature of each symbol (variable or parameter) is indicated at the top of the table.

Finally, we will presumably need our definition of "fraction of equilibrium":

$$F_{eq} = Q_{tot}/Q_{tot.eq} \qquad (13\text{-}1)$$

All of these equations may now be written in a tabular array (Table 13-3) as described in Chapter 12. Notice that Q_{tot}, $Q_{tot.eq}$, and dQ_{tot} must each be allotted a separate column because each has a different meaning.

At this point you may well ask whether there are not other elementary equations which should be included in Table 13-3, or, for that matter, whether we do not already have more equations than we need. Although either a deficit or an excess of equations is possible, neither is serious. If an essential equation is missing, not only the need for, but very likely also the nature of the missing relationship will become evident when further analysis is blocked by the lack of some necessary connection between two equations. For example, if we had forgotten to include Equation 13-17, we should soon notice that there was no way of eliminating the unwanted term, C_{tot}, since it would then be peculiar to Equation 13-19. On the other hand, if a superfluous equation is included, no harm is done, since it will only remain unused. For example, we would not spoil the subsequent analysis by including equations stating that $C_{art}\dot{V}_B = \dot{Q}_{art}$, $C_{ven}\dot{V}_B = \dot{Q}_{ven}$, etc., even though the information contained therein has already been fully utilized in Equations 13-14 and 13-16.

Are all of the elementary equations simultaneously true? For example, can Equations 13-18 and 13-19, which are rather similar, both be true at the same time? A moment's thought will convince you that there is no conflict between these two equations. For at equilibrium, when $t = \infty$ and the partial pressures of the inert gas are by definition everywhere equal, $C_{tot} = C_{ven} = C_{art} = C_I S_B$. Thus Equation 13-18, which applies *only* at equilibrium, is merely a special case of Equation 13-19, which is true at *all* times including $t = \infty$.

13-8. Step 4. Combining the Elementary Equations to Attain the Desired Objective

Our objective is "to find a general equation for F_{eq} as a function of time and other important variables." Which of the symbols heading the columns of Table 13-3 represent important variables and which do not? It has already been pointed out that the concentrations of inert gas are subsidiary to other factors and cannot themselves influence the rate of approach to equilibrium. Let us therefore begin by eliminating the terms for concentration.

TABLE 13-3

Equation No.	Equation	C_I Par*	C_A (t)*	C_{art} (t)	C_{ven} (t)	C_{tot} (t)	V_A Par	V_B Par	(Vdist) Par	S_B Par	Q_{tot} (t)	$Q_{tot\cdot eq}$ Par	dQ_{tot} (t)	dt Independent Variable	P_{eq} (t)	dP_{eq} (t)
13-14a	$\dot{V}_A(C_I - C_A) = \dot{V}_B(C_{art} - C_{ven})$	✓	✓	✓	✓		✓	✓								
13-15	$C_{art} = S_B C_A$		✓	✓						✓						
13-16a	$dQ_{tot}/dt = \dot{V}_B(C_{art} - C_{ven})$			✓	✓			✓					✓	✓		
13-17	$C_{tot} = C_{ven}$				✓	✓										
13-18	$Q_{tot.eq} = (Vdist)C_I S_B$	✓							✓	✓		✓				
13-19	$Q_{tot} = (Vdist)C_{tot}$					✓			✓		✓					
13-1	$F_{eq} = Q_{tot}/Q_{tot.eq}$										✓	✓			✓	
13-20 (14a and 15)	$\dot{V}_A\left(C_I - \dfrac{C_{art}}{S_B}\right) = \dot{V}_B(C_{art} - C_{ven})$	✓	×	✓	✓		✓	✓		✓						
13-21 (17 and 19)	$Q_{tot} = C_{ven}(Vdist)$		×		✓	×			✓		✓					
13-22 (16a and 20)	$\dot{V}_A\left(C_I - \dfrac{dQ_{tot}}{dt}\cdot\dfrac{1}{\dot{V}_B S_B} - \dfrac{C_{ven}}{S_B}\right) = \dfrac{dQ_{tot}}{dt}$	✓	×	×	✓	×	✓	✓		✓			✓	✓		

Eq. no.	Equation	1	2	3	4	5	6	7	8	9	10	11	12
13-23 (21 and 22)	$\dot{V}_A\left(C_I - \dfrac{dQ_{tot}}{dt}\cdot\dfrac{1}{\dot{V}_B S_B} - \dfrac{Q_{tot}}{(V\mathrm{dist})S_B}\right) = \dfrac{dQ_{tot}}{dt}$	✓	×	×	×	×	✓	✓	✓	✓	✓		✓	✓
13-24 (18 and 23)	$\dot{V}_A\left(\dfrac{Q_{tot\text{-}eq}}{(V\mathrm{dist})S_B} - \dfrac{dQ_{tot}}{dt}\cdot\dfrac{1}{\dot{V}_B S_B} - \dfrac{Q_{tot}}{(V\mathrm{dist})S_B}\right) = \dfrac{dQ_{tot}}{dt}$	×	×	×	×	×	✓	✓	✓	✓	✓	✓	✓	
13-24a	$\dfrac{dQ_{tot}}{dt} = \dfrac{\dot{V}_A \dot{V}_B}{(V\mathrm{dist})(\dot{V}_A + S_B \dot{V}_B)}\,(Q_{tot\text{-}eq} - Q_{tot})$	×	×	×	×	×	✓	✓	✓	✓	✓	✓	✓	
13-25	$dQ_{tot} = Q_{tot\text{-}eq}\,dF_{eq}$	×	×	×	×	×				✓	✓			✓
13-26 (24a and 25)	$Q_{tot\text{-}eq}\left(\dfrac{dF_{eq}}{dt}\right) = \left[\dfrac{\dot{V}_A \dot{V}_B}{(V\mathrm{dist})(\dot{V}_A + S_B \dot{V}_B)}\right](Q_{tot\text{-}eq} - Q_{tot})$	×	×	×	×	×	✓	✓	✓	✓	✓	×	✓	✓
13-27 (1 and 26)	$\dfrac{dF_{eq}}{dt} = \dfrac{\dot{V}_A \dot{V}_B}{(V\mathrm{dist})(\dot{V}_A + S_B \dot{V}_B)} - \left[\dfrac{\dot{V}_A \dot{V}_B}{(V\mathrm{dist})(\dot{V}_A + S_B \dot{V}_B)}\right](F_{eq})$	×	×	×	×	×	✓	✓	✓	✓	×	×	✓	✓

* Nature of quantity symbolized: Par = parameter, (t) = a time-dependent variable.

The order in which the equations are combined is immaterial so long as we bear in mind which terms are to be eliminated. The particular sequence illustrated in Table 13-3 is but one of many possible routes to the same final equation. The successive steps used should be obvious from the table. For example, the first row below the seven elementary equations shows that Equation 13-20 was obtained by combining Equations 13-14a and 13-15 with the elimination of C_A. The symbols in Equation 13-20 are indicated by check marks in the proper boxes, and the elimination of C_A is indicated by placing an $\times$ in the appropriate box, both in the row for Equation 13-20 and in all subsequent rows. According to Rule 5 (Chap. 12), Equation 13-20 entirely replaces Equations 13-14a and 13-15 which may therefore be deleted. This can best be done by covering up each unwanted row with a narrow strip of paper. Then if you make a mistake and have to retrace your steps, you can restore the deleted equations simply by uncovering them.

Equation 13-24, which represents the useful residue of six of the elementary equations, can be written more concisely by collecting terms and solving for dQ_{tot}/dt (Equation 13-24a). We now have a choice between integrating Equation 13-24a and then combining it with Equation 13-1, or combining the two equations first and then integrating. If we choose the latter, we will need to use Equation 13-1 in the differential form given in Equation 13-25. (This may, if you wish, be regarded as an example of requiring an additional elementary equation which was not included in the original set.) Now by eliminating dQ_{tot}, Equations 13-24a and 13-25 may be combined to give Equation 13-26. Equation 13-26 may, in turn be combined with Equation 13-1, thereby eliminating both $Q_{tot.eq}$ and Q_{tot}, and yielding Equation 13-27. (Notice that the simultaneous disappearance of $Q_{tot.eq}$ and Q_{tot} is in accordance with Rule 3 (Chap. 12); for if we divide both sides of Equation 13-26 by $Q_{tot.eq}$, the only term containing Q_{tot} which remains is the ratio $Q_{tot}/Q_{tot.eq}$. This ratio can be wholly replaced by F_{eq} both in Equation 13-1 and in Equation 13-26.)

Equation 13-27 is an explicit equation for "the rate of approach to equilibrium" as defined earlier in this chapter. It is of the general form "$dy/dt = a - by$" which we have encountered before (e.g., Equation 7-43). It may therefore be integrated to the exponential form:

$$F_{eq} = 1 - \exp\left[-\frac{\dot{V}_A \dot{V}_B t}{(V\text{dist})(\dot{V}_A + S_B \dot{V}_B)} \right] \qquad (13\text{-}28)$$

If an expression for the *time* needed to attain a given fraction of equilibrium is desired, Equation 13-28 may be written in logarithmic form and solved for t:

$$t = [-\ln (1 - F_{eq})] \frac{(V\text{dist})(\dot{V}_A + S_B\dot{V}_B)}{\dot{V}_A\dot{V}_B} \qquad (13\text{-}29)$$

To show that the alveolar ventilation, $\dot{V}_A$, is itself a function of three other variables, Equation 13-28 may be combined with Equation 13-13:

$$F_{eq} = 1 - \exp\left[-\frac{\dot{N}(V_{tid} - V_D)\dot{V}_B t}{(V\text{dist})[\dot{N}(V_{tid} - V_D) + S_B\dot{V}_B]} \right] \qquad (13\text{-}30)$$

With the derivation of Equation 13-30, we have reached our objective: "a general equation for F_{eq} as a function of time and other important variables." Equation 13-30 describes *exactly* how equilibrium is approached when an inert gas is inhaled by the simplified model of Figure 13-5. It will *not*, in general, describe accurately how equilibrium is approached when an inert gas is inhaled by a living animal. In particular, the failure of the equation to take into account the peculiarities of individual organs and tissues is a serious defect, as was made evident by Equation 13-12. But even though Equation 13-30 is not fully applicable to the living system, it will at least indicate correctly the general magnitude of the influence, and certainly the direction of the influence of the several factors which it does take into account. For example, the equation states unequivocally that the more soluble an inert gas is in blood and tissues, the more slowly is equilibrium approached. This effect of solubility upon the rate of equilibration (by no means intuitively obvious!) explains why recovery from anesthesia produced by the highly soluble diethyl ether ($S_B = 15$) is very much slower than recovery from anesthesia after the administration of a relatively insoluble anesthetic such as cyclopropane ($S_B = 0.46$).

No mathematical analysis of the sort discussed in this chapter is complete until the correctness of the final equation has been meticulously verified. How to check equations will be discussed in the next chapter.

EXERCISES. CHAPTERS 12 AND 13

Exercise 1

Criticize each of the following statements about general anesthetic agents. The statements are quoted from the designated editions of textbooks of pharmacology.

  A. *Goodman and Gilman, 2nd Edition, 1955, page 49 (49):* "Circulatory velocity in the lungs is not particularly important under most circumstances, except in regard to rapidly acting anesthetics such as nitrous oxide, ethylene, and cyclopropane. Even then it determines only the speed of induction."

B. *Wilson and Schild, 8th Edition of Clark, 1952, page 203 (117):*
"Cyclopropane has a high oil-water distribution coefficient and is
accordingly a powerful anaesthetic. Owing to its high solubility in
plasma (45 vol. per cent.) it is rapidly taken up by the blood stream
and produces anaesthesia in one to three minutes."

C. *Krantz and Carr, 5th Edition, 1961, page 469 (66):* "The solubility
of a gas in a liquid is proportional to its pressure."

Page 471: ". . . with the removal of ether the following equilibria
shifts occur in rapid succession.

1. The partial pressure of ether in the inhaled air is reduced to
   zero.
2. Therefore the arterial blood carries no ether.
3. The tissues are bathed with blood containing no anesthetic."

D. *Davison, 3rd Edition, 1944, page 254 (30):* "Divinyl ether is
more volatile than diethyl ether, hence its onset is more rapid and the
recovery more quickly accomplished."

E. *Gaddum, 5th Edition, 1958, page 125 (43):* "Ether is less active
than chloroform, in the sense that the anaesthetic concentration in the
blood is 2–4 times as high and correspondingly larger total quantities
are required for an operation of any given duration. Because of this
difference in activity, the induction of anaesthesia with ether is slower
than with chloroform."

Exercise 2

Using the tabular method described in the text,

A. Derive Equation 11-12.
B. Derive the equation for thyroid blood flow which was discussed in
   Exercise 6, Chapter 9.
C. Derive the equation for fraction of S bound to plasma protein
   which was discussed in Exercise 9, Chapter 10.

Exercise 3

Assuming that $C_I = 0$, use Equation 13-14a and 13-15 to show that the
"ventilation/perfusion ratio," $\dot{V}_A/\dot{V}_B$, is equal to the concentration ratio,
C_{ven}/C_A minus the solubility.

Exercise 4

Using only the upper (pulmonary) portion of the model depicted in
Figure 13-5, derive a general equation for the pulmonary clearance of an
inert gas from the mixed venous blood entering the lungs.

Exercise 5

Assuming that there is an inert gas present in the blood but not in the inspired air, use the equation for the pulmonary clearance derived in Exercise 4 above to find an equation for the solubility, S_B, when $\partial(\dot{V}\text{cl})/\partial\dot{V}_A = \partial(\dot{V}\text{cl})/\partial\dot{V}_B$, in other words, when a very small change in ventilation has the same effect upon pulmonary clearance as an equally small change in perfusion.

Exercise 6

An investigator wants to determine the concentration of a gas X in whole blood. Gas chromatography, which requires X to be present in a gas phase, is the only method available for the analysis of X. The investigator therefore equilibrates an aliquot of blood of volume V_B with air at constant temperature in a tonometer whose total volume is V_{ton}. The concentration of X in the gas phase, $C_{air.eq}$, after equilibrium between gas and blood has been established is then measured in the gas chromatograph. If S is the solubility of gas X in blood at the temperature of equilibration, derive an equation for calculating the concentration originally present in the blood sample from $C_{air.eq}$ and the parameters of the system.

Exercise 7

X is a substance which is not appreciably stored in nor metabolized by any organs or tissues of the body. X does not enter red blood cells, nor is it appreciably bound to plasma protein. It can be recovered quantitatively from the urine, the only route of excretion.

X was infused continuously at a rate of 42 mg. per hour into the leg vein of a dog whose right kidney had been removed. Blood samples could be collected from the femoral artery and from the left renal vein. After the concentration of X in arterial blood had become essentially constant, the following data were obtained:

$(\dot{V}\text{cl})$ = renal plasma clearance of X = 50 ml. per minute
$\dot{V}_U$ = urine flow = 4.0 ml. per minute
C_{ven} = concentration of X in renal venous plasma = 10 mg. per liter

Calculate the renal plasma flow.

Exercise 8

Suppose that at time zero a normal human subject takes a *single* breath of room air to which has been added a tracer quantity of a biochemically inert, radioactive gas not previously present in the inspired air. The subject holds the breath for several seconds.

Derive an equation for the concentration of the inert gas in the pulmonary venous blood leaving the lungs at any time during the period of breath holding.

The following is a list of some of the quantities which you will want to consider. (You are at liberty to add any others which you find convenient.)

V_I = the volume of air inspired

V_D = the volume of the physiological dead space

V_{FRC} = the functional residual capacity, *i.e.*, the volume of air in the alveoli before the breath was inspired

V_A = the volume of air in the alveoli (after the inspiration)

V_L = the volume of lung tissue including blood but excluding air

$\dot{V}_B$ = the pulmonary blood flow

C_I = the concentration of the inert gas in the inspired air

$C_{A.t}$ = the concentration of the inert gas in alveolar air at time t

$C_{B.t}$ = the concentration of the inert gas in the pulmonary vein blood leaving the lungs at time t

t = time measured from the time of taking the breath

S = the solubility of the inert gas in blood, *i.e.*, the ratio of its concentration in blood to its concentration in air at equilibrium. (We will assume that the same ratio applies to the distribution of the inert gas between air and lung tissue.)

We will *assume* that all of these quantities remain constant except $C_{A.t}$, $C_{B.t}$, and t.

In your derivation, use the following steps:

1. State the additional simplifying assumptions which you find it desirable to make concerning
   a. the temperature and the humidity of the inspired air
   b. the rate at which the air is drawn into the lungs during inspiration
   c. the distribution of the inspired air and of blood flow to various parts of the lung
   d. the rate of diffusion of inert gas between alveolar air and lung tissue and between lung tissue and pulmonary blood
   e. the return of the inert gas to the lungs in the mixed venous blood
2. On the basis of these assumptions, draw a diagram of the model system which you are going to describe mathematically.
3. Write down all of the equations you can think of to describe various aspects of the model system.
4. Combine the individual equations so as to obtain a differential equation for dC_B/dt entirely in terms of the *independent* parameters

of the system. (For example, V_A is *not* an independent parameter because if either V_I or V_{FRC} changed, the value of V_A would necessarily change. But V_I *is* an independent parameter because its value will not be altered by any changes which it is possible to make in any other factor in the system.)

5. Integrate the differential equation so as to obtain an explicit expression for $C_{B.t}$ as a function of t.

6. Check the equation which you have derived.

14

CHECKING THE VALIDITY OF EQUATIONS

To a conscientious scientist, few things are more embarrassing than publishing an erroneous equation. You should therefore learn to regard every newly derived equation with the deepest distrust until it has survived the simple but rigorous tests described in this chapter. You should use the same tests to check any published equation which puzzles you. By so doing, you can usually either clarify its meaning in your own mind or prove the equation false.

14-1. A Sequence of Steps for Checking

The steps to be used in checking the validity of an equation are listed below, roughly in the order in which they are most conveniently applied.

I. Making sure that the equation is suitably arranged for checking

    A. *Is the equation written in its simplest form?* Are there still terms which could be collected or canceled, complex fractions which should be simplified, etc.?

        Examples: Compare Equation 12-29e with Equation 12-29 and Equation 13-24 with Equation 13-24a in Table 13-3.

    B. *Is the equation written in terms of primary factors?* If not, you may be checking only part of the derivation, or concealing an identity. This may produce an artificial correlation.

        Example: In the last chapter, $\dot{V}_A$ in Equation 13-28 is not a primary factor, but a composite of three other primary variables. Theoretically, therefore, it would be better to check Equation 13-30 in which the three factors are explicitly written in place of $\dot{V}_A$. For a discussion of artificial and inevitable correlations, see Sections 4-10 and 4-11.

    C. *Can the equation be further simplified*, without significant loss of accuracy, *by using approximations?* (See Section 3-18 and Appendix C.) If so, the simplified version must be checked for validity just as carefully as the original.

II. Testing for internal consistency

    A. *Has every symbol been precisely defined and consistently used?* (See Chapter 1.)

    B. *Is the equation dimensionally correct?* (See Chapter 2.)

    C. *Has the same set of units been used throughout?* If not, have appropriate numerical conversion factors been included in the equation, and clearly explained in the derivation? (See Chapter 2.)

    D. *Test the equation by letting each appropriate variable approach first zero, then infinity.* Do these special cases make sense? If not, what are the limits within which the equation does hold? Such limits should be clearly specified.

    E. *Test the equation at least twice by numerical substitution.* Using the method described in this chapter, make up a set of arbitrary but reasonable and internally consistent values for the variables in the equation. Is the equation correct when these values are substituted into it?

III. Testing for external consistency

    A. *Compare the equation with any similar equations previously derived.* Are the differences of the sort to be expected from using different models? Is there at least qualitative agreement about the effect of varying different factors?

    B. *Compare the equation with actual experimental data.* Is the equation able to predict the trend of the data reasonably well? If not, is the disagreement what you might expect in view of the simplifying assumptions which had to be made?

The steps up to and including II-C have already been discussed or exemplified, as indicated in the above list by references to other chapters. The remaining items will now be illustrated by using them to check the equations for inert gas accumulation which were derived in the preceding chapter.

14-2. II-D. Testing the Validity of an Equation at the Limits of Each Important Variable

Equation 13-28

$$F_{eq} = 1 - \exp\left[-\frac{\dot{V}_A \dot{V}_B t}{(V\text{dist})\,(\dot{V}_A + S_B \dot{V}_B)}\right] \qquad (13\text{-}28)$$

contains five independent variables, each of which we can allow to approach zero or infinity while holding the others constant. For example, as t approaches zero, the exponent approaches zero, the exponential approaches unity, and F_{eq} approaches zero as it should. When t approaches infinity, the exponential approaches zero, and F_{eq} approaches unity (*i.e.*, complete equilibrium) again as it should. To choose another example, as S_B approaches zero, the denominator of the exponent approaches $(V\mathrm{dist})\dot{V}_A$, and the entire exponent therefore approaches $\dot{V}_B t/(V\mathrm{dist})$. In other words, the equation would have us believe that with gases of extremely low solubility the fraction of equilibrium attained at any time, t, is no longer influenced by alveolar ventilation. Does this make sense? If a gas is practically insoluble, so little of it will be distributed into the blood from alveolar air that the concentration in alveolar air will be virtually undiminished. Under these circumstances (once initial "lung washout" has been completed) the concentration of inert gas in the alveoli will practically equal the inspired concentration whether the ventilation of the alveoli be large or small. The limiting factor in the absorption of gas from the lungs will then be the rate at which the blood flows through the lungs to carry the gas away. Thus the equation *does* seem to make sense when we allow S_B to approach zero.

The values approached by F_{eq} as, one by one, the several factors are allowed to approach zero or infinity, are given in Table 14-1. Note that, oddly enough, as $\dot{V}_A$ becomes very large and approaches infinity, it ceases to influence F_{eq}! For as $\dot{V}_A$ approaches infinity, $S_B \dot{V}_B$ becomes negligibly small compared with $\dot{V}_A$, and the exponent therefore approaches $\dot{V}_A \dot{V}_B t/\dot{V}_A(V\mathrm{dist})$, or $\dot{V}_B t/(V\mathrm{dist})$. You should be able to verify the remainder of the table and to convince yourself that in each instance the equation cor-

TABLE 14-1

Testing the validity of Equation 13-28 at the extremes of the variables

Variable	Value Approached by F_{eq} When the Variable Shown at the Left Approaches	
	Zero	Infinity
t	0	1
$\dot{V}_A = \dot{N}(V_{\mathrm{tid}} - V_D)$	0	$1 - \exp\left[-\dfrac{\dot{V}_B t}{(V\mathrm{dist})}\right]$
$\dot{V}_B$	0	$1 - \exp\left[-\dfrac{\dot{V}_A t}{(V\mathrm{dist})S_B}\right]$
S_B	$1 - \exp\left[-\dfrac{\dot{V}_B t}{(V\mathrm{dist})}\right]$	$1 - \exp\left[-\dfrac{\dot{V}_A t}{(V\mathrm{dist})S_B}\right]$
$(V\mathrm{dist})$	1	0

rectly predicts what ought to happen as the variable assumes very large or very small values.

14-3. II-E. Testing the Validity of an Equation by Numerical Substitution

This is the most critical test of all and should never be neglected. In brief, it consists of giving numerical values to the variables used in the derivation so that the final equation will reduce to a numerical identity only if it has been properly derived from the elementary equations.

Consider an equation where y is a function of $n - 1$ other variables. Including y, such an equation will have a total of n variables. Now we may assign whatever arbitrary values we please to any $n - 1$ of the variables. But having done so, *the possible values of the remaining variable are completely determined by the functional relationship specified by the equation.*

Example

Let

$$y^2 = (a - b + \tfrac{1}{2}c)/4\,d^2 \tag{14-1}$$

Assign the following arbitrary values to the first four variables: $y = 7$, $a = 11$, $b = 3$, and $c = 4$. Then d can have only the values 0.2258, or -0.2258.

Similarly for *each* equation used in a derivation, the value of one of the variables becomes fixed, and can no longer be assigned arbitrarily.

Now consider Table 14-2 in which the elementary equations of Table 13-3 again appear. Equation 13-13 has been added for completeness. The only other change from Table 13-3 has been to treat each time derivative as a single unit because a derivative can be given a definite numerical value, whereas an isolated differential cannot. Accordingly, Equation 13-25 has been divided through by dt to obtain Equation 13-25a. As in Table 13-3, each symbol has been allotted its own column. Equation 14-2, below the 9 elementary equations from which it is derived, is the same as Equation 13-27 except that $\dot{V}_A$ has been replaced by $\dot{N}(V_{\text{tid}} - V_D)$ from Equation 13-13. We shall now check the derivation of Equation 14-2.

There are 17 variables listed across the top of Table 14-2. We can assign arbitrary values to only 8 of these, since for each of the 9 elementary equations used in the derivation 1 of the 17 variables must assume a fixed value. To check the internal consistency of Equation 14-2, we want to have as many of its terms as possible fixed by the constraints of the elementary equations. Let us accordingly assign arbitrary values to 8 of the "unwanted" variables, *i.e.*, those which do *not* appear in the final equation. We may as well choose values which are biologically reasonable, although, in fact, *any*

TABLE 14-2

Checking the derivation of Equation 14-2 by numerical substitution

The calculated values are in **bold face**. See Table 14-3 for the sequence of steps used to obtain the numerical values in this table.

Equation No.	Equation	C_I	C_A	C_{art}	C_{ven}	C_{tot}	$\dot{V}_A$	$\dot{V}_B$	$(V\text{dist})$	V_{tid}	V_D	$\dot{N}$	S_B	Q_{tot}	$Q_{tot.eq}$	dQ_{tot}/dt	F_{eq}	dF_{eq}/dt
13-14a	$\dot{V}_A(C_I - C_A) = \dot{V}_B(C_{art} - C_{ven})$	440	380	300	230		**6.00**	**5.14**										
13-15	$C_{art} = S_B C_A$		380	**300**									0.790					
13-16a	$dQ_{tot}/dt = \dot{V}_B(C_{art} - C_{ven})$			300	230			5.14								**360**		
13-17	$C_{tot} = C_{ven}$				230	**230**												
13-18	$Q_{tot.eq} = (V\text{dist})C_I S_B$	440							69.1				0.790		**24,000**			
13-19	$Q_{tot} = (V\text{dist})C_{tot}$					230			69.1					**15,900**				
13-1	$F_{eq} = Q_{tot}/Q_{tot.eq}$													15,900	24,000		**0.663**	
13-13	$\dot{V}_A = \dot{N}(V_{tid} - V_D)$						6.00			0.500	0.150	**17.1**						
13-25a	$dQ_{tot}/dt = Q_{tot.eq}(dF_{eq}/dt)$														24,000	360		**0.0150**
14-2	$$\frac{dF_{eq}/dt}{} = \left[\frac{\dot{N}(V_{tid} - V_D)\dot{V}_B}{(V\text{dist})\dot{N}(V_{tid} - V_D) + S_B\dot{V}_B}\right](1 - F_{eq})$$							5.14	69.1	0.500	0.150	17.1	0.790				0.663	**0.0150**

$$0.0150 = \frac{17.1(0.5 - 0.15)5.14}{(69.1)17.1(0.5 - 0.15) + 0.79(5.14)}(1 - 0.663) = 0.0149 \text{ (slide rule calculations)}$$

342

numerical values would serve to check the correctness of the derivation. (Zero, unity, and infinity should, of course, be avoided.)

The proper procedure is best explained by describing how to obtain the particular numerical values exemplified in Table 14-2. First, we assign an arbitrary value of 440 μmoles per liter to C_I. (Any other value and any other "unwanted" variable would have served equally well as a starting point.) We write "440" in the two checked boxes of the column for C_I. By itself, this value does not fix the value of any of the other variables, hence we are free to assign another arbitrary value to any other "unwanted" variable, for example, 380 μmoles per liter to C_A, entering this value in the checked boxes of the C_A column. Again there are no variables whose values can be calculated, so we assign an arbitrary 300 μmoles per liter to C_{art}. Now the value of S_B is fully determined because it is the only remaining unevaluated term in Equation 13-15. *Before assigning any further arbitrary values* we must therefore calculate the value of S_B, write it in the three proper boxes, and see whether the evaluation of S_B makes it possible to calculate the values of still more variables. The remainder of the table is completed in the same manner, always taking care at each step to calculate as many values as possible before assigning another arbitrary value. All of the steps used in the example of Table 14-2 are summarized in Table 14-3.

We now have a complete set of synthetic data which are known to be consistent with the elementary equations. Substitution of these synthetic values into Equation 14-2 should yield a numerical identity, as indeed it does (see bottom line of Table 14-2). If it did not, the equation would be wrong, and no doubt about it! However, if only a *single* set of made-up values is used, obtaining a numerical identity does not prove conclusively that the equation is correct. There is always a remote chance that the particular set of values chosen will satisfy both the correct equation and a wrong equation. It is therefore important to test the equation with at least *two* sets of consistent values which differ markedly from each other. Fortuitous satisfaction of an erroneous equation by both sets of synthetic data is exceedingly unlikely.

The procedure described above does not even require a separate table. You can write the synthetic data directly upon a sheet of tracing paper placed over the original Table 13-3. Better still, make up a set of values *before* undertaking the original derivation. Then if the need should arise, you are ready to carry out a numerical check of the intermediate steps as well as of the final equation.

We have checked Equation 14-2 for internal consistency, but what about its integrated form, Equation 13-30? If we were to substitute our numerical values into the logarithmic form of Equation 13-30 we could solve for t. However, this does not test the correctness of the integration because we

TABLE 14-3

An outline of the steps which were used to obtain the numerical values in Table 14-2

An Arbitrary Value of	Was Assigned to	Necessitating Calculation of
440 μmoles/L.	C_I	
380 μmoles/L.	C_A	
300 μmoles/L.	C_{art}	$S_B = 0.790$
230 μmoles/L.	C_{ven}	$C_{\text{tot}} = 230$ μmoles/L.
6 L./min.	$\dot{V}_A$	$\dot{V}_B = 5.14$ L./min. $dQ_{\text{tot}}/dt = 360$ μmoles/min.
24,000 μmoles	$Q_{\text{tot.eq}}$	(Vdist) $= 69.1$ L. $dF_{\text{eq}}/dt = 0.0150$/min. $Q_{\text{tot}} = 15,900$ μmoles $F_{\text{eq}} = 0.663$
0.5 L.	V_{tid}	
0.15 L.	V_D	$\dot{N} = 17.1$/min.

have no way of calculating t from the original equations. We can best check Equation 13-30 by showing that, upon differentiating it with respect to time, we obtain Equation 14-2.

14-4. III-A. Comparing the Equation with Similar Equations Previously Derived

Few equations are wholly original. Frequently a new equation is actually a generalized version or a specialized version of a previous equation. It is then often possible to show that, under restricted conditions, the old and the new reduce to the same equation. A comparison with similar equations previously derived is valuable both as a means of checking the validity of the newly derived equation and as a way of bringing out clearly its novel features.

As an illustration, consider Equations 13-12 and 13-28. Both are equations for F_{eq} derived quite independently of each other. Equation 13-12 is valid only when C_{art} is assumed to be constant throughout the entire course of equilibration. Equation 13-28 is valid only when the entire body is assumed to consist of a single homogeneous tissue. Now let us make *both* of these assumptions for each equation in turn, and see whether they reduce to the same equation.

First consider Equation 13-12. If the body consists of a single homogeneous tissue, Equation 13-12 reduces to

$$F_{eq} = \left[\frac{Q_{tot.eq}}{Q_{tot.eq}}\right]\left[1 - \exp\left(\frac{\dot{V}_BS_Bt}{(V\text{dist})S_B}\right)\right]$$
$$= 1 - \exp\left(-\frac{\dot{V}_Bt}{(V\text{dist})}\right)$$

(14-3)

Next consider Equation 13-28. If C_{art} remains constant throughout, it must always be equal to its equilibrium value, C_IS_B. But we are also assuming (Equation 13-15) that C_{art} is at all times equal to C_AS_B. Hence C_A must equal C_I. Now C_A can equal C_I only when a negligibly small fraction of the gas in alveolar air is removed by the blood. This will be true if the solubility of the gas in blood is very small, or if the air flow is very large, or if the blood flow is very small. But we have already shown (Table 14-1) that as S approaches zero or as $\dot{V}_A$ approaches infinity, Equation 13-28 also reduces to Equation 14-3. And if blood flow is allowed to approach zero, both Equation 13-12 and Equation 13-28 approach zero. Hence, when the two equations are subjected to the same restrictions, they become identical. But the importance of this demonstration goes beyond the mere proof that the two equations are compatible. For if S_B is so small or V_A is so large that C_{art} is, in fact, practically equal to C_IS_B, then Equation 13-12 (or a more realistic equation of like form) will be by all odds the better and more general equation.

Further comparison of Equation 13-28 with other published equations (64, 89) will be left to the interested reader. But one general point is worth making here. It is always wise to translate a published equation into the system of symbols with which you are familiar. This avoids the confusion likely to result from trying to work with two different sets of symbols simultaneously. Moreover, it forces you to make sure you understand the exact meaning of every term in both equations.

14-5. III-B. Comparing the Equation with Experimental Data

In the tests so far discussed the validity of the simplified model has not really been under scrutiny at all. We have simply been making certain that the equations being checked were properly derived from the model. But if the model itself is unrealistic, equations based upon it are hardly likely to be useful for explaining physiological phenomena or for advancing biological theory. In short, one can never be satisfied with a theoretical equation unless it can fit experimental data. And if the data to test it are not available, the equation points straight to the laboratory bench.

In the present example, we can compare the theory with some of Haggard's experimental observations on the accumulation of diethyl ether (53). In his "Experiment 1," Haggard allowed a 10-kg. dog, under deep sedation with morphine, to breathe for 2 hr. a constant concentration of 0.2 gm. of ether per liter of air (BTPS). The quantity of ether in the dog at any time, Q_{tot}, was calculated as the difference between the total quantity of ether inspired and the total quantity expired up to that time. The cardiac output, $\dot{V}_B$, was estimated at 0.681 L. per minute. The respiratory dead space was 30.5 per cent of the tidal volume. Although Haggard gave values for only the total respiratory minute volume, i.e., $\dot{N}V_{tid}$, we can estimate the alveolar ventilation by assuming that throughout the experiment the dead space continued to represent 30.5 per cent, and the alveolar ventilation $100 - 30.5 = 69.5$ per cent of the total respiratory minute volume. Let us also assume that both cardiac output and alveolar ventilation remained constant, although actually ventilation tended to decrease somewhat toward the end of the 2-hr. experiment (Table 14-4). Finally let us assume that the volume of distribution of ether was equal to the total body volume, i.e.,

TABLE 14-4

Testing the agreement between theory and experiment

Time (t)	Tidal Volume per Minute ($\dot{N}V_{tid}$)	Calculated Alveolar Ventilation per Minute ($\dot{V}_A = 0.695\dot{N}V_{tid}$)	Total Ether Absorbed (Q_{tot})	Fraction of Equilibrium (F_{eq})	
				From experimental data	From Equation 13-28
min.	*L./min.*	*L./min.*	*gm.*		
0	1.57	1.09	0.0	0.000	0.000
10	1.51	1.05	1.8	0.060	0.059
25	1.60	1.11	4.2	0.140	0.142
40	1.50	1.04	6.5	0.217	0.217
55	1.50	1.04	8.5	0.283	0.286
70	1.44	1.00	10.2	0.340	0.349
85	1.35	0.94	12.0	0.400	0.406
100	1.28	0.89	13.0	0.433	0.458
120	1.28	0.89	14.5	0.483	0.520

roughly 10 L. in a 10-kg. dog. Then at equilibrium

$$Q_{tot.eq} = (V\text{dist})C_I S_B \tag{13-18}$$

or, for this experiment,

$$Q_{tot.eq} = (10)(0.2)(15) = 30 \text{ gm. of ether}$$

We can now calculate an "experimental" value for F_{eq} at each of the times given in Table 14-4. We can also calculate "theoretical" values for F_{eq} by means of Equation 13-28 and the following data:

$\dot{V}_A$ = 1.01 L. per minute (mean of the nine values in Table 14-4)

$\dot{V}_B$ = 0.681 L. per minute

$(V\text{dist})$ = 10.0 L.

S_B = 15.0 (for ether in blood at 37°C.)

The experimental and theoretical values of F_{eq} are given in the last two columns of Table 14-4 and are compared graphically in Figure 14-1. The agreement between experiment and theory is remarkably close — closer, indeed, than we have any right to expect! Even the tendency of the last two points to fall below the theoretical line is in accord with the decrease

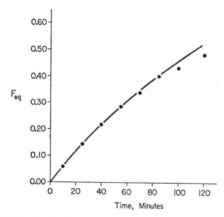

FIG. 14-1. Testing the Agreement between Theory and Experiment

The *solid line* is the fraction of equilibrium attained at various times as predicted from the theoretical equation, Equation 13-28. The *points* represent actual values derived from the experimental data of Haggard (53). The points and the corresponding theoretical values are compared in Table 14-4.

in alveolar ventilation towards the end of the experiment. Now whenever the agreement between two sets of values is as close as this, one should suspect a hidden identity leading, perhaps, to an artificial or an inevitable correlation (see Sections 4-10 and 4-11). While there is nothing of the sort here to account for the remarkable fit of the experimental points to the theoretical line, it is worth noting that of the four variables (other than t) used to calculate the theoretical values, two, (Vdist) and S_B, were *also* used to calculate the experimental values. For this reason, the ratio

experimental F_{eq}/theoretical F_{eq}

is rather insensitive to small errors in the value assumed for (Vdist) (Fig. 14-2). Furthermore, the cardiac output, which Haggard estimated by a rather crude method, has comparatively little influence upon the rate of approach to equilibrium with a gas as highly soluble as diethyl ether. Hence the two biological variables whose values are most in doubt ((Vdist) and $\dot{V}_B$) have the least influence upon the agreement between theory and experiment (Fig. 14-2). This makes the close agreement actually observed somewhat less surprising.

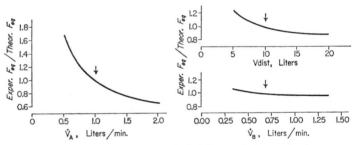

Fig. 14-2. How Much Do Various Biological Factors Influence the Agreement between Theory and Experiment Illustrated by Figure 14-1?

According to Table 14-4, at 85 min. the fraction of equilibrium predicted from Equation 13-28 (0.406) was practically identical with the fraction calculated from Haggard's experimental data (0.400), so that ratio between the experimental fraction and the theoretical fraction is, to all intents and purposes, unity. But might this good agreement still be good if we were to use in our calculations values which were very far from the true values? If so, the close agreement we have observed would be meaningless. To rule out this possibility, in each of the graphs of this figure one of the biological variables has been allowed to assume erroneous values greater than or less than the true value indicated by an *arrow*. A rather small departure of alveolar ventilation (*graph at left*) from its true value spoils the agreement between theory and experiment. This is less true of the volume of distribution (*top right*) and still less true of blood flow (*bottom right*) which, with a gas as soluble as ether, has little influence upon rate of equilibration.

The ability of Equation 13-28 to fit Haggard's data gives some assurance that the equation is reasonable, at least for gases of high solubility. Further testing of the equation with similar data for gases of intermediate and low solubilities would certainly be worthwhile.

EXERCISES. CHAPTER 14

In each of the following exercises, the equation cited is wrong. Identify the error or errors by checking the equation, using the method outlined in the text. Then correct the equation if possible.

Exercise 1

Brownell (18) gives the following equation for a single compartment of some substance, let us say S, labeled at time zero by a tracer dose of an isotope:

$$u = (100/M)e^{-kt}$$

where

u is the total quantity of S entering (and, steady state being assumed, leaving) the compartment per unit time (gm./sec.)

M is the total quantity of S present in the compartment (gm.)

100 is the total dose of isotope (*i.e.*, 100 per cent) present in the compartment at time zero

t is time, measured from the time of giving the tracer dose (sec.)

k is the rate constant, defined as u/M (proportion per sec.)

Exercise 2

The following is quoted from Houssay's physiology textbook, 2nd edition, 1955, page 482 (60).

"Calcium ion depresses excitability, but its effect depends upon the concentration of other ions. Monovalent ions (Na^+ and K^+) have the opposite effect of bivalent ions (Ca^{++} and Mg^{++}); hydrogen ion has the same effect as bivalent ions. Excitability depends on an equilibrium between these ions, which can be expressed as follows:

$$K = \frac{(Na^+) + (K^+)}{(Ca^{++}) + (Mg^{++}) + (H^+)}$$

Excitability is depressed when the value of the constant K diminishes and it becomes higher as K increases."

Exercise 3

In reviewing various procedures for estimating the volume of extracellular fluid, White and Rolf (115) state that "... following a single injection of substances which are completely recovered in the urine . . . and whose falling plasma curve is a straight semi-log line . . ." the volume of distribution of the substances may be estimated as ". . . renal clearance in a period/ change in natural log of plasma level in that period."

Write this statement in the form of an equation, and then check the equation.

Exercise 4

The following passages are quoted from Smith, *The Kidney*, 2nd edition, 1951 (98).

Pages 144 to 145: "If the plasma concentration of X is indicated by P_x, the unbound fraction by F, the water content of the plasma by W, and the filtration rate by C_F, then the rate of filtration of X in mg./ min. will be

$$P_x F W C_F$$"

Page 146: "The rate of tubular excretion, T_x, is the difference between the total rate of excretion; $U_x V$, and the filtration rate of X, *i.e.*,

$$T_x = U_x V - C_F P_x F W = \left(\frac{C_x}{C_F} FW\right) P_x$$"   (Equation 3)

U_x is the concentration of X in urine; V is urine flow; and C_x is the renal plasma clearance of X. (In order to check and correct this equation you may find it desirable to rewrite it, using the kind of symbols which have been employed in the present text.)

Exercise 5

The following two equations are taken from a paper (96) as outstanding for the implausibility of the experimental method used as for the erroneous mathematical analysis of the resulting "data" (square brackets supplied for the sake of clarity):

$$F_c = \frac{A_c}{T_c \left[\int_0^a \Delta C_c\right] S_c} \quad \text{and} \quad F_c = \frac{I_c \left[\dfrac{1}{\int_0^a \Delta C_c}\right] S_c}{A_c}$$

where

F_c is coronary blood flow (presumably in ml./sec.).

A_c is an area (sq. in.) measured from a recording of counts-per-minute (cpm) over the precordial region in man as a function of time after the sudden intravenous injection of a dose of radioactive iodinated human albumin or Diodrast.

(The authors believe that the area chosen for measurement represents radioactive material in the coronary bed.)

T_c is the "transit time" for the coronary circulation, *i.e.*, the interval between the time when the first radioactivity is believed to be detected in the coronary circulation, and the time when a curve (which is supposed to bound the "coronary peak") "seems to end" as it is extrapolated to the base line.

S_c is the sensitivity of the recording equipment (inches deflection per cpm).

I_c is the total quantity of tracer traversing the coronary circulation during its first circulation (cpm).

$\int_0^a \Delta C_c$ is best described in the authors' own words: "The concentration between (*sic*) the tracer substance within the blood bolus is not known as a continuous dilution is occurring. Let the integral …

$$\int_0^a \Delta C_c$$

at the coronary level express this changing concentration."

Exercise 6

Forster has published the following form of "the alveolar air equation" (38):

$$\text{alveolar } P_{O_2} = \text{inspired } P_{O_2} - \frac{\text{arterial } P_{CO_2}}{\text{inspired } P_{CO_2}} (\text{inspired } P_{O_2} - \text{expired } P_{O_2})$$

where P is the partial pressure of the designated gas.

Exercise 7

Crawford *et al.* (27) studied the effect of variations in the total rate of solute excretion upon the osmolar concentration of the urine of water-loaded human subjects who were receiving continuous intravenous infusions of the antidiuretic hormone "Pitressin," at various constant rates. Careful reading of their abstract indicates that they obtained the following results:

1. If the rate of infusion of Pitressin was less than 0.17 mU. per square meter of body surface per minute, the concentration of the urine was less than 154 mOs. per liter when the rate of solute excretion was small but tended to increase toward a limiting value of about 154 mOs. per liter as the rate of solute excretion increased.

2. If the rate of infusion of Pitressin was greater than 0.17 mU. per

square meter of body surface per minute, the concentration of the urine was greater than 154 mOs. per liter when the rate of solute excretion was small, but tended to decrease toward the same limiting value of about 154 mOs. per liter as the rate of solute excretion increased.

3. It is thus *implied* that with an infusion of 0.17 mU./M^2-min. the concentration of the urine would be about 154 mOs. per liter regardless of the rate of solute excretion.

The authors summarize their results in the following general equation which they claim is valid at any rate of infusion of Pitressin (presumably the absolute values of the parameters a and b would be different for different rates):

$$\text{``Log } \Delta K = a + b \text{ Log } X$$

where ΔK is the difference between observed urine water concentration and the constant K; a is the intercept constant; b, the slope constant and X, the output rate of osmotically active solute in urine.''

At all rates of Pitressin infusion from zero to a maximally effective rate of 0.75 mU./M^2-min., their data were best fitted when K had a constant value of 6.5. As defined by these authors, the peculiar term "water concentration" means the reciprocal of the osmolar concentration expressed in millimoles per milliliter. For example, if the concentration of the urine were 154 mOs. per liter, the "water concentration" would be $1000/154 = 6.5$ ml. per milliosmole.

Exercise 8

Riggs (88) attempted to analyze iodine metabolism according to the following model:

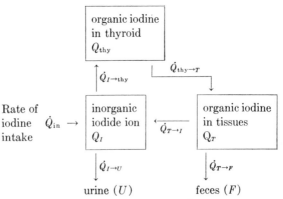

If we assume that the system is in a steady state,

$$(1) \quad \dot{Q}_{\text{in}} + \dot{Q}_{T \to I} = \dot{Q}_{I \to U} + \dot{Q}_{I \to \text{thy}}$$

$$(2) \quad \dot{Q}_{I \to \text{thy}} \qquad = \dot{Q}_{\text{thy} \to T}$$

$$(3) \quad \dot{Q}_{\text{thy} \to T} \qquad = \dot{Q}_{T \to I} + \dot{Q}_{T \to F}$$

Now suppose we completely block the uptake of iodine by the thyroid gland by giving adequate doses of such a drug as propylthiouracil. Because the compartment of inorganic iodide is very small and its contained iodide therefore "turns over" much more rapidly than is true for the other compartments, it may be legitimate to assume that it soon reaches a new steady state:

$$(4) \quad \dot{Q}_{\text{in}} + \dot{Q}_{T \to I_{\text{bl}}} = \dot{Q}_{I \to U_{\text{bl}}}$$

where the subscript "bl" signifies "when the thyroid is blocked." We shall also assume that both $\dot{Q}_{\text{in}}$ and $\dot{Q}_{T \to F}$ (which is comparatively small anyway) remain constant and that

$$(5) \quad \dot{Q}_{\text{thy} \to T_{\text{bl}}} = \dot{Q}_{T \to I_{\text{bl}}} + \dot{Q}_{T \to F}$$

With these assumptions, the five equations listed above can be combined to give

$$\dot{Q}_{\text{thy} \to T_{\text{bl}}} = \dot{Q}_{I \to U_{\text{bl}}} - \dot{Q}_{I \to U}$$

Because both of the quantities on the right are measurable, the rate of secretion of iodine from the blocked gland can be calculated.

Next let us suppose that the organic iodine in compartments thy and T had been previously labeled with radioactive iodine and that the specific activities of these two compartments are equal. The following erroneous equation (Equations 56 and 57 in the original paper) was said to permit estimation of the quantity of iodide in the thyroid gland, symbols for radioactive iodine being designated by asterisks:

$$\text{Specific activity} = \frac{Q^*_{\text{thy}}}{Q_{\text{thy}}} = \frac{\dot{Q}^*_{\text{thy} \to T_{\text{bl}}}}{\dot{Q}_{\text{thy} \to T_{\text{bl}}}} = \frac{\dot{Q}^*_{I \to U_{\text{bl}}} - \dot{Q}^*_{I \to U}}{\dot{Q}_{I \to U_{\text{bl}}} - \dot{Q}_{I \to U}}$$

or, solving for Q_{thy} :

$$Q_{\text{thy}} = \frac{Q^*_{\text{thy}} (\dot{Q}_{I \to U_{\text{bl}}} - \dot{Q}_{I \to U})}{(\dot{Q}^*_{I \to U_{\text{bl}}} - \dot{Q}^*_{I \to U})}$$

Accepting all of the numerous assumptions given above, write the equation as it *should* be.

Exercise 9

In an interesting attempt to analyze a hypothetical mechanism which might be responsible for certain periodic phenomena in biological systems,

Spangler and Snell (103) have postulated a symmetrical "cross-coupled catalytic system" which we may symbolize as follows:

$$A + E \underset{k_{-1}}{\overset{k_1}{\rightleftharpoons}} B \overset{k_2}{\longrightarrow} E + P; \qquad P \overset{k_3}{\longrightarrow}$$

$$+$$

$$\alpha P'$$

$$k_4 \Big\downarrow\Big\uparrow k_{-4}$$

$$I$$

and

$$A' + E' \underset{k'_{-1}}{\overset{k_1'}{\rightleftharpoons}} B' \overset{k_2'}{\longrightarrow} E' + P'; \qquad P' \overset{k_3'}{\longrightarrow}$$

$$+$$

$$\alpha P$$

$$k_4' \Big\downarrow\Big\uparrow k'_{-4}$$

$$I'$$

where the symbols A, E, B, etc., and A', E', B', etc., all refer to different compounds, and α is the number of molecules of P' (or of P) combining with one molecule of the enzyme, E, (or E') to form the inactive complex, I (or I').

Assume that both sequences of reactions diagramed above are in the steady state, *i.e.*, that the flow of molecules into A (or A') is equal to the flow of molecules out of P (or P'). (Note here, by the way, the use of the common and convenient symbology in which if k_i is the rate constant for a particular reaction, k_{-i} is the rate constant for the reverse reaction.)

Spangler and Snell give the following equation derived from the system outlined above:

$$"\left(\frac{k_3 P}{k_2 E_t - k_3 P}\right)$$

$$\left(1 + K_4\left[\frac{k_1' k_2' E_t' A'}{k_3'(k_2' + k'_{-1})(1 + k_4' P^\alpha) + k_1' k_3' A'}\right]^\alpha\right) - \frac{k_1 A}{k_2 + k_{-1}} = 0$$

$$K_4 \equiv \frac{k_4}{k_{-4}} \qquad K_4' \equiv \frac{k_4'}{k'_{-4}}$$

where the species symbols are used to denote concentrations ... E_t and E_t' are the total catalyst concentrations which are conserved."

Check this equation dimensionally, and try to use your common sense in correcting it. (As a much more rigorous alternative, derivation of the equation by the tabular method is perfectly straightforward but somewhat tedious.)

Exercise 10

Malvin *et al.* (73) have attempted to calculate the concentration of sodium in the fluid being reabsorbed from the lumen of the proximal tubules of the kidney while flow along the tubule is temporarily arrested (or at least slowed) by occluding the ureter during a massive osmotic diuresis ("stop-flow" technique). The urine which flows out when the ureter is unclamped is collected in small serial samples, some of which are believed to represent "proximal" samples, *i.e.*, samples which remained in the proximal convoluted tubule during the stoppage of flow and whose composition was therefore modified by the abnormally prolonged operation of the reabsorptive processes of the proximal tubule during the stoppage of flow. Malvin *et al.* measured the concentrations of sodium and of creatinine in such "proximal samples" obtained after restoration of flow, and they compared these concentrations with the concentrations in "free-flow" urine collected just before stoppage of flow and with the concentrations in plasma. They *assume* that creatinine is neither absorbed by, nor secreted by the renal tubules and that its concentration at any point along the nephron (compared to its concentration in plasma) therefore indicates how much water has been reabsorbed from the glomerular filtrate. They further *assume* that sodium absorption from the stop-flow sample has occurred in three stages: $Q_{Na.before.s}$, taking place prior to stoppage of flow between the glomerulus and the point at which flow is arrested, $Q_{Na.during.s}$, taking place while flow is stopped, and $Q_{Na.after.s}$, taking place after restoration of flow through those segments of the nephrons which are distal to the point at which the "proximal sample" was temporarily arrested. (The subscript, s, designates the "stop-flow" sample.)

$$Q_{Na.tot.s} = Q_{Na.before.s} + Q_{Na.during.s} + Q_{Na.after.s}$$

They further *assume* that in free-flow samples of urine collected before stopping flow, the reabsorption of sodium differs from the stop-flow sample *only* in its lack of the "$Q_{Na.during}$" component of reabsorption. In other words (with the subscript, f, to designate the free-flow sample), it is assumed that

$$Q_{Na.tot.f} = Q_{Na.before.s} + Q_{Na.after.s}$$

Combining these two equations and solving for $Q_{Na.during.s}$,

$$Q_{Na.during.s} = Q_{Na.tot.s} - Q_{Na.tot.f}$$

Exactly the same assumptions are made for water reabsorption. Using the argument presented above, the authors then calculate the quantity of sodium reabsorbed per milliliter of urine, and the quantity of water reabsorbed per milliliter of urine, during stoppage of flow. Their equation is

$C_{\text{Na.reabs.}s}$

$$= \frac{[(C_{\text{cr.}U.s}C_{\text{Na.}P}/C_{\text{cr.}P}) - C_{\text{Na.}U.s}] - [(C_{\text{cr.}U.f}C_{\text{Na.}P}/C_{\text{cr.}P}) - C_{\text{Na.}U.f}]}{[(C_{\text{cr.}U.s}/C_{\text{cr.}P}) - 1] - [(C_{\text{cr.}U.f}/C_{\text{cr.}P}) - 1]}$$

where

$C_{\text{Na.reabs.}s}$ is the concentration of sodium in the hypothetical "reabsorbate" of the proximal tubules during stoppage of flow. In other words, it is the calculated quantity of sodium reabsorbed, $Q_{\text{Na.during.}s}$, divided by the similarly calculated volume of water reabsorbed, per milliliter of urine.

cr   means "of creatinine"

U means "in the urine sample"

P means "in plasma"

and the other subscripts and symbols are already familiar. (The symbols used here are not the ones used by the authors.)

REFERENCES

(*Note:* References 14, 50, 55, 111, 112, and 119 were deleted in proof.)

1. ABRAMSON, A. B.: *Dimensional Analysis for Students of Medicine.* Josiah Macy, Jr., Foundation, New York, 1950.
2. ADOLPH, E. F.: *Physiological Regulations.* Jaques Cattell Press, Lancaster, Pa., 1943.
3. ALBERT, A.: Ionization, pH, and biological activity. Pharmacol. Rev., 4: 136–167, 1952.
4. ALBERT, A.: *Selective Toxicity,* Ed. 2. Methuen & Co., Ltd., London, 1960.
5. ALLELA, A., WILLIAMS, F. L., BOLENE-WILLIAMS, C., AND KATZ, L. N.: Interrelation between cardiac oxygen consumption and coronary blood flow. Amer. J. Physiol., 183: 570–582, 1955.
6. ARIËNS, E. J., VAN ROSSUM, J. M., AND SIMONIS, A. M.: Affinity, intrinsic activity and drug interactions. Pharmacol. Rev., 9: 218–236, 1957.
7. BAKER, N., SHIPLEY, R. A., CLARK, R. E., AND INCEFY, G. E.: C^{14} studies in carbohydrate metabolism: glucose pool size and rate of turnover in the normal rat. Amer. J. Physiol., 196: 245–252, 1959.
8. BELL, P. H., AND ROBLIN, R. O. JR.: Studies in chemotherapy. VII. A theory of the relation of structure to activity of sulfanilamide type compounds. J. Amer. chem. Soc., 64: 2905–2917, 1942.
9. BERGLUND, E., BORST, H. G., DUFF, F., AND SCHREINER, G. L.: Effect of heart rate on cardiac work, myocardial oxygen consumption, and coronary blood flow in the dog. Acta physiol. scand., 42: 185–198, 1958.
10. BERKSON, J.: Are there two regressions? J. Amer. statist. Ass., 45: 164–180, 1950.
11. BERLINER, R. W., LEVINSKY, N. G., DAVIDSON, D. G., AND EDEN, M.: Dilution and concentration of the urine and the action of antidiuretic hormone. Amer. J. Med., 24: 730–744, 1958.
12. BERLINER, R. W., AND ORLOFF, J.: Carbonic anhydrase inhibitors. Pharmacol. Rev., 8: 137–174, 1956
13. BERRY, R. N.: Quantitative relations among vernier, real depth, and stereoscopic depth acuities. J. exp. Psychol., 38: 708–721, 1948.
15. BRINK, F.: The role of calcium ions in neural processes. Pharmacol. Rev., 6: 243–298, 1954.
16. BRODIE, B. B., AND HOGBEN, C. A. M.: Some physico-chemical factors in drug action. J. Pharm., Lond., 9: 345–380, 1957.
17. BRONNER, F., BENDA, C. E., HARRIS, R. S., AND KREPLICK, J.: Calcium metabolism in a case of gargoylism studied with the aid of radiocalcium. J. clin. Invest., 37: 139–147, 1958.
18. BROWNELL, G. L.: Isotopes: radioactive; tracer experiments, theory. In *Medical Physics,* edited by O. Glasser, Vol. 3, pp. 353–364. Year Book Publishers, Inc., Chicago, 1960.
19. BROWNELL, G. L., CAVICCHI, R. V., AND PERRY, K. E.: Electrical analog for analysis of compartmental biological systems. Rev. sci. Instrum., 24: 704–710, 1953.

357

20. Burton, A. C.: The basis of the principle of the master reaction in biology. J cell. comp. Physiol., **9**: 1–14, 1936.

21. Cander, L.: Solubility of inert gases in human lung tissue. J. appl. Physiol., **14**: 538–540, 1959.

22. Carlson, L. D.: Gas exchange and transportation. In *Medical Physiology and Biophysics*, Ed. 18 of Howell's *Textbook of Physiology*, edited by T. C. Ruch and J. F. Fulton. W. B. Saunders Company, Philadelphia, 1960.

23. Cole, L. C.: Biological clock in the unicorn. Science, **125**: 874–876, 1957.

24. Cowgill, G. R., and Drabkin, D. L.: Determination of a formula for the surface area of the dog together with a consideration of formulae available for other species. Amer. J. Physiol., **81**: 36–61, 1927.

25. Crank, J.: *The Mathematics of Diffusion.* Oxford University Press, London, 1956.

26. Crawford, B. H.: The dependence of pupil size upon external light stimulus under static and variable conditions. Proc. roy. Soc. Ser. B., **121**: 376–395, 1937.

27. Crawford, J. D., Cushman, A. N., Parisi, A., and Terry, M. L.: The influence of pitressin and osmolar excretion on urine water concentration in normal man. J. clin. Invest., **36**: 880, 1957.

28. Danielli, J. F.: Theory of the penetration of a thin membrane. In *Permeability of Natural Membranes* by H. Davson and J. F. Danielli, Appendix. Cambridge University Press, London, 1943.

29. Daniels, F.: *Mathematical Preparation for Physical Chemistry.* McGraw-Hill Book Company, Inc., New York, 1928. (Republished in a paperbound edition.)

30. Davison, F. R.: *Synopsis of Materia Medica, Toxicology, and Pharmacology.* Ed. 3. C. V. Mosby Company, St. Louis, 1944.

31. Davson, H.: *A Textbook of General Physiology*, Ed. 2. J. & A. Churchill, Ltd., London, 1959.

32. Denton, J. E., and Beecher, H. K.: New analgesics. J. Amer. med. Ass., **141**: 1051–1057; 1146–1153, 1949.

33. Dominguez, R.: Kinetics of elimination, absorption and volume of distribution in the organism. In *Medical Physics*, edited by O. Glasser, Vol. 2, pp. 476–489. Year Book Publishers, Inc., Chicago, 1950.

34. Dominguez, R., Goldblatt, H., and Pomerene, E.: Kinetics of the elimination of substances injected intravenously (experiments with creatinine). Amer. J. Physiol., **114**: 240–254, 1935–1936.

35. DuBois, E. F.: *Basal Metabolism in Health and Disease*, Ed. 3. Lea & Febiger, Philadelphia, 1936.

36. Dwight, H. B.: *Mathematical Tables of Elementary and Some Higher Mathematical Functions*, Ed. 2. Dover Publications, New York, 1958.

37. Finney, D. J.: *Probit Analysis: A Statistical Treatment of the Sigmoid Response Curve*, Ed. 2. Cambridge University Press, London, 1952.

38. Forster, R. E.: Exchange of gases between alveolar air and pulmonary capillary blood: pulmonary diffusing capacity. Physiol. Rev., **37**: 391–452, 1957.

39. Fox, I. J., and Wood, E. H.: Circulatory system: methods; blood flow measurement by dye-dilution technics. In *Medical Physics*, edited by O. Glasser, Vol. 3, pp. 155–163. Year Book Publishers, Inc., Chicago, 1960.

40. Furchgott, R. F.: The pharmacology of vascular smooth muscle. Pharmacol. Rev., **7**: 183–265, 1955.

41. Gaddum, J. H.: Lognormal distributions. Nature, **156**: 463–466, 1945.

42. GADDUM, J. H.: Theories of drug antagonism. Pharmacol. Rev., **9**: 211–218, 1957.
43. GADDUM, J. H.: *Pharmacology*, Ed. 5. Oxford University Press, London, 1959.
44. GERBER, G. B., GERBER, G., AND ALTMAN, K. I.: Studies of collagen turnover in lathyritic rats. Arch. Biochem., **96**: 601–604, 1962.
45. GILBERT, D. L., AND FENN, W. O.: Calcium equilibrium in muscle. J. gen. Physiol., **40**: 393–408, 1957.
46. GLASSER, O. (Editor): *Medical Physics*, Vol. 1 (1944), Vol. 2 (1950), Vol. 3 (1960). Year Book Publishers Inc., Chicago.
47. GOLDMAN, S.: Cybernetic aspects of homeostasis. In *Mineral Metabolism*, edited by C. L. Comar and F. Bronner, Vol. 1, Part A, Chapter 3. Academic Press, Inc., New York, 1960.
48. GOLDSTEIN, A.: The interactions of drugs and plasma proteins. Pharmacol. Rev., **1**: 102–165, 1949.
49. GOODMAN, L. S., AND GILMAN, A.: *The Pharmacological Basis of Therapeutics*, Ed. 2. The Macmillan Company, New York, 1955.
51. GREEN, N. M., AND LOWTHER, D. A.: Formation of collagen hydroxyproline *in vitro*. Biochem. J., **71**: 55–66, 1959.
52. GRODINS, F. S., GRAY, J. S., SCHROEDER, K. R., NORINS, A. L., AND JONES, R. W.: Respiratory responses to CO_2 inhalation. A theoretical study of a nonlinear biological regulator. J. appl. Physiol., **7**: 283–308, 1954.
53. HAGGARD, H. W.: The absorption, distribution, and elimination of ethyl ether. II. Analysis of the mechanism of absorption and elimination of such a gas or vapor as ethyl ether. J. biol. Chem., **59**: 753–770, 1924.
54. HAMMOND, E. C.: The effects of smoking. Sci. Amer., **207**: 39–51, 1962.
56. HILL, A. V.: Diffusion of O_2 and lactic acid through tissues. Proc. roy. Soc. Ser. B., **104**: 39–96, 1929.
57. HITCHCOCK, D. I.: Diffusion in liquids. In *Physical Chemistry of Cells and Tissues*, edited by R. Höber, Chapter 1. The Blakiston Company, Philadelphia, 1945.
58. HODGMAN, C. D. (Editor-in-Chief): *Handbook of Chemistry and Physics*. Chemical Rubber Publishing Co., Cleveland, Ohio. (Frequently revised. The edition used in preparing this book has been the 41st., 1959–1960.)
59. HOFSTEE, B. H. J.: Plotting titration data. Science, **131**: 1068, 1960.
60. HOUSSAY, B. A.: *Human Physiology*, Ed. 2. McGraw-Hill Book Company, Inc., New York, 1955.
61. JACOBS, M. H.: Diffusion processes. Ergebn. Biol., **12**: 1–160, 1935.
62. JOHNSON, L. H.: *Nomography and Empirical Equations*. John Wiley & Sons, Inc., New York, 1952.
63. JONES, H. B.: Respiratory system: nitrogen elimination. In *Medical Physics*, edited by O. Glasser, Vol. 2, pp. 855–871. Year Book Publishers, Inc., Chicago, 1950.
64. KETY, S. S.: The theory and applications of the exchange of inert gas at the lungs and tissues. Pharmacol. Rev., **3**: 1–41, 1951.
65. KETY, S. S., AND SCHMIDT, C. F.: The nitrous oxide method for the quantitative determination of cerebral blood flow in man; Theory, procedure, and normal values. J. clin. Invest., **27**: 476–483, 1948.
66. KRANTZ, J. C., JR., AND CARR, C. J.: *The Pharmacologic Principles of Medical Practice*, Ed. 5. The Williams & Wilkins Company, Baltimore, 1961.
67. LACROIX, A., AND RAGOT, C. L.: *A Graphic Table Combining Logarithms and Antilogarithms*. The Macmillan Company, New York, 1925.

68. LANGHAAR, H. L.: *Dimensional Analysis and Theory of Models*. John Wiley & Sons, Inc., New York, 1951.

69. LEWIS, W. H., JR., AND ALVING, A. S.: Changes with age in the renal function in adult men. Amer. J. Physiol., 123: 500-515, 1938.

70. LLOYD, B. B., JUKES, M. G. M., AND CUNNINGHAM, D. J. C.: The relation between alveolar oxygen pressure and the respiratory response to carbon dioxide in man. Quart. J. exp. Physiol., 43: 214-227, 1958.

71. LLOYD, B. B., AND TAYLOR, K. B.: Genesis of human cerebrospinal fluid. J. appl. Physiol., 14: 401-404, 1959.

72. LOEWE, S.: Relationship between stimulus and response. Science, 130: 692-695, 1959.

73. MALVIN, R. L., WILDE, W. S., VANDER, A. J., AND SULLIVAN, L. P. Localization and characterization of sodium transport along the renal tubule. Amer. J. Physiol., 195: 549-557, 1958.

74. MELLOR, J. W.: *Higher Mathematics for Students of Chemistry and Physics*. Dover Publications, Inc., 1955. (Paperbound reprint of the 4th edition, originally published 1912.)

75. MILNE, M. D., SCRIBNER, B. H., AND CRAWFORD, M. A.: Non-ionic diffusion and the excretion of weak acids and bases. Amer. J. Med., 24: 709-729, 1958.

76. MOKOTOFF, R., ROSS, G., AND LEITER, L.: Renal plasma flow and sodium reabsorption and excretion in congestive heart failure. J. clin. Invest., 27: 1-9, 1948.

77. MOORE, W. J.: Chemical kinetics. In *Physical Chemistry*, Ed. 3, Chapter 17, Longmans, Green, & Co., New York, 1957.

78. NEUMAN, W. F., AND NEUMAN, M. W.: *The Chemical Dynamics of Bone Mineral*. University of Chicago Press, Chicago, 1958.

79. NORRIS, W. P., TYLER, S. A., AND BRUES, A. M.: Retention of radioactive bone-seekers. Science, 128: 456-462, 1958.

80. ORLOFF, J., AND BERLINER, R. W.: The mechanism of the excretion of ammonia in the dog. J. clin. Invest., 35: 223-235, 1956.

81. OSMOND, H., AND HOFFER, A.: *Pro domo sua*. A brief account of the Saskatchewan research in psychiatry. J. Neuropsychiat., 2: 287-291, 1961.

82. PAPPENHEIMER, J. R. (Chairman): Standardization of definitions and symbols in respiratory physiology. Fed. Proc., 9: 602-605, 1950.

83. PAPPENHEIMER, J. R., RENKIN, E. M., AND BORRERO, L. M.: Filtration, diffusion, and molecular sieving through peripheral capillary membranes. A contribution to the pore theory of capillary permeability. Amer. J. Physiol., 167: 13-46, 1951.

84. PRICE, H. L., AND HELRICH, M.: Significance of the competence index in the measurement of myocardial contractility. J. Pharmacol., 115: 199-205, 1955.

85. RALL, D. P., STABENAU, J. R., AND ZUBROD, C. G.: Distribution of drugs between blood and cerebrospinal fluid: general methodology and effect of pH gradients. J. Pharmacol., 125: 185-193, 1959.

86. REINER, J. M.: *Behavior of Enzyme Systems*. Burgess Publishing Company, Minneapolis, 1959.

87. RICHARDS, D. W., JR.: Respiratory system: external respiration. In *Medical Physics*, edited by O. Glasser, Vol. 2, pp. 836-845. Year Book Publishers, Inc., Chicago, 1950.

88. RIGGS, D. S.: Quantitative aspects of iodine metabolism in man. Pharmacol. Rev., 4: 284-370, 1952.

89. Riggs, D. S., and Goldstein, A.: Equation for inert gas exchange which treats ventilation as cyclic. J. appl. Physiol., 16: 531–537, 1961.

90. Riker, W. F.: Cholinergic drugs. In *Pharmacology in Medicine*, Ed. 2, edited by V. A. Drill. McGraw-Hill Book Company, Inc., New York, 1958.

91. Ritchie, J. M., and Greengard, P.: On the active structure of local anesthetics. J. Pharmacol., 133: 241-245, 1961.

92. Ross, J. M., Fairchild, H. M., Weldy, J., and Guyton, A. C.: Autoregulation of blood flow by oxygen lack. Amer. J. Physiol., 202: 21-24, 1962.

93. Roughton, F. J. W.: Diffusion and chemical reaction velocity in cylindrical systems of physiological interest. Proc. roy. Soc. Ser. B., 140: 203–229, 1952.

94. Sapirstein, L. A., and Ogden, E.: Theoretic limitations of the nitrous oxide method for the determination of regional blood flow. Circulat. Res., 4: 245–249, 1956.

95. Scott, J. C., and Balourdas, T. A.: The interpretation of "spurious" correlations in coronary flow literature. Circulat. Res., 7: 169–172, 1959.

96. Sevelius, G., and Johnson, P. C.: Myocardial blood flow determined by surface counting and ratio formula. J. Lab. clin. Med., 54: 669–679, 1959.

97. Skinner, S. M., Clark, R. E., Baker, N., and Shipley, R. A.: Complete solution of the three-compartment model in steady state after single injection of radioactive tracer. Amer. J. Physiol., 196: 238–244, 1959.

98. Smith, H. W.: *The Kidney. Structure and Function in Health and Disease.* Oxford University Press, New York, 1951.

99. Snedecor, G. W.: *Statistical Methods, Applied to Experiments in Agriculture and Biology*, Ed. 5. Iowa State College Press, Ames, 1956.

100. Snell, F. M., Shulman, S., Spencer, R. P., and Moos, C.: *Biophysical Principles of Structure and Function.* University of Buffalo, Buffalo, 1961.

101. Solomon, A. K.: Equations for tracer experiments. J. clin. Invest., 28: 1297–1307, 1949.

102. Solomon, A. K.: Compartmental methods of kinetic analysis. In *Mineral Metabolism*, edited by C. L. Comar and F. Bronner, Vol. 1, Part A, Chapter 5. Academic Press, Inc., New York, 1960.

103. Spangler, R. A., and Snell, F. M.: Sustained oscillations in a catalytic chemical system. Nature, Lond., 191: 457–458, 1961.

104. Stacy, R. W.: Computers: analog. In *Medical Physics*, edited by O. Glasser, Vol. 3, pp. 193-201. Year Book Publishers, Inc., Chicago, 1960.

105. Stacy, R. W.: Computers: digital. In *Medical Physics*, edited by O. Glasser, Vol. 3, pp. 201–208. Year Book Publishers, Inc., Chicago, 1960.

106. Stacy, R. W., Williams, D. T., Worden, R. E., and McMorris, R. O.: *Essentials of Biological and Medical Physics.* McGraw-Hill Book Company, Inc., New York, 1955.

107. Stanbury, J. B., Brownell, G. L., Riggs, D. S., Perinetti, H., Itoiz, J., and del Castillo, E. B.: *Endemic Goiter. The Adaptation of Man to Iodine Deficiency.* Harvard University Press, Cambridge, 1954.

108. Stark, L.: Vision: servoanalysis of pupil reflex to light. In *Medical Physics*, edited by O. Glasser, Vol. 3, pp. 702-719. Year Book Publishers, Inc., Chicago, 1960.

109. Stephenson, R. P.: A modification of receptor theory. Brit. J. Pharmacol., 11: 379–393, 1956.

110. Straus, O. H., and Goldstein, A.: Zone behavior of enzymes. J. gen. Physiol., 26: 559–585, 1943.

113. von Euler, C.: Physiology and pharmacology of temperature regulation. Pharmacol. Rev., **13**: 361–398, 1961.

114. West, E. S.: *Textbook of Biophysical Chemistry*, Ed. 2. The Macmillan Company, New York, 1956.

115. White, H. L., and Rolf, D.: Comparison of various procedures for determining sucrose and inulin space in the dog. J. clin. Invest., **37**: 8–19, 1958.

116. Wilbrandt, W., and Rosenberg, T.: The concept of carrier transport and its corollaries in pharmacology. Pharmacol. Rev., **13**: 109–183, 1961.

117. Wilson, A., and Schild, H. O.: *Clark's Applied Pharmacology*, Ed. 8. The Blakiston Company, New York, 1952.

118. Wilson, E. B., Jr.: *An Introduction to Scientific Research*. McGraw-Hill Book Company, Inc., New York, 1952. (Available in a paperbound edition.)

120. Wolf, A.: History of philosophy. In *The Encyclopaedia Britannica*, Vol. 17, p. 750. Encyclopaedia Britannica, Inc., Chicago, 1955.

121. Wolf, A. V.: *The Urinary Function of the Kidney*. Grune & Stratton, Inc., New York, 1950.

122. Wollenberger, A., and Krayer, O.: Experimental heart failure caused by central nervous system depressants and local anesthetics. J. Pharmacol., **94**: 439–443, 1948.

123. Wu, H., Sendroy, J., Jr., and Bishop, C. W.: Interpretation of urinary N^{15}-excretion data following administration of an N^{15}-labeled amino acid. J. appl. Physiol., **14**: 11–21, 1959.

124. Zilversmit, D. B., Entenman, C., and Fishler, M. C.: On the calculation of "turnover time" and "turnover rate" from experiments involving the use of labeling agents. J. gen. Physiol., **26**: 325–331, 1942-1943.

APPENDIX A. DISCUSSION OF EXERCISES

CHAPTER 1

Exercise 1

A. "Per cent of doses" and "per cent of subjects" would be the same only if 1) each single dose of the particular size being tested had the same effect in a given subject, so that there would be no question about whether the patient did or did not obtain relief of pain, and 2) the same number of single doses were given to each subject. These conditions were not fulfilled in Denton and Beecher's study. In fact, (page 1057) "the number of doses of a given drug from which a patient got no relief was limited to two, whereupon another drug was substituted. But the number of doses of a given drug from which a patient got complete relief was limited only by the frequency with which that patient required medication during the first three post-operative days." As the authors themselves recognize, this curious convention seriously distorts the results, since the AD 90 per cent was in fact actually calculated on the basis of the total number of *doses* given (see C below).

B. "Per cent of the trials." "Trials" might refer either to patients or to doses. In view of the ambiguous definitions offered, it is impossible, without further analysis, to tell which is meant.

C. The table shows that the percentages of pain relief *must* have been calculated from the total number of doses, not the number of patients. Since patients cannot be split, 87.5 per cent relief could not have been calculated from 27 *patients*, the nearest possible values being $24/27 = 0.889$ or 88.9 per cent and $23/27 = 0.852$ or 85.2 per cent. In contrast, $56/64$ of the *doses* would give precisely 87.5 per cent. Other values listed in the same table confirm the conclusion that per cent of the number of *doses*, not patients, was actually used.

In passing, we may note the artlessly frank admission of bias made by the authors on page 1052. "Occasionally, when low doses were given to determine the shape of the dose-effect curve, *if data on a dozen patients fitted the curve*, they were accepted as adequate." (Italics supplied.)

Exercise 2

A. According to the *first passage* quoted, the tolerance is to be measured by the percentage of a large randomly selected population which exhibits an effect of a particular magnitude when all individuals in the population are given the same dose of the drug. Tolerance would thus be a characteristic of the population, not of the individuals composing the population. According to the *second passage* quoted, tolerance is clearly an individual attribute. But we are told that it must be expressed as an *ordinal* quantity which can be specified only with reference to other individuals whose tolerance has also been measured (how?). This is like saying, "If we arrange 100 individuals in order of increasing height, we shall define the height of the 69th individual as 69." Probably what the author means is that if the tolerances of a very large number of individuals, that is to say, a population, are measured (again, how?), the tolerance of any individual will be expressed as the per cent of the population with a smaller tolerance.

B. As indicated above, the author does not say how the tolerance of an individual is to be determined. Indeed, the definitions offered suggest that an individual tolerance can be measured only by ranking many individuals with respect to their tolerance so as to see where the individual falls in the group.

C. The most obvious and straightforward way of defining tolerance is as follows: The tolerance of an individual for a particular drug is directly proportional to the dose of the drug per kilogram of body weight needed to produce a particular effect. This could be written as an equation merely by defining a proportionality constant, k, such that

$$T = kD \qquad \text{(effect constant)}$$

Having thus defined tolerance, we are perfectly free to deal with the frequency distribution of tolerances in a population if we wish. By not giving us any such clear definition, the author leaves us bewildered.

Exercise 3

The definitions offered are far too vague. The rate of production of ultrafiltrate, u, is certainly an actual flow of fluid, *e.g.*, milliliters per minute. Since a is "an immediate reabsorption operating on the ultrafiltrate," we may surmise that a is the volume of ultrafiltrate reabsorbed per unit of time, *i.e.*, like u, an actual flow. In contrast, the "transfer constants" s and f which refer to "secretion of solute" and "diffusion of solute" presumably denote quantity of solute transferred per unit of time, *e.g.*, millimoles per minute. But these interpretations are incompatible with the text of the paper. For example, the authors say that the flow, u, is the same throughout the system. Hence a cannot represent reabsorption of the ultrafiltered fluid. Furthermore, scrutiny of the equations presented shows that a, s, and f can be added to or subtracted from u, and hence *must* be measured in the same units as u. The final conclusion from this detective work is that a, s, and f are all *clearances* rather than actual flows or quantities of solute per unit of time. Had the authors chosen better symbols and defined them more clearly, we would have been spared this frustrating search for the real meanings of the symbols used. For example:

$\dot{V}_{\text{filt}}$ = volume of cerebrospinal fluid formed per unit time by ultrafiltration

$(\dot{V}_f\text{cl})_{\text{abs}}$ = clearance by immediate absorption of solute from the ultrafiltrate, f, before it mixes with the ventricular fluid

$(\dot{V}_v\text{cl})_{\text{sec}}$ = clearance by secretion of solute out of the ventricular fluid, v

$(\dot{V}_l\text{cl})_{\text{dif}}$ = clearance by diffusion of solute from the "lumbar" cerebrospinal fluid, l, to plasma

Exercise 4

A. Brink does not explicitly define either of the E's in the equation quoted. However, the form of the equation makes it clear that the second E has exactly the same significance as the more familiar symbol f, and simply means "is a function of." Letting the symbol for the dependent variable mean "is a function of" is common notation, used particularly when many different functions of different variables are to be discussed. As for the first E, the author implies that it means "the stability of the membrane." But since the stability of the membrane cannot be measured directly, we may assume that E actually means "excitability" rather than stability. The equation would then state that the excitability of the membrane is a function of potential difference, L-fraction, and so forth.

B. As written, Riker's equation states that excitation is directly proportional to the rate at which the concentration of acetylcholine changes, *i.e.*, if the concentration changes slowly, excitation will be small. If the concentration changes 10 times as rapidly, excitation will be 10 times as great. But for many cells (*e.g.*, skeletal muscle cells) excitation is an *all-or-none event* which either does or does not occur. What the author probably meant was not excitation itself but some *graded phenomenon* which might be proportional to the rate of change of acetylcholine concentration, and which, if sufficiently intense, would lead to excitation.

CHAPTER 2

Exercise 1

Not definable mathematically

8. Excitation, an event
15. Senility, an ill-defined condition
17. Turbulence, patterns of flow too variable, complex, and chaotic to describe mathematically
20. Zygote, a biological object, not a mathematical entity

Dimensionless numbers

 2. Avogadro's number

 9. Log$_{10}$ of the injected dose

 12. pH

 13. Q_{10} (the ratio of the rates of some process measured at two temperatures 10°C. apart)

Dimensional quantities

 1. Absolute zero, $[\theta]$

 3. Barometric pressure, $[ML^{-1}T^{-2}]$

 4. Cardiac output, $[L^3T^{-1}]$

 5. Clearance, $[L^3T^{-1}]$

 6. "Counts per minute," $[T^{-1}]$

 7. $-dN/dt$, $[T^{-1}]$

 10. Neonatal period, $[T]$

 11. Molarity, $[L^{-3}]$ (Notice that molarity is *number* of moles per liter, not grams per liter)

 14. R, the ideal gas law constant, $[ML^2T^{-2}\theta^{-1}]$

 16. Stroke volume, $[L^3]$

 18. $\dot{V}_w \int_{t_1}^{t_2} C_{x.\text{in}}\, dt$, $[M]$

 (Notice that had concentration been given in millimoles per liter, the expression would have been dimensionless.)

 19. White blood cell count, $[L^{-3}]$

Exercise 2

Since an angle is fundamentally determined by the ratio of an arc length to the radius of the arc, it is a dimensionless number. An *angular* acceleration, such as radians per sec^2, therefore, has the dimensions, $[T^{-2}]$ and should not be listed among units with the dimensions $[LT^{-2}]$.

Exercise 3

The transport rate of a substance S would presumably be expressed as quantity of S transported per unit time. Letting $[Q]$ be the dimensions of quantity (however expressed), v would have the dimensions, $[Q][T^{-1}]$. The concentrations have the dimensions $[Q][L^{-3}]$. The dimensional equivalent of the equation is thus

$$[Q][T^{-1}] = [D'][Q][L^{-3}]$$

Solving for $[D']$, the dimensions of D', we find $[D'] = [L^3T^{-1}]$. D', therefore, has the dimensions of a flow or a clearance. But in transport of a substance S by a carrier C no actual fluid flow is involved. Moreover, we cannot relate the transport of S to the concentration of a different entity, CS, by a clearance in the ordinary sense. It is better, therefore, to regard D' merely as a proportionality constant which specifies the rate of transport of S per unit difference in concentration of carrier-substrate complex between the two sides of the membrane.

While this analysis is logical enough, it is probably not what the authors actually had in mind. For, as we shall see in Section 7-7, the term "permeability constant" is commonly used to mean a proportionality constant with the dimensions $[LT^{-1}]$ which expresses the rate of transfer of a substance across a membrane *per unit of membrane area* and per unit of concentration difference across the membrane (see Equation 7-22). *If* this is what the authors intended, they should have made it clear that v is not "quantity of S transported per unit of time," but rather "quantity of S transported per unit of time across one unit of area" which would have the dimensions $[Q][T^{-1}L^{-2}]$.

Exercise 4

The "equation" is preposterous. "Duration of illness" is the only quantity symbolized which can be given a rigorous mathematical definition. To this interval of time, we are asked to add a number of ill-assorted terms, many of which are exceedingly hazy. By trying to present their notions in a pseudoscientific mathematical equation, the authors have only succeeded in making themselves ridiculous. *Moral:* If you must be vague, don't try to be mathematical.

Exercise 5

The statement in the footnote is incorrect. The constants, c, (or C) must all have the same dimensions as the quantity a in the equation, which is defined as isotope concentration, and which the authors express as the dimensionless ratio: excess atoms of N^{15} per hundred atoms of nitrogen. λ is clearly a rate constant with the dimension $[T^{-1}]$. But the expression $C_1 e^{-\lambda t} + C_2 t e^{-\lambda t} + C_3 t^2 e^{-\lambda t}$ is the sum of three terms which successively have the dimensions (none), $[T]$, and $[T^2]$. A sum of this kind is impossible.

Exercise 6

The dimensions of $U\sqrt{V}/B$ are $[L^{1.5}T^{-0.5}]$. This quantity is not obtained by multiplying UV/B by $\sqrt{V}$, as the author erroneously states, but rather by dividing UV/B by $\sqrt{V}$.

Exercise 7

We may surmise that the equation contains an unwritten dimensional constant, numerically equal to unity, which would make it dimensionally correct. If the relationship were written as a general equation, it would be

$$t = a + bL_{H_2O}$$

where a and b are to be found from the data. In this equation, a would have the dimension $[T]$, and b would also have the dimension $[T]$ since L_{H_2O} is expressed as per cent of body weight, a dimensionless ratio. On fitting this general equation to specific data, it was evidently found that when t was measured in hours, the numerical value of b was unity. Had time been expressed in minutes, the *same* equation would have been

$$t = 78 + 60L_{H_2O}$$

thus making the dimensional coefficient of L_{H_2O} obvious. But it could (and should) have been made equally obvious in the equation for time in hours simply by writing the number "one" as the coefficient of L_{H_2O}. This was, in fact, done in Adolph's original equation (2, page 90): "The times in hours required to complete the diuretic response are related to the volumes administered (ΔW_c) by the equation

$$t_c = 1.3 + 1.0\Delta W_e"$$

Even with the misprint of W_e in place of W_c, this original equation is less confusing than Wolf's version in which the dimensional constant is omitted.

Exercise 8

Since $\overline{BN}$ has the dimensions $[ML^{-3}]$, so also must 7.56. Since A has the dimension $[T]$, its coefficient, 0.1119, must have the dimensions $[ML^{-3}T^{-1}]$ so that the term as a whole will have the same dimensions as the other terms in the equation.

Exercise 9

According to the definitions given, the dimensions of the several quantities are V, $[T^{-1}]$; a, $[L^{-3}]$; k, $[T^{-1}]$; and x, (none). Dimensionally, the equation is thus impossible. Either a must be defined as number of moles (dimensionless) or V must be defined as change in concentration per unit time, and x as number of moles per liter which have been changed into products.

Exercise 10

A. For dimensional correctness, the function of body length and body volume must have the same dimensions as surface area, A, namely, $[L^2]$. Let a be the exponent of body length, l. Since volume, V, already has the dimensions, $[L^3]$, the requirement for dimensional correctness will be satisfied by any equation of the general form

$$A = kl^a(V^{1/3})^{2-a} \quad \text{or,} \quad A = kl^a V^{(2-a)/3}$$

where k is a dimensionless constant of proportionality.

B. The problem is to estimate a and k from the observed values of A,

l, and V. Since one of the unknowns, a, occurs in the exponents, we cannot solve the general equation directly for a. Under these circumstances, a logarithmic transformation is always worth trying. Taking logs on both sides of the equation,

$$(\log A) = (\log k) + a(\log l) + (2 - a)(\log V)/3$$

Now this equation is a simple linear equation in two unknowns, a and $(\log k)$. It can be rearranged, therefore, so that one of the unknowns will be the slope, and the other the intercept of a straight line, *i.e.*, so that we have an equation of the form

$$y = a + (\log k)x$$

or of the form

$$y' = (\log k) + ax'$$

where y, y', x, and x' are appropriate functions of the known quantities, $(\log A)$, $(\log l)$, and $(\log V)$. We can then calculate y and x, (or y' and x'), plot one against the other on graph paper, fit a straight line to the points, and estimate a and $(\log k)$ from the slope and intercept of the line. (*Caution.* Some of the pitfalls in fitting straight lines which are involved in this procedure will not be discussed until Chapter 3.) The two straight-line equations are

$$\left[\frac{3(\log A) - 2(\log V)}{3}\right] = (\log k) + \left[\frac{3(\log l) - (\log V)}{3}\right] a$$

and

$$\left[\frac{3(\log A) - 2(\log V)}{3(\log l) - (\log V)}\right] = a + \left[\frac{3}{3(\log l) - (\log V)}\right] (\log k)$$

The first of these is the easier to work with.

C. The following table gives the values of

$$y = \left[\frac{3(\log A) - 2(\log V)}{3}\right] \text{ and } x = \left[\frac{3(\log l) - (\log V)}{3}\right]$$

calculated from the data of Cowgill and Drabkin (24):

Dog no.	x	y
4	0.531	1.012
6	0.550	1.031
1	0.624	1.091
7	0.545	1.034
2	0.579	1.084
5	0.529	1.017
3	0.508	1.023

The straight line which I happened to fit by eye to these points has the equation

$$y = 0.573 + 0.849\,x$$

Whence: $a = 0.849$, and $k = $ antilog $0.573 = 3.741$.
These values yield

$$A = 3.741\ l^{0.849} V^{0.384}$$

for the original equation.

Exercise 11

A. 60 (mi./hr.) = 60 (5280 ft./3600 sec.) = 88 ft./sec.
B. 15 (lb./in.²) = 15 (4.45 × 10⁵ dynes/2.54 × 2.54 cm.²) = 6.90 × 10⁴ dynes/cm.²
C. 186,000 (mi./sec.) = 186,000 (2.54 × 12 × 5280 cm./sec.) = 2.99 × 10¹⁰ cm./sec.
D. 32 (ft./sec.) per sec. = $32 \left(\dfrac{1}{5280}\ \text{mi.} \Big/ \dfrac{1}{3600}\ \text{hr.} \right)$ per $\dfrac{1}{60}$ min. = $32 \left(\dfrac{3600}{5280} \text{mi./hr.} \right) \Big/ \dfrac{1}{60}$ min. = $32 \left(\dfrac{[3600 \times 60]}{5280} \right)$ (mi./hr.) per min. = 1309 (mi./hr.) per min. = 1309 mi. hr.⁻¹ min.⁻¹

Exercise 12

Assuming that the number 0.0313 in this equation is a dimensionless unit conversion factor, we find that the dimensional equivalent of the equation as written is

$$ML^2T^{-3} = MLT^{-1}$$

which is wrong. It is obvious that the right side of the equation must be multiplied by an acceleration (dimensions, LT^{-2}) to make it correct. Presumably the author defined m as "mass of subject in pounds" inadvertently. He meant "weight of subject in pounds," which would make the equation dimensionally correct.

Since this equation is valid only for the particular units used, we are obliged to check the numerical value also. In other words, we must find the factor by which the product of pound weight and miles per hour must be multiplied to convert it to kcal./min. Using the method outlined in the text,

$$\frac{\text{mile-pound}}{\text{hour}} = \frac{(5280\ \text{ft.})(\text{lb.})}{60\ \text{min.}} = 88 \left(\frac{\text{ft.-lb.}}{\text{min.}} \right)$$

The *Handbook of Chemistry and Physics* (58) tells us that 1 ft.-lb. per minute equals 3.239 (10⁻⁴) kcal. per minute.

$$88 \left(\frac{\text{ft.-lb.}}{\text{min.}} \right) = 88 \ (0.0003239 \text{ kcal./min.}) = 0.0285 \text{ kcal./min.}$$

Hence the unit conversion factor by which (mile-pound) per hour must be multiplied to convert to kilogram-calories per minute is *not* 0.0313 but 0.0285. The equation was wrong both dimensionally and numerically.

Exercise 13

The equation as written is in dimensional chaos. Although it appears to be a general equation, it is, in fact, applicable only to water. Since it is an equation for temperature, all of the addends on the right must have the dimensions of temperature. But as it stands, the second term has the dimensions $[ML^{-1}T^{-2}]$, and the third term cannot be examined dimensionally because the dimensions of α are not defined.

To correct the dimensions of the second term, we must multiply it by some factor, or factors, with the dimensions $[M^{-1}LT^2\theta]$. If the problem has to do with water, whose characteristics are often used as a standard of comparison for other substances, it is possible that a dimensional quantity *numerically equal to unity* has been omitted from the equation. Careful analysis of the problem will reveal not one but two such quantities, *thermal capacity* (with the dimensions $[L^2T^{-2}\theta^{-1}]$) which *for water* is about 1 small calorie per gram-degree, and *density* (dimensions, $[ML^{-3}]$) which *for water* is about 1 gm. per milliliter. The product of these two dimensional quantities has the dimensions $[ML^{-1}T^{-2}\theta^{-1}]$, which is just the reciprocal of the dimensions of the multiplier required to correct the second term. We may conclude, therefore, that to make the equation general (and more easily understood), the second term should be divided by the thermal capacity and by the density of the flowing fluid.

Finally, if the third term on the right is to have the dimensions of temperature,

$$\left[\frac{\alpha H X}{V^2} \right] = [\theta]$$

where the square brackets mean "the dimensions of." Solving for the dimensions of α,

$$[\alpha] = [\theta] \left[\frac{V^2}{HX} \right] = \left[\frac{\theta L^6 T^{-2}}{ML^2T^{-3}L} \right] = [M^{-1}L^3T\theta]$$

CHAPTER 3

Exercise 1

    A. $a = 70$ gm.; $b = -10$ gm./hr.
    B. $a = 344$ L./min., $b = -20.7$ L./min. per mm. Hg,
       $c = 0.3125$ L., min.$^{-1}$, (mm. Hg)$^{-2}$

This second-order polynomial happened to fit a particular set of data very well. But notice how preposterous it would be to extrapolate it beyond the observations. For example, it predicts a ventilation of 344 L./min. when the partial pressure of CO_2 in the alveoli is zero!

C. $a = -0.1$ msec., $b = 23$ mv., $c = 16.7$ mv.-msec.

D. For convenience, let $E_{max} = a$, and let $Q_{0.5}^n = b$. Then, with a slight rearrangement of the original equation, the three simultaneous equations are

$$10b + 10(10^{-1})^n = (10^{-1})^n a$$
$$55b + 55(10)^n = (10)^n a$$
$$120b + 120(10^3)^n = (10^3)^n a$$

Step 1. Solve the first equation for a, and substitute this expression for a in the second and third equations thus obtaining two equations in two unknowns.

Step 2. Solve each of these two equations for b.

Step 3. If we now set these two expressions for b equal to each other, we have an equation in the single unknown n, which can be reduced to

$$6.5(10^{2n})^2 - 60.5(10^{2n}) + 54 = 0$$

This equation is a quadratic equation in the quantity (10^{2n}). Solving for 10^{2n}, and rejecting the root, 1 (which would correspond to $n = 0$), we find $10^{2n} = 8.31$, whence, by using logarithms, $n = 0.46$. By appropriate substitution in earlier expressions, we then easily obtain $Q_{0.5} = 27.5$ μg. and $E_{max} = 142$ mm. Hg. Note, however, that this solution was possible only because the three doses differed successively from each other by a simple power of 10. A more general solution would be to plot each of the two expressions for b which were obtained in Step 2 against n on a single sheet of graph paper. The point of intersection of these two curves would have as coordinates the desired values of b and of n.

Exercise 2

A. The values on the scale of ordinates *decrease* from below upwards. This makes it appear as though all of the lines had positive slopes, whereas in fact the slopes are negative.

B. The values on the scale of abscissas *decrease* from left to right. Again, this makes the lines appear to have a slope opposite in sign to their true slope.

C. Each of the curves in these graphs is based upon only four experimental points. Yet the authors have used a wide variety of curves (presumably fitted by eye) to express the trend of the data. In no instance is any statistical justification given for fitting a curved line rather than a straight line to the four points, and since at least three points are needed to determine a curved line, at most only one degree of freedom is available for estimating residual error variation. Consequently, few, if any, of the curved lines can be justified objectively. This is particularly unfortunate in Figure 3, from which the authors draw a conclusion based upon a supposed difference in the shape of the curves. Remember that there are no more degrees of freedom available for fitting a curve by eye than for fitting it mathematically, and that sweeping curves drawn through four points are apt to provoke at best, ridicule, at worst, a suspicion of bias.

Exercise 3

If we take antilogs on both sides, we obtain

$$Y' = 8.47X^{0.986}$$

which, for the range of data analyzed, is scarcely different from

$$Y' = 8.47X$$

The law of parsimony suggests that the latter equation is preferable. In addition it has the virtue of being dimensionally correct.

Exercise 4

All sums are to be taken from $i = 1$ through $i = n$.

A-1. Deviation $= y_i - (k/x_i)$

Deviation$^2 = y_i^2 - 2k(y_i/x_i) + k^2(1/x_i^2)$

$SS = (\sum y_i^2) - 2k[\sum (y_i/x_i)] + k^2[\sum (1/x_i^2)]$

$dSS/dk = -2[\sum (y_i/x_i)] + 2k[\sum (1/x_i^2)]$

Normal equation: $k \sum (1/x_i^2) = \sum (y_i/x_i)$

Solving for k: $k = \sum (y_i/x_i)/\sum (1/x_i^2)$

A-2. $SS = (\sum y_i^2) - 2a(\sum y_i) - 2b(\sum x_iy_i) + na^2 + 2ab(\sum x_i) + b^2(\sum x_i^2)$

Normal equations:

$$-(\sum y_i) + na + b(\sum x_i) = 0$$
$$-(\sum x_iy_i) + a(\sum x_i) + b(\sum x_i^2) = 0$$

Parameters:

$a = \bar{y} - b\bar{x}$ (bar above symbol indicates the mean value of the quantity symbolized)

$$b = \frac{(\sum x_i y_i) - [(\sum x_i \sum y_i)/n]}{(\sum x_i^2) - [(\sum x_i)^2/n]}$$

Note carefully the distinction between two such symbols as $\sum x_i y_i$ which means "the sum of the individual products," and $\sum x_i \sum y_i$ which means "the product of the sums."

A-3. If the line *must* pass through the origin, the intercept, a, must be zero, and the equation is simply $y = bx$. The least-squares value of b is: $b = \sum x_i y_i / \sum x_i^2$

B.　For A-1 there are $n - 1$ degrees of freedom for variation about the curve.

For A-2 there are $n - 2$ degrees of freedom for variation about the line.

For A-3 there are $n - 1$ degrees of freedom for variation about the line.

C.　$a = (\sum y_i)/n = \bar{y}$. In other words, the least-squares solution for the parameter a of the equation $\bar{y}_{est} = a$ is simply the arithmetic mean of the series of observations.

Exercise 5

Since the old method had 8 times the error of the new, each of the new analyses is worth 64 times as much as a single analysis by the crude method. The entire set of 247 old analyses should therefore be given only about as much weight as 4 analyses by the new method. The investigator would be well advised to discard the old analyses entirely, thus avoiding the nuisance of weighting and the trouble of describing two methods, one of which is practically worthless.

Exercise 6

In the *original* data, the error is proportional to the absolute difference between $y - \Delta y$ and $y + \Delta y$, which is simply $2\Delta y$. The weight is therefore inversely proportional to $(2\Delta y)^2$ which in this problem is independent of y. But in the *transformed* data, the error is proportional to the difference between $1/(y - \Delta y)$ and $1/(y + \Delta y)$ which is $[1/(y - \Delta y)] - [1/(y + \Delta y)] = 2\Delta y/[y^2 - (\Delta y)^2]$. The weight to be given is therefore inversely proportional to the square of this quantity, which is

$$4(\Delta y)^2/[y^4 - 2y^2(\Delta y)^2 + (\Delta y)^4]$$

Since Δy is constant, this entire expression is inversely proportional to y^4 provided y is enough larger than Δy for the term $-2y^2(\Delta y)^2$ to be negligible in comparison with y^4. The *weight* will then be *directly* proportional to y^4.

CHAPTER 4

Exercise 1

An equation of this sort merely *lists* the independent variables which influence the dependent variable, and cannot indicate either the direction or the magnitude of the effects of the independent variables.

Exercise 2

A. $1 \dashrightarrow 2$ E. $1 \dashleftarrow 2$
B. $1 \leftrightarrow 2$ F. Inversely correlated, not functionally related
C. $1 \rightarrow 2$ G. $1 \leftrightarrow 2$
D. $1 \leftrightarrow 2$ H. $1 \leftrightarrow 2$

> *Comment on C.* No one denies the *existence* of a strong association between cigarette smoking and lung cancer. As for the *nature* of this association, the law of parsimony unequivocally *requires* us to accept the functional relationship shown above until it is proved false. The only alternative would be for us to jettison the law of parsimony altogether, and to entertain upon equal terms all manner of complex hypotheses, however preposterous they may be (54).
>
> *Comment on D.* Note that *either* variable could be altered independently. We could determine the duration of current flow needed to stimulate for various independently controlled voltages, or we could determine the voltage needed to stimulate for various independently controlled durations of current flow.
>
> *Comment on F.* One example of an infinitude of inevitable correlations. *Any* two unrelated variables showing a trend with time over the same period are inevitably correlated because they are both functions of the same third variable, time. These "nonsense" correlations are genuine correlations all right, but they are utterly meaningless and uninteresting.
>
> *Comment on H.* The two variables listed are functionally related by the law of mass action:
>
> $$(C_{HPO_4^-})(C_{H^+})/(C_{H_2PO_4^-}) = K$$
>
> which involves a third variable, C_{H^+}, as well. Under these circumstances, one can specify the relation between two of the variables only by assuming that the third remains constant. The *entire* functional relationship could be diagrammed as

$$C_{H^+} \longleftrightarrow C_{HPO_4^-}$$
$$\nwarrow \quad \nearrow$$
$$C_{H_2PO_4^-}$$

Exercise 3

A given value of n ("x") may happen to be paired with any value of n' from 0 to 10,000, each of the 10,001 possible pairs being equally probable. Consequently, if n happens to be small, say 10, the values of "y" (*i.e.*, nn') will, on the average, be evenly distributed between 0 and 100,000; whereas if n happens to be large, say 9,550, the values of y will be evenly distributed between 0 and 95,500,000. Since each value of x between 0 and 10,000 is also equally likely, the points will tend to be evenly scattered throughout the triangular area bounded by the three lines $y = 0$, $x = 10,000$, and $y = 10,000x$. The least-squares line fitting the points will have the equation, $y = 5,000x$.

Exercise 4

A.1. In advance of the experiment, there is nothing which makes a relationship between coronary flow and oxygen consumption by the heart inevitable in spite of the fact that *by definition* $\dot{Q}_{O_2} = \dot{V}_{cor}(C_{O_2 \cdot art} - C_{O_2 \cdot ven})$. *A priori*, an increased consumption of oxygen might be accompanied by an increased extraction of oxygen from the blood (*i.e.*, by an increased arteriovenous difference in the concentration of oxygen) with little or no change in blood flow; or by an increase in blood flow, with little or no change in oxygen extraction; or by an increase in *both* extraction *and* flow. But once it has been demonstrated that over a substantial range of oxygen consumption there is very little change in arteriovenous oxygen difference, a close relationship between $\dot{Q}_{O_2}$ and $\dot{V}_{cor}$ becomes inevitable.

A.2. The conclusion quoted is perfectly valid. The question of what type of relationship exists between $\dot{Q}_{O_2}$ and $\dot{V}_{cor}$ can be answered only on physiological, not mathematical grounds, and there is abundant physiological evidence that normally coronary flow is dependent upon oxygen consumption, not vice versa. This conclusion is in no way altered by the authors' having *calculated* oxygen consumption from blood flow.

B. The criticism offered by Scott and Balourdas is completely unjustified. Let us apply their line of reasoning to a very simple example. Consider the size of personal accounts in savings banks throughout the United States. This variable will have a very wide range. Consider also the various annual rates of interest offered by these banks. This variable will have a very small range. If we now plot the number of dollars of interest paid yearly on an account against the size of that account, we will find a strikingly good relationship which is, of course, a true functional relationship between size of account and amount of interest paid. Now we would find very much the same strikingly good relationship if we were to assign random values (with the same range and variation) to size-of-account and to rate-of-interest. But this fact does not invalidate the evidence for a functional relation between amount of interest paid and size of account.

Scott and Balourdas imply that a high correlation between two variables which do *not* have a common element can be taken as evidence of a functional relationship. This is wrong on two counts. *First* two variables may be highly correlated without being functionally related (Category 4). *Second*, in the final analysis, two variables, A and B, *cannot* be correlated unless they *do* have a common element. For either there is a functional relationship between them, let us say $A = f(B)$, in which case B is an element common to both A and itself; or both are functions of a third variable, C, in which case C is an element common to both A and B.

Exercise 5

In this example, the maximum tension is identically equal to the product of time of rise, and mean rate of rise of tension:

$$(\text{Ten})_{\max} = t_{\text{rise}}(\text{mean rate of rise})$$

Furthermore, the investigator showed that

$$(\text{mean rate of rise}) = k \ (d(\text{Ten})/dt)_{\max}$$

where k is a proportionality constant. Hence,

$$(\text{Ten})_{\max} = k t_{\text{rise}}(d(\text{Ten})/dt)_{\max}$$

Consequently, once it is demonstrated that t_{rise} and $(d(\text{Ten})/dt)_{\max}$ are independent of each other, a "correlation" (actually a functional relation) between $(\text{Ten})_{\max}$ and t_{rise} becomes inevitable and need not be illustrated graphically.

Exercise 6

$$\dot{N}_{\text{beat}} \dashrightarrow V_{\text{beat}}$$
$$\searrow \quad \swarrow$$
$$\dot{V}_{\text{syst}}$$

Neglecting coronary flow, by definition

$$\dot{V}_{\text{syst}} = N_{\text{beat}} V_{\text{beat}}$$

However, we have no right to conclude from this equation that systemic output is inevitably directly proportional to heart rate. For though an increase in heart rate does tend to *increase* systemic output directly (as is stated by the equation), it simultaneously tends to *decrease* systemic output by decreasing stroke volume. As a result, in a competent heart, the heart rate has remarkably little effect upon systemic output over a wide range where these two tendencies counterbalance each other. Only at low heart rates where the direct effect predominates, or at excessive heart rates where the effect upon stroke volume predominates is there a marked net effect of heart rate upon systemic output.

Exercise 7

By changing the tube, we could change either the radius or the length. By changing the fluid or the temperature, we could change the viscosity. We could also change either the inflow pressure or the end-tube pressure. Each of these variables could be altered independently of each other. But the only way flow can be altered is by changing one of the other variables. Therefore, flow is the dependent variable in this system, all others being independent. Even a superficial familiarity with the system allows us to specify the direction of the change in flow produced by each independent variable. We may therefore write

where ΔP is the difference in pressure between the two ends of the tube and is partly independent, partly dependent.

Poiseuille's law as given in the *Handbook of Chemistry and Physics* (58) is

$$\dot{V} = \frac{\pi \Delta P r^4}{8 l \eta}$$

which agrees with the qualitative relationships shown in the symbol-and-arrow diagram.

CHAPTER 5

Exercise 1

A simple buffer solution is not a feedback mechanism and cannot be treated as one. No two feedback equations exist. The behavior of all of the components of the buffer system is completely specified by a *single* equation, the Henderson-Hasselbalch equation, and accordingly the relationships among the variables belong to Category 2-b, not to Category 3.

Exercise 2

In the first place, it is most confusing to talk about the "gain" of an autoregulatory system whose purpose is to minimize, not to amplify, changes. In the second place, by basing their estimates on *total* changes from the normal control, the authors obtain values which still indicate a positive "gain" at a time when the system is no longer able to compensate at all (see the three points on the right of their Fig. 5). This could be avoided by comparing the *slopes* of the open-loop and closed-loop curves, *i.e.*, by calculating the homeostatic index as described in the text. But even if we were to calculate the homeostatic index, we would still have to explain the apparent "overcompensation" which occurred when the hemoglobin oxygen saturation was but slightly decreased. Now as a matter of fact *overcompensation in the steady state is impossible.* (Why?) Since the data reported in this paper were observed during the steady state, we must look for some other explanation.

Let us begin by trying to identify the interrelationships among the factors which may be concerned in the regulation of local blood flow in response to anoxia. First, applying "Ohm's law" to fluid flow, blood flow, $\dot{V}_B$, must be equal to the quotient obtained by dividing the pressure difference across the vascular bed ($P_{art} - P_{ven}$) by the resistance of the vascular bed, R:

$$(1) \qquad \dot{V}_B = (P_{art} - P_{ven})/R$$

Also,

$$(2) \qquad \dot{Q}_{O_2 \cdot in} = C_{O_2 \cdot art}\dot{V}_B$$

where $\dot{Q}_{O_2 \cdot in}$ is the quantity of oxygen entering the tissue per unit of time and therefore has exactly the same significance as the authors' "oxygen available to the tissues"

$C_{O_2 \cdot art}$ is the concentration of oxygen in the arterial blood

Conservation ("the Fick principle") demands that

$$(3) \qquad C_{O_2 \cdot ven} = (\dot{Q}_{O_2 \cdot in} - \dot{Q}_{O_2 \cdot cons})/\dot{V}_B$$

where $C_{O_2 \cdot ven}$ is the concentration of oxygen in the effluent venous blood

$\dot{Q}_{O_2 \cdot cons}$ is the rate of oxygen consumption by the tissue

Now it is reasonable to suppose that the pressure difference remains essentially constant, and that the increased blood flow is due to decreased resistance, presumably caused by relaxation of the smooth muscle of the small blood vessels in response to anoxia. For simplicity, let us assume that the concentration of oxygen in the venous effluent is the same as it is in the tissues, including the vascular smooth muscle. Although we cannot write an explicit equation for the effect of anoxia on resistance to flow, we can at least indicate that the relationship which is being assumed is

$$(4) \qquad R = f(C_{O_2 \cdot ven})$$

Next, let us draw a symbol-and-arrow diagram summarizing the relations specified by these four equations.

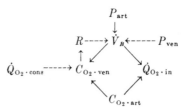

According to this diagram, an increase in blood flow might *increase* $C_{O_2 \cdot \text{ven}}$ through its effect upon $\dot{Q}_{O_2 \cdot \text{in}}$, but might *decrease* $C_{O_2 \cdot \text{ven}}$ by a direct effect. This makes no sense at all, because increasing the flow of blood through the tissues could not possibly decrease the oxygen concentration therein. We are thus led to re-examine Equation 3 above which induced us to put the broken diagonal arrow into the diagram. And we now notice that $Q_{O_2 \cdot \text{in}}$ is not a single separate entity (as, for example, oxygen consumption is) but actually contains blood flow as an integral factor (Equation 2). Let us accordingly rewrite Equation 3 as

$$C_{O_2 \cdot \text{ven}} = C_{O_2 \cdot \text{art}} - (\dot{Q}_{O_2 \cdot \text{cons}} / \dot{V}_B)$$

which tells a less devious story! The corresponding symbol-and-arrow diagram

shows at a glance that $\dot{Q}_{O_2 \cdot \text{in}}$, the variable chosen by Ross *et al.* as an indicator of regulatory effectiveness *is not even a member of the feedback loop*; it is merely a wholly dependent (and wholly irrelevant) variable which might better have been omitted from the analysis entirely. If you are interested, you can now choose sensible values for the several parameters such as P_{art}, $\dot{Q}_{O_2 \cdot \text{cons}}$, etc., and analyze the authors' data correctly, calculating the homeostatic index from the changes which must have occurred in the values of $\dot{V}_B$ and $C_{O_2 \cdot \text{ven}}$ which *are* members of the feedback loop. If you do this, you will find no evidence of "overcompensation."

Exercise 3

Merely apply the same procedure as outlined in the text to each of the other five pairs of variables in turn.

Exercise 4

The rate at which the substance is filtered will be

$$\dot{Q}_{filt} = C_P \dot{V}_{filt}$$

where $\dot{V}_{filt}$ is the glomerular filtration rate. The rate of excretion will be

$$\dot{Q}_U = C_P \dot{V}_{filt} - \dot{Q}_{abs \cdot max}$$

The clearance will be

$$(\dot{V}cl) = \dot{Q}_U/C_P = \dot{V}_{filt} - (\dot{Q}_{abs \cdot max}/C_P)$$

$$d(\dot{V}cl)/dC_P = \dot{Q}_{abs \cdot max}C_P^{-2}$$

By definition of clearance

$$C_P = \dot{Q}_U/(\dot{V}cl)$$

But in the steady state

$$\dot{Q}_{in} = \dot{Q}_U$$

Hence,

$$C_P = \dot{Q}_{in}/(\dot{V}cl)$$

$$dC_P/d(\dot{V}cl) = -\dot{Q}_{in}(\dot{V}cl)^{-2}$$

The negative product of these first derivatives is

$$\dot{Q}_{in}\dot{Q}_{abs \cdot max}/(\dot{V}cl)^2 C_P^2$$

But since

$$(\dot{V}cl)C_P = \dot{Q}_U = \dot{Q}_{in}$$

the negative product of these first derivatives is

$$\text{Homeostatic index} = \dot{Q}_{abs \cdot max}/\dot{Q}_{in}$$

The homeostatic index is greatest when the threshold is barely exceeded, approaching ∞ as $\dot{Q}_U = \dot{Q}_{in}$ approaches zero. At first this may seem anomalous, but according to the terms of our problem, even when $\dot{Q}_{in}$ is infinitesimally small, a steady state can be achieved only by having $\dot{Q}_{filt}$ exceed $\dot{Q}_{reabs}$ by the same infinitesimal amount. In the face of a steady infusion, however small, a *little* excretion (closed-loop response) is infinitely better than *no* excretion (open-loop response), for if $(\dot{V}cl)$ did not increase from zero to a small positive value when the threshold plasma concentration is reached, the concentration in plasma would continue to increase without limit.

Exercise 5

The first part of the analysis is very similar to Exercise 4 and, indeed, leads to an exactly analogous expression for the product of first derivatives, but with opposite sign:

Product of first derivatives $= \dot{Q}_{tub \cdot max}/\dot{Q}_{in}$

By Equation 5-49,

$$(\text{Magnification} - 1) = \frac{\dot{Q}_{tub \cdot max}/\dot{Q}_{in}}{1 - (\dot{Q}_{tub \cdot max}/\dot{Q}_{in})} = \frac{\dot{Q}_{tub \cdot max}}{\dot{Q}_{in} - \dot{Q}_{tub \cdot max}}$$

But since in the steady state $\dot{Q}_{in} = \dot{Q}_U = \dot{Q}_{filt} + \dot{Q}_{tub \cdot max}$, we may rewrite the expression as

$$(\text{Magnification} - 1) = \dot{Q}_{tub \cdot max}/\dot{Q}_{filt}$$

CHAPTER 6

Exercise 1

Your ancestor's dollar has been undergoing simple exponential growth expressed by the general equation,

$$N_t = N_0 e^{kt}$$

where N_t is the number of dollars present at time t measured from the time of the original deposit

N_0 is the number of dollars in the original deposit at time zero

k is the proportional rate of increase per year, *i.e.*, the rate of interest

Applying this general equation to the specific problem at hand,

$$N_{1962} = (1)e^{(0.02)(1962)}$$

or,

$$\ln (N_{1962}) = (0.02)(1962) = 39.24$$

or,

$$\log_{10}(N_{1962}) = (0.4343)(39.24) = 17.042$$

or,

$$N_{1962} = \text{antilog}_{10} (17.042) = \text{about } 1.10 \times 10^{17}$$

or about 110 *quadrillion* dollars!

Exercise 2

Since N_0 is a constant, $dN/dt = N_0 \, d(e^{-kt})/dt$. In a table of differentials (for example, in the *Handbook of Chemistry and Physics* (58)) we find that $d(e^{au}) = ae^{au} \, du$. Hence, $dN/dt = -N_0 k e^{-kt}$. But by our original equation, $N_0 e^{-kt} = N_t$. Hence, $dN/dt = -kN_t$.

Exercise 3

Begin by writing an equation for the *broken line* representing the asymptote. For example, in Figure 6-24*d*, the asymptote is

$$y_{\text{asymp}.t} = a + bt$$

where a is the intercept and b is the slope of the *broken line*. From this equation, it is obvious that at time zero

$$y_{\text{asymp}.0} = a$$

Substituting these values for $y_{\text{asymp}.t}$ and $y_{\text{asymp}.0}$ into Equation 6-39a,

$$y_t = a + bt + (y_0 - a)e^{-kt}$$

which is the specific variant of Equation 6-39a for the *solid line* of graph *d*. The other equations are similarly found. They are

$$
\begin{aligned}
&\underline{a}. \ \ y_t = y_0 e^{-kt} \\
&\underline{b}. \ \ y_t = (y_0 + a)e^{-kt} - a \\
&\underline{c}. \ \ y_t = a + (y_0 - a)e^{-kt} \\
&\underline{e}. \ \ y_t = bt + y_0 e^{-kt} \\
&\underline{f}. \ \ y_t = a - bt + (y_0 - a)e^{-kt}
\end{aligned}
$$

Exercise 4

Under the circumstances specified, the number of photons absorbed per second and per square centimeter while passing through an infinitesimal layer lying x cm. below the surface will be directly proportional to the number of photons per second per square centimeter incident upon the layer, $\dot{N}_{\text{phot}.x}/A$, and directly proportional to the thickness, dx, of the layer:

$$\dot{N}_{\text{phot.abs}.x}/A = k(\dot{N}_{\text{phot}.x}/A)\,dx$$

where k is an appropriate constant of proportionality. But *absorption* of photons in the layer dx thick is equivalent to the *decrease* in the number of photons which occurs in that layer. Hence,

$$\left(\frac{\dot{N}_{\text{phot.abs}.x}}{A}\right)\Big/ dx = -d\left(\frac{\dot{N}_{\text{phot}.x}}{A}\right)\Big/ dx = k\left(\frac{\dot{N}_{\text{phot}.x}}{A}\right)$$

or, integrating,

$$\dot{N}_{\text{phot}.x}/A = (\dot{N}_{\text{phot}.0}/A)e^{-kx}$$

This is equivalent to *Bouguer's law* (more commonly known as *Lambert's law*):

$$I_x = I_0 e^{-kx}$$

where I signifies *intensity* of a beam of light (which is directly proportional to the photon flux, $\dot{N}_{\text{phot.}}/A$, used above) and k is the *absorption coefficient* for the particular wave length of monochromatic photons being used.

In solutions of colored material where the solvent is nonabsorbing, k is proportional to the molar concentration, C, of the absorbing material:

$$k = \alpha C$$

where α is the *molar absorption coefficient*. Then,

$$I_x = I_0 e^{-\alpha C x}$$

which is *Beer's law*.

Just to make simple matters as complicated as possible, when these laws are expressed in terms of common logarithms (to the base 10), $0.4343k$ is called the *extinction coefficient*, and 0.4343α is called the *molar extinction coefficient*.

Exercise 5

Since C_D decreases exponentially with time,

$$C_{D.t} = C_{D.0} e^{-kt}$$

By the terms of the problem, $C_{D.0} = aQ_D$. Furthermore, by Equation 6-10,

$$-k = -0.693/t_{1/2}$$

Combining these equations, we obtain

$$C_{D.t} = aQ_D \exp(-0.693t/t_{1/2})$$

But when $C_{D.t} = C_{\min}$, $t = t_{\text{eff}}$. Hence,

$$C_{\min} = aQ_D \exp(-0.693\, t_{\text{eff}}/t_{1/2})$$

or, taking logarithms,

$$\ln C_{\min} = \ln a + \ln Q_D - (0.693\, t_{\text{eff}}/t_{1/2})$$

Solving for t_{eff},

$$t_{\text{eff}} = \frac{t_{1/2}}{0.693} \left[\ln \left(\frac{aQ_D}{C_{\min}} \right) \right]$$

Exercise 6

You should inject 111 mg.

Exercise 7

A. It is impossible to fit the points by a single exponential.

B. An equation such as,

$$C_t = 96.1 - 27.1 \, e^{-2.77t} - 48.5 \, e^{-0.120t}$$

would fit the points very closely. But such an equation, with five parameters calculated from seven points would be completely unwarrantable, deserving only our scorn. As the data stand, all we can safely conclude is that potassium accumulation was not a single exponential process.

Exercise 8

Let Y be the quantity which is decreasing exponentially. Then, by Equation 6-56, the logarithmic mean of Y, $\bar{Y}_{1\text{-}2}$, between t_1 and t_2 is

$$\bar{Y}_{1\text{-}2} = \frac{Y_1 - Y_2}{\ln (Y_1/Y_2)}$$

where the subscripts 1 and 2 mean "at time 1" and "at time 2." But, in general (see Section 6-11), the quantity decreasing exponentially, Y, is

$$Y = y - y_{\text{asymp}}$$

Hence,

$$\overline{(y - y_{\text{asymp}})}_{1\text{-}2} = \frac{y_1 - y_{\text{asymp}.1} - y_2 + y_{\text{asymp}.2}}{\ln \left(\dfrac{y_1 - y_{\text{asymp}.1}}{y_2 - y_{\text{asymp}.2}} \right)}$$

If $y_{\text{asymp}} = $ a constant, c, this becomes

$$\overline{(y - c)}_{1\text{-}2} = \frac{y_1 - y_2}{\ln \left(\dfrac{y_1 - c}{y_2 - c} \right)} \quad \text{or,} \quad \bar{y}_{1\text{-}2} = \frac{y_1 - y_2}{\ln \left(\dfrac{y_1 - c}{y_2 - c} \right)} + c$$

Exercise 9

If equilibrium is instantaneously attained, then at all times

$$C_{\text{art}} = S_B C_A$$

and therefore,

$$\bar{C}_{\text{art}} = S_B \bar{C}_A \text{ or } 0.11 \text{ mM./L.} = 3.0 \, \bar{C}_A \text{ or } \bar{C}_A = 0.0367 \text{ mM./L.}$$

Since there is no X in blood returning to the lungs, $C_{A.\text{asymp}} = 0$. Hence, by Equation 6-56,

$$\bar{C}_{A.1\text{-}2} = \frac{C_{A.1} - C_{A.2}}{\ln (C_{A.1}/C_{A.2})} = 0.0367 \text{mM./L.}$$

Also, since k, the proportional rate of decrease of C_A, is 0.05 per second and the interval between t_1 and t_2 is 3 sec.

$$C_{A.2} = C_{A.1} e^{-0.05 \, (3.0)} = 0.861 \, C_{A.1}$$

We can now solve these two equations simultaneously:

$$0.0367 = \frac{C_{A.1}(1 - 0.861)}{\ln (1/0.861)} = \frac{0.139 C_{A.1}}{2.303 \,(\log_{10} 1.1615)} = \frac{0.139 C_{A.1}}{0.1497} = 0.928 C_{A.1}$$

$$C_{A.1} = \frac{0.0367}{0.928} = 0.03955 \text{mM./L.} \quad \text{Hence, } C_{A.2} = (0.861 C_{A.1})$$

$$= 0.03406 \text{mM./L.}$$

Exercise 10

The equation which I fitted to these data was

$$y_t = 17e^{-0.0885t} + 42e^{-0.0194t}$$

Your equation should be at least vaguely similar.

Exercise 11

It is patently ridiculous to derive so elaborate an equation from so few points unless the points are errorless. These points are not. Notice that only 7 points are used in fitting the triple exponential, the point at 250 min. being disregarded (or, more correctly, being joined to the "calculated" line by an arbitrary curve). Whenever you encounter an equation with 6 parameters calculated from 7 points, you may safely dismiss the analysis as worthless.

This is but one of many examples in which an unjustifiably elaborate multiple exponential is used to fit a small number of points. As a second example, Gilbert and Fenn (45) fitted the equation (Equation 3, p. 399 of their paper):

$$C = -38.9e^{-0.114t} - 41.7e^{-0.0107t} - 14.4e^{-3.58t} + 95.0$$

to a series of highly scattered points for calcium uptake by frog sartorius muscle (see their Figs. 3 and 4). As a final example, study Figure 1, page 140, in a paper by Bronner *et al.* (17). Here is a pathetically overblown *quadruple* exponential containing 8 parameters calculated from 17 points, subject to considerable error, and obtained from a single moribund patient.

CHAPTER 7

Exercise 1

As a general statement, the first sentence of the passage quoted is incorrect, since the solubility of, let us say, a series of sugars, has nothing to do with the rate at which they diffuse through a liquid in accordance with a given concentration gradient. Even for gases (where the term "solubility" has a different meaning) the statement is confusing and misleading. Gases diffusing through liquids follow *Graham's law*. Therefore, the ratio of diffusivities of the two gases, D_{O_2}/D_{CO_2}, will be inversely proportional to the square roots of their molecular weights:

$$D_{O_2}/D_{CO_2} = \left(\frac{44}{32}\right)^{1/2} = 1.17$$

Accordingly, if we ran a "diffusion race" between, say, 10^6 molecules of carbon dioxide and 10^6 molecules of oxygen, both "teams" of molecules being liberated at the same instant and at the same point in an aqueous solution, and if we awarded the prize to the team which first got 10,000 of its molecules at least 1 mm. away from the starting point, oxygen would win. If we confine our attention entirely to the process of diffusion through a given liquid, this is the only fair kind of comparison to make, and we must conclude that oxygen diffuses a little faster than carbon dioxide.

The statement can be justified only when it is applied to the rather peculiar *diffusion constant* for gases defined by Equation 7-26. The Bunsen solubility coefficient for carbon dioxide in water, α_{CO_2}, is about 23.6 times that for oxygen, α_{O_2}, at body temperature. Hence,

$$\frac{(\text{diffusion constant})_{CO_2}}{(\text{diffusion constant})_{O_2}} = \frac{\alpha_{CO_2}D_{CO_2}}{\alpha_{O_2}D_{O_2}} = \frac{23.6}{1.17} = 20$$

Notice that *diffusivity* is proportional to the rate of diffusion across unit area of an infinitesimal layer of liquid per unit of concentration gradient across that layer. But the *diffusion constant* is proportional to the rate of diffusion across unit area of an infinitesimal layer of liquid per unit of concentration gradient across the corresponding layer in an imaginary gas phase in equilibrium with the liquid phase. This seems an odd way to express rate of diffusion!

Exercise 2

The statement is grossly erroneous. For convenience, let us consider a gas, G, for which $\alpha = 1$ so that the diffusion constant is numerically equal to the diffusivity. The statement quoted implies that in a given interval of time, the fraction of equilibrium attained 1 meter away from the source of the diffusing gas in air would be the same as the fraction of equilibrium attained 1 μ away from the source of the diffusing gas in liquid. Therefore, $K_{F_{eq}}$ and t in Equation 7-10 must be the same for the gas phase and for the liquid phase. But in the present situation, $D_{G.\text{gas}} = 1,000,000 \, D_{G.\text{liq}}$. Let us accordingly write Equation 7-10 for the gas phase:

$$K_{F_{eq}} = \frac{x_{G.\text{gas}}}{2\sqrt{1,000,000 \, D_{G.\text{liq}}t}}$$

and for the liquid phase:

$$K_{F_{eq}} = \frac{x_{G.liq}}{2\sqrt{D_{G.liq}t}}$$

Setting these two expressions equal to each other and solving for the ratio of distances, $x_{G.gas}/x_{G.liq}$,

$$\frac{x_{G.gas}}{x_{G.liq}} = \sqrt{1,000,000} = 1000$$

Therefore, under comparable conditions, the same quantity of gas might be expected to diffuse across 1 *millimeter* of "perfectly still air" as across 1 micron of "pulmonary membrane."

Exercise 3

The conditions described *approximate* the geometrical arrangement assumed for Equation 7-3, except that the final equilibrium concentration being approached is not $C_{S.0}/2$ but $C_{S.0}$, the *constant* concentration of oxygen maintained at one face of the tissue.

Since $t = 5$ min. and $D_{O_2} = 4.5 (10^{-4})$, we may rewrite Equation 7-9 as

$$1 - F_{eq} = \text{erf}[x/9.48(10^{-2})]$$

By means of this equation and a table of the error function we may calculate the values of F_{eq} for various values of x. (For example, at $x = 0.65$ mm. $= 0.065$ cm., $F_{eq} = 0.332$ at 5 min.) If we now plot F_{eq} against x (Fig. A-1), we find that the curve between $F_{eq} = 1.00$ at $x = 0$ and $F_{eq} = 0.136$ at $x = 1.0$ mm. can be *approximated* by a straight line (*broken line* in Fig. A-1) whose equation is

$$F_{eq.x} = 1.00 - 1.00x$$

For this line, the mean value of F_{eq} between $x = 0$ mm. and $x = 1.00$ mm. is obviously 0.500. This is very close to the (presumably accurate) value of 0.534 calculated by Hill. (By what method could the accurate value be calculated?)

Exercise 4

By Equation 7-11, for a *particular* D_S and x, the factor (x^2/D_S) is a constant. All we need do, therefore, is to prove that $t_{F_{eq}}$ is approximately proportional to

$$\left(\frac{1}{1 - F_{eq}}\right)^2$$

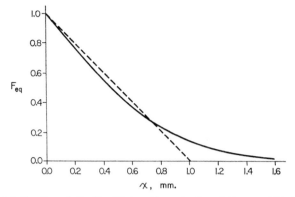

Fig. A-1. Fraction of the Equilibrium Concentration of Oxygen at Various Distances from the Initial Boundary at Zero Millimeters.

The *solid line* illustrates the situation 5 min. after the start of diffusion. The *straight broken line* roughly approximates the true curve between the limits of 0 and 1.0 mm., the thickness of the tissue in Exercise 3, Chapter 7.

"over a considerable range" of values of F_{eq}. The easiest way to do this is to let $(1 - F_{eq})$ assume various values from 0 to 1, and then to calculate from a table of the error function corresponding values for $(K_{F_{eq}})$. These may then be substituted into Equation 7-11 for calculation of $t_{F_{eq}}$. (In doing this *any* constant may be used for (x^2/D_S) so we may as well be lazy and let $(x^2/D_S) = 1$). Now if we plot $t_{F_{eq}}$ against the corresponding value of

$$\left(\frac{1}{1 - F_{eq}}\right)^2,$$

we will obtain a curve which approximates a straight line "over a considerable range."

CHAPTER 8

Exercise 1

A. $dQ_A/dt = -\dot{Q}_{A\to out}$ $Q_{A.t} = Q_{A.0} - \dot{Q}_{A\to out}t$
   (*Example:* Ethyl alcohol for which, in man, $\dot{Q}_{A\to out}$ due to metabolism is about 167 mg. per minute or 10 gm. per hour.) Note that with a constant absolute rate of loss, $Q_{A.t}$ will reach zero at a finite time when $t = Q_{A.0}/\dot{Q}_{A\to out}$.

B. $dQ_A/dt = -k_{A\to out}Q_{A.t}$ $Q_{A.t} = Q_{A.0} \exp[-k_{A\to out}t]$

C. $dQ_A/dt = \dot{Q}_{in \to A}$ $Q_{A.t} = \dot{Q}_{in \to A}t$

D. $dQ_A/dt = \dot{Q}_{in \to A} - k_{A \to U}Q_{A.t}$

$Q_{A.t} = (\dot{Q}_{in \to A}/k_{A \to U})(1 - \exp[-k_{A \to U}t])$

(*Example:* Continuous infusion of a substance into a single compartment from which it is lost by a constant rate of clearance.) Note carefully that the *proportional rate at which a steady state is approached during continuous infusion is independent of the rate of infusion and depends solely upon the proportional rate at which the substance is lost from the compartment.* In other words, under like conditions the half-time for accumulation is the same as the half-time for disappearance.

E. $dQ_A/dt = \dot{Q}_{in \to A} - k_{A \to U}Q_{A.t} - k_{A \to liv}Q_{A.t} = \dot{Q}_{in \to A} - k_A Q_{A.t}$

where $k_A = k_{A \to U} + k_{A \to liv}$

$Q_{A.t} = (\dot{Q}_{in \to A}/k_A)(1 - e^{-k_A t})$

F. $dQ_A/dt = (\dot{Q}_{in \to A} - \dot{Q}_{A \to out}) - k_{A \to U}Q_{A.t}$

$Q_{A.t} = [(\dot{Q}_{in \to A} - \dot{Q}_{A \to out})/k_{A \to U}](1 - \exp[-k_{A \to U}t])$

(*Example:* Such an equation might describe a situation in which a substance secreted by the renal tubules was being infused at a rate greater than the "transfer maximum" of the tubules. $\dot{Q}_{A \to out}$ would then be the transfer maximum, while $k_{A \to U}$ would then represent only that portion of urinary loss which was due to filtration of the substance through the glomeruli. The equation would be applicable only when the concentration of the substance in plasma was already high enough to "saturate" the tubular secretory mechanism.)

G. $dQ_A/dt = k_{B \to A}Q_{B.t}$

But by the terms of the problem, $Q_{B.t} = Q_{tot} - Q_{A.t}$

Hence, $dQ_A/dt = k_{B \to A}Q_{tot} - k_{B \to A}Q_{A.t}$

and, $Q_{A.t} = Q_{tot}(1 - \exp[-k_{B \to A}t])$

Exercise 2

The data are most easily analyzed by using Equation 8-18 which may be rewritten in the form

$$(k_{I \to U}/k_I)Q_{tot} - Q_{U.t} = (k_{I \to U}/k_I)Q_{tot}e^{-k_I t}$$

which shows that if the proper asymptotic value, $(k_{I \to U}/k_I)Q_{tot}$, is chosen, the difference between that value and the observed cumulative amounts in the urine, $Q_{U.t}$, should give a straight line on semilog graph paper. (Note that the *time* for a given point is the time at the *end* of the collection period, not at its middle). My estimated asymptote was 91 per cent of the dose in the urine, leaving only 9 per cent for the thyroid. (Rather low for a normal subject, but this particular individual had been taking iodide supplements.) The ratio of thyroid clearance to renal clearance was therefore 9/91, so that with a renal clearance of 34 ml. per minute, the thyroid clearance was 3.36 ml. per minute. From the slope of the straight line, the half-time was 7.23 hr.

Exercise 3

It is obvious from the general trend of the data that $Q_{A.t}$ is decreasing toward some asymptote. By trial and error I found that the data could be fitted reasonably well by assuming an asymptote of $Q_{A.\infty} = 20$ mg. In other words, $Q_{A.t} - 20$ plotted against time on semilog paper gave a straight line. The slope of this line indicated a half-time for equilibration of 14 min., or a total rate constant of 0.0495 per minute, equal to the sum, $k_{A \to B} + k_{B \to A}$. The intercept of the line on the time = 0 axis was 84 mg. which must therefore have been the Q_{tot} dissolved in Compartment A at time zero. $Q_{B.\infty} = Q_{tot} - Q_{A.\infty} = 84 - 20 = 64$ mg. Now at equilibrium (time infinity) the quantity of S transferred from A to B must equal the quantity transferred in the opposite direction. Hence,

$$k_{A \to B} Q_{A.\infty} = k_{B \to A} Q_{B.\infty} \quad \text{or,} \quad 20 k_{A \to B} = 64 k_{B \to A}$$

Moreover, we found previously that

$$k_{A \to B} + k_{B \to A} = 0.0495$$

Solving these two equations simultaneously, we obtain

$$k_{A \to B} = 0.0377, \qquad k_{B \to A} = 0.0118.$$

Exercise 4

Each successive dose will add an amount, Q_D, of drug to whatever remains from previous doses so that there will be an accumulation of drug in the body. Accumulation will continue more and more slowly, until, at infinite time, the quantity disappearing exponentially during the t hours between doses will precisely equal the amount, Q_D, added by a single dose. Let $Q_{0_\infty} = Q_{max}$ be the maximum quantity then present at time zero just after a dose, and let Q_{t_∞} be the quantity present just before the next dose, t hours later. Now with simple exponential disappearance, F, the fraction of the quantity present which disappears during the t hours following *any* single dose remains constant throughout. Hence,

$$Q_{t_\infty} = F Q_{max}$$

But at time infinity the difference between Q_{max} and Q_{t_∞} is equal to Q_D:

$$Q_{max} - Q_{t_\infty} = Q_{max} - F Q_{max} = Q_{max}(1 - F) = Q_D$$

Dividing both sides of this last equation by V, the constant volume of the compartment,

$$(Q_{max}/V)(1 - F) = Q_D/V$$

But $Q_{max}/V = C_{max}$, and $Q_D/V = C_0$, the peak concentration after the first dose. Hence,

$$C_{max}(1 - F) = C_0, \quad \text{or,} \quad C_{max} = C_0/(1 - F)$$

which is the expression required.

Exercise 5

If each datum for rate of urinary excretion is subtracted from an assumed asymptote of 60 mg. per minute, the resulting series of differences gives a moderately good straight line when plotted against time on semilog graph paper, the half-time calculated from my line being 47 min. This indicates that the substance is probably distributed through a single compartment, A, from which it is lost at a constant proportional rate of $k_A = 0.693/t_{1/2} = 0.693/47 = 0.01475$ per minute. But the asymptote for rate of *urinary* loss, 60 mg. per minute, accounts for only 0.706 of the total 85 mg. being infused into A (and hence also being lost from A) when a steady state has been reached. The remainder, 25 mg. per minute or 0.294 of the total, must be accounted for by extrarenal routes.

Now at 4.0 hr., when 58.5 mg. per minute is being excreted into the urine, the plasma concentration is 0.836 mg. per milliliter. The renal clearance is therefore $58.5/0.836 = 70.0$ ml. per minute. Since this 70.0 ml. per minute accounts for only 0.706 of the total, the total clearance must be $70.0/0.706 = 99.2$ ml. per minute, extrarenal clearance thus being 29.2 ml. per minute.

Finally, since but one compartment, A, is involved, $(V\text{dist}) = V_A$. This allows us to use the very convenient relationship of Equation 8-5 to calculate the volume:

$$(V\text{dist}) = V_A = (\dot{V}\text{cl})/k_A = 99.2/0.01475 = 6730 \text{ ml. or } 6.73 \text{ L.}$$

(*N.B.* Because of considerable scatter in the points, your estimates may differ substantially from the values given here. It is also quite obvious that the number of significant figures used to express the above results is *not* justified by the accuracy of the original data!)

Note again that the half-time of approach to a steady state during a continuous infusion is determined solely by the exponential rate of *loss* from the compartment. This rate of loss can therefore be calculated just as well from the rate of approaching the steady state during the infusion as from the rate of disappearance from the compartment when the infusion is discontinued.

Exercise 6

My estimates were

A. $C_{A.t} = 0.86e^{-0.0473t} + 0.35e^{-0.00425t}$

(Concentration in grams per liter; time in minutes)

B. $k_{A \to B} = 0.0228$ $k_{A \to U} = 0.0121$ $k_{B \to A} = 0.0167$

$V_A = 5.50$ L. $(V\text{dist})_{B_A} = 7.52$ L. $(V\text{dist}) = 13.0$ L. Renal plasma clearance of creatinine $= 0.0121(5.50) = 0.0664$ L. per minute $= 66.4$ ml. per minute.

Exercise 7

Most of the proof is elementary. Note that Z reduces to $k_{A \to B} + k_{B \to A}$. We can then easily obtain

$$Q_{A.t} - \frac{k_{B \to A}Q_{\text{tot}}}{k_{A \to B} + k_{B \to A}} = \frac{k_{A \to B}Q_{\text{tot}}}{k_{A \to B} + k_{B \to A}} \exp\left[-(k_{A \to B} + k_{B \to A})t\right]$$

Now note that the coefficient of the exponential on the right may be written

$$\frac{k_{A \to B}Q_{\text{tot}}}{k_{A \to B} + k_{B \to A}} = Q_{\text{tot}} - \left(\frac{k_{B \to A}}{k_{A \to B} + k_{B \to A}}\right) Q_{\text{tot}}$$

and that according to the boundary conditions assumed for Equation 8-25, $Q_{\text{tot}} = Q_{A.0}$. Hence,

$$\left(Q_{A.t} - \frac{k_{B \to A}Q_{\text{tot}}}{k_{A \to B} + k_{B \to A}}\right) = \left(Q_{A.0} - \frac{k_{B \to A}Q_{\text{tot}}}{k_{A \to B} + k_{B \to A}}\right) \exp\left[-(k_{A \to B} + k_{B \to A})t\right]$$

which is Equation 8-23.

Exercise 8

(For uniformity, we will change "W" and "Z" of Equation 7-52 to "A" and "B".) First, let us show that the asymptotes are equivalent. Since Q_{tot} is a factor common to both, it suffices to prove that

$$\frac{V_A}{V_A + R_{B/A}V_B} = \frac{k_{B \to A}}{k_{B \to A} + k_{A \to B}}$$

Or, cross-multiplying, that

$$k_{A \to B}V_A = k_{B \to A}R_{B/A}V_B$$

By definition, at equilibrium, the rate of transfer of solute via the two pathways must be equal:

$$(dQ/dt)_{A \to B.\text{eq}} = (dQ/dt)_{B \to A.\text{eq}}$$

Replacing these derivatives by their equivalents from Equation 8-2 and solving for $(\dot{V}\text{cl})_{A \to B}$,

$$(\dot{V}\text{cl})_{A\rightarrow B} = \frac{(\dot{V}\text{cl})_{B\rightarrow A}C_{B.\text{eq}}}{C_{A.\text{eq}}} = (\dot{V}\text{cl})_{B\rightarrow A}R_{B/A}$$

Eliminating the clearances from this equation by combining it with Equation 8-5,

$$k_{A\rightarrow B}V_A = k_{B\rightarrow A}R_{B/A}V_B$$

which was to be proved

Second, we must show that the exponents are equivalent, *i.e.*, that

$$\frac{D_M A_M}{\Delta x_M}\left(\frac{R_{M/A}}{V_A} + \frac{R_{M/B}}{V_B}\right) = k_{A\rightarrow B} + k_{B\rightarrow A}$$

To do this easily, let us consider the rate of transfer by diffusion from A to B *only*, disregarding any diffusion in the opposite direction from B to A. For this unidirectional diffusion Equation 7-33 may be written

$$(dQ/dt)_{A\rightarrow B} = \frac{D_M A_M R_{M/A}}{\Delta x_M} C_{A.t}$$

Eliminating

$$\frac{(dQ/dt)_{A\rightarrow B}}{C_{A.t}}$$

by combining with Equation 8-2, and then eliminating $(\dot{V}\text{cl})_{A\rightarrow B}$ by combining with Equation 8-5,

$$k_{A\rightarrow B} = D_M A_M R_{M/A}/\Delta x_M V_A$$

Precisely the same steps applied to the pathway from B to A give

$$k_{B\rightarrow A} = D_M A_M R_{M/B}/\Delta x_M V_B$$

Summing these two rate constants,

$$\frac{D_M A_M}{\Delta x_M}\left(\frac{R_{M/A}}{V_A} + \frac{R_{M/B}}{V_B}\right) = k_{A\rightarrow B} + k_{B\rightarrow A}$$

which was to be proved.

CHAPTER 9

Exercise 1

    A. $\dot{V}_{\text{art}}dtC_{\text{art}} + \dot{V}_{\text{port}}dtC_{\text{port}} = dC_{\text{liv}}V_{\text{liv}} + \dot{V}_{\text{ven}}dtC_{\text{ven}} + \dot{V}_{\text{bile}}dtC_{\text{bile}}$
      where the subscript "port" refers to portal vein blood
    B. $\dot{V}_{\text{art}} + \dot{V}_{\text{port}} = \dot{V}_{\text{ven}} + \dot{V}_{\text{bile}}$

Exercise 2

The proportional rate of washout will be (Equation 9-1) (5.0 L. per minute)/(3.0 L.) = 1.67 per minute. The half-time will therefore be 0.693/1.67 = 0.415 min.

Exercise 3

Assume that *all* of the Diodrast present in the "effective renal plasma flow" is removed from the plasma. Then $C_{ven} = C_{out} = 0$.

Assume that Diodrast is not accumulated in the tissue of the kidney. Then $dC_{kid} = dC_W = 0$.

With these simplifications, Equation 9-6 may be written

$$\dot{V}_{ERPF}dtC_{art.t} = \dot{Q}_{v.t}dt$$

where

$\dot{V}_{ERPF}$ = the effective renal plasma flow
$C_{art.t}$ = the concentration of Diodrast in arterial plasma at time t
$\dot{Q}_{v.t}$ = the rate of Diodrast excretion during the interval dt

Now *assume* that C_{art} remains constant during the time of interest. Then we may drop the restriction that we are confining our attention to the infinitesimal interval, dt:

$$\dot{V}_{ERPF}C_{art} = \dot{Q}_v$$

But by definition, $(\dot{V}cl) = \dot{Q}_v/C_{art}$. Hence, $\dot{V}_{ERPF} = (\dot{V}cl)_{Diodrast}$.

Exercise 4

It is meaningless to speak of the time of disappearance of indicator when it is assumed to be approaching zero concentration asymptotically. The value for "mean concentration" is therefore equally meaningless, since it will be determined very largely by the time at which one arbitrarily chooses to regard the indicator as having "disappeared."

Exercise 5

The problem is simply to integrate the exponential from the time of its commencement (which we may call time zero) to time infinity. The general equation for the curve is

$$C_{art.t} = C_{art.0}e^{-kt}$$

The desired integral, or area, is therefore

$$\text{Area} = \int_0^\infty C_{\text{art.0}}e^{-kt}\,dt = C_{\text{art.0}}\int_0^\infty e^{-kt}\,dt = \left[C_{\text{art.0}}\,e^{-kt}/-k\right]_0^\infty$$

$$= (C_{\text{art.0}}e^{-\infty}/-k) - (C_{\text{art.0}}e^{-0}/-k) = C_{\text{art.0}}/k$$

Exercise 6

Let us make the simple assumption that we are dealing with two separate, well-mixed compartments, thyroid and body, forming a closed system and able to exchange iodide with each other only via thyroid blood flow. Let us further assume that the radioactive iodide in the thyroid bloodstream comes into complete and instantaneous equilibrium with the iodide ion pool in the thyroid. The "volume" of these two compartments will be proportional to the quantity of iodide ion they contain, and this in turn will be proportional to the fractions of the dose of radioactive iodide they contain at equilibrium:

$$V_{\text{body}}/V_{\text{thy}} = (1 - F_{\text{thy}})/F_{\text{thy}}$$

Furthermore, the rate constant for the approach to equilibrium in a closed two-compartment system is the sum of the two rate constants for transfer between the compartments. In the present system, the rate constant for transfer from body to thyroid will be $\dot{V}_{\text{thy}}/V_{\text{body}}$, whereas the rate constant for transfer from thyroid to body will be $\dot{V}_{\text{thy}}/V_{\text{thy}}$. The half-time for approaching equilibrium will therefore be

$$t_{1/2} = 0.693/\dot{V}_{\text{thy}}\left(\frac{1}{V_{\text{body}}} + \frac{1}{V_{\text{thy}}}\right)$$

Lastly, since

$$C^*_{\text{body}}V_{\text{body}} = Q^*_{\text{body}} \quad \text{(the asterisks indicating radioactive iodide)}$$

and

$$Q^*_{\text{body.eq}} = (1 - F_{\text{thy}})Q^*_{\text{tot}}$$

we may write

$$V_{\text{body}} = (1 - F_{\text{thy}})Q^*_{\text{tot}}/C^*_{\text{body.eq}}$$

Combining these various equations to eliminate all unknowns except $\dot{V}_{\text{thy}}$,

$$\dot{V}_{\text{thy}} = 0.693Q^*_{\text{tot}}F_{\text{thy}}(1 - F_{\text{thy}})/t_{1/2}C^*_{\text{body.eq}}$$

(If it is not obvious how to combine the equations, try the method described in Chapter 13.)

I am not aware that anyone has attempted to measure thyroid blood flow in this way, but it probably would not be very satisfactory because during the very rapid initial approach to equilibrium between the two compartments, it is unlikely that the compartment of body iodide would actually behave as a single well-stirred compartment. However, perhaps the method suggested here is worth exploring further.

Exercise 7

    A. Every pair illustrated
    B. Every pair except A-B
    C. Every pair except A-E

Exercise 8

Since the observations illustrated by the figure were made after a single dose of sulfanilamide, it is possible to explain them by postulating that one or more compartments intervene between the plasma compartment and the compartment of cerebrospinal fluid actually sampled.

Exercise 9

According to the curves illustrated, the specific activity in Compartment 2 continues to rise for some time after it has surpassed the specific activity in its immediate precursor compartment, Compartment 1. This is impossible. Similar anomalies appear elsewhere in Figure 3. [Dr. Shipley informs me that the fault lies in the labelling of the graphs. The ordinates of the curve plotted for the ith compartment actually represent q_i/Q_1 rather than specific activity which would have been q_i/Q_i.]

Exercise 10

The statement is not true. A constant ratio of unity should *not* be expected unless the "precursor" and the "product" are parts of the same metabolic pool. But if they are parts of the same pool, they cannot be precursor and product!

CHAPTER 10

Exercise 1

Let N_W be the total number of molecules of W present. Let us name the *individual* molecules W_1, W_2, W_3, W_4, $\cdots$, W_N. Next, let us identify all possible collisions by making a square array as follows:

	W_1	W_2	W_3	W_4	$\cdots$	W_N
W_1		x	x	x		x
W_2			x	x		x
W_3				x		x
W_4						x
$\vdots$						
W_N						

In such a square there are N^2 boxes. But the N of these which lie on the corner-to-corner diagonal from top left to bottom right represent impossible "collisions" between a particular molecule and itself. This leaves $N^2 - N$ boxes. But only half of these (marked by x) represent *different* opportunities for collision, since a collision of W_1 with W_2 is the same as a collision of W_2 with W_1. Hence, the total number of different collisions is $(N^2 - N)/2 = N(N - 1)/2$. Now usually we deal with numbers of molecules so large that $(N - 1)$ is insignificantly different from N. Hence we may take the number of different opportunities for collision as $N^2/2$.

Exercise 2

Let k_T be the rate of reaction at temperature T (absolute). Then by definition, $(Q_{10}) = k_{T+10}/k_T$, and $(Q_{\Delta T}) = k_{T+\Delta T}/k_T$. For convenience, let us define $(E_{act}/R) = C$. Then Equation 10-20, applied to the three temperatures of present interest becomes

$$\ln k_T = \ln A - (C/T)$$

$$\ln k_{T+\Delta T} = \ln A - [C/(T + \Delta T)]$$

$$\ln k_{T+10} = \ln A - [C/(T + 10)]$$

Hence,

$$\ln (Q_{\Delta T}) = \ln k_{T+\Delta T} - \ln k_T = \frac{C}{T} - \frac{C}{T + \Delta T}$$

Similarly,

$$\ln (Q_{10}) = \ln k_{T+10} - \ln k_T = \frac{C}{T} - \frac{C}{T + 10}$$

So that

$$\ln (Q_{10}) - \ln (Q_{\Delta T}) = \ln \left(\frac{(Q_{10})}{(Q_{\Delta T})} \right) = C \left(\frac{1}{T + \Delta T} - \frac{1}{T + 10} \right)$$

Taking antilogarithms and rearranging suitably:

$$(Q_{10}) = (Q_{\Delta T}) \exp \left[\frac{E_{act}}{R} \left(\frac{10 - \Delta T}{(T + \Delta T)(T + 10)} \right) \right]$$

Exercise 3

Let k = the rate constant, and let A, B, C, etc., be the reactants. Then we may write the rate of reaction as

$$dC_A/dt = kC^\alpha_A C^\beta_B C^\gamma_C \cdots$$

Let $[C]$ be the dimensions of molar concentration. Then the corresponding dimensional equation is

$$[C]T^{-1} = [C]^{-1}T^{-1}[C]^n$$

where $\alpha + \beta + \gamma + \cdots = n$ is the order of the reaction. Solving for $[C]^n$,

$$[C]^n = [C]^2 \quad \text{or} \quad n = 2$$

Hence the reaction was a second-order reaction.

Exercise 4

Water acts as an "acid" when it becomes negatively charged by dissociating a hydrogen ion:

$$HOH \rightleftarrows H^+ + OH^-$$

In pure water, equal numbers of H^+ and of OH^- are present, each in a concentration of 10^{-7} mole per liter at 24°C. Since the density of water is essentially 1.00, the molar concentration of pure water is

$$(1000 \text{ gm./L.})/(18.00 \text{ gm./mole}) = 55.5 \text{ moles/L.}$$

Substituting these values into the Henderson-Hasselbalch equation (Equation 10-34) we have

$$7.0 = pK_a + \log_{10}(10^{-7}/55.5)$$

or, solving for the pK_a,

$$pK_a = 7.0 - (-7.0 - 1.744) = 7.0 + 8.744 = 15.744$$

For comparison, the pK_a of the extremely weak acid, glucose, is about 13.

A very similar line of argument shows that when water acts as a "base," *i.e.*, when it becomes positively charged by associating a hydrogen ion.

$$HOH + H^+ \rightleftarrows H_3O^+$$

the Henderson-Hasselbalch equation becomes

$$7.0 = pK_a + \log_{10}(55.5/10^{-7})$$

or,

$$pK_a = 7.0 - (1.744 + 7.0) = 7.0 - 8.744 = -1.744$$

For comparison, the pK_a of the very weak base, *p*-nitroaniline, is about 1.

Exercise 5

Applying the Henderson-Hasselbalch equation to the tertiary nitrogen, we find that, at pH 7.4, 13.6 per cent of the molecules do not carry a positive charge. Applying the Henderson-Hasselbalch equation to the phenolic hydroxyl, we find that, at pH 7.4, 99.7 per cent of the molecules do not carry a negative charge. Hence, $(0.997)(0.136) = 0.136$ or 13.6 per cent of the molecules are wholly uncharged. Similar reasoning for pH $= 9.05$ shows that 87.6 per cent of the nitrogen groups and 87.6 per cent of the phenolic hydroxyl groups are uncharged. Hence, $(0.876)(0.876) = 0.767$ or 76.7 per cent of the molecules carry no charge at all.

Exercise 6

By further consulting the same *Handbook*, we find that the vapor pressure of water at 25°C. is 24 mm. Hg, so that the partial pressure of CO_2 would be 736 mm. Hg. We find also that the Bunsen solubility coefficient for CO_2 in water at 25°C. is 0.759 ml. (STPD) of CO_2 per milliliter of solution per atmosphere of partial pressure of CO_2. Hence, the total amount of "CO_2" (presumably including dissolved CO_2, H_2CO_3, and HCO_3^-) present in solution will be $(736/760)(0.759) = 0.735$ ml. of CO_2 (STPD) per milliliter of solution or 0.735 L. of CO_2 (STPD) per liter of solution. Now the density of CO_2 (STPD) (again, the *Handbook*!) is 1.977 gm. per liter (STPD). Hence there are $(0.735)(1.977) = 1.452$ gm., or $1.452/44 = 0.0330$ mole of "total" CO_2 per liter of solution at 25°C.

Now in pure water, there must be 1 mole of bicarbonate ion formed for each mole of H^+. Since the pH is given as 3.8, the concentration of HCO_3^- must be $10^{-3.8}$ M per liter, or 0.00016 mole per liter. By difference, therefore, the sum of dissolved CO_2 and H_2CO_3 must be 0.0328 mole per liter. The ratio of $C_{HCO_3^-}$ to $C_{H_2CO_3} + C_{CO_2}$ is therefore about 0.0049. Substituting the value thus obtained into Equation 10-44 we have

$$3.8 = pK'_{a.\text{H}_2\text{CO}_3} + \log_{10}(0.0049) = pK'_{a.\text{H}_2\text{CO}_3} - 2.31$$

Solving for $pK'_{a.\text{H}_2\text{CO}_3}$

$$pK'_{a.\text{H}_2\text{CO}_3} = 3.8 + 2.3 = 6.1$$

Exercise 7

The statement is not true. By "total carbon dioxide" Richards presumably means the sum of CO_2, H_2CO_3, and HCO_3^-, of which sum roughly 5 per cent is CO_2 plus H_2CO_3 at the pH of plasma. But since only about $1/800$ of the CO_2 plus H_2CO_3 is actually H_2CO_3, Richards' statement *should* be, "About 0.006 per cent of the sum of dissolved CO_2, bicarbonate, and carbonic acid is in the form of carbonic acid."

Exercise 8

A. In both compartments, the pH of the solution is very much lower than the pK_a of the acid. Accordingly there will be negligible traces of the ionized form of the acid present, and the entire problem reduces to calculating the rate of approach to equilibrium across a membrane equally permeable from either side to a single constituent. We are told that initially, when 10 mg. of Y are in the 1.0 L. of A, and no Y is in B, the concentration of Y in A decreases by 0.01 mg. per minute or $0.01/10 = 0.001$ proportion per minute, so that $k_{A \to B} = 0.001$. But at equilibrium, the concentration of Y in both A and B will be 2.5 mg. per liter, the original 10 mg. having been evenly distributed throughout the entire 4-L. volume of the system. At equilibrium, therefore, A will contain 2.5 mg. while B will contain 7.5 mg. But by definition, at equilibrium the rate of transfer of Y must be the same in both directions. Hence,

$$k_{A \to B} Q_{A.\text{eq}} = k_{B \to A} Q_{B.\text{eq}}, \quad \text{or,} \quad 0.001 \ (2.5) = k_{B \to A}(7.5)$$

Solving for $k_{B \to A}$ we find $k_{B \to A} = 0.000333$ per minute. Since this is a closed two-compartment system, the rate constant for approach to equilibrium will be the sum of the two unidirectional rate constants, or 0.001333. The half-time of approach to equilibrium will thus be $t_{1/2} = 0.693/0.001333 = 520$ minutes.

B. To calculate the ratio $C_{\text{tot}.A.\text{eq}}/C_{\text{tot}.B.\text{eq}}$ for X, we need an equation for bases analogous to Equation 10-53 for acids. Applying to bases the same steps as were used in deriving Equation 10-53, we obtain

$$C_{\text{tot}.A.\text{eq}}/C_{\text{tot}.B.\text{eq}} = \frac{1 + \text{antilog}_{10}(pK_a - \text{pH}_A)}{1 + \text{antilog}_{10}(pK_a - \text{pH}_B)}$$

Substituting into this equation the values given in the problem, we find

$$C_{tot.A.eq}/C_{tot.B.eq} = 0.0688$$

Exercise 9

In addition to Equation 10-55 for the mass action equilibrium,

$$(C_S C_{Pr}/C_{SPr}) = (Keq)_{SPr}$$

we need only the following three simple equations which are really nothing but definitions:

$$C_{S.tot} = C_S + C_{SPr}$$

$$C_{Pr.tot} = C_{Pr} + C_{SPr}$$

$$F_{bound} = C_{SPr}/C_{S.tot}$$

Since we want an equation for F_{bound} in terms of $C_{S.tot}$, $C_{Pr.tot}$, and $(Keq)_{SPr}$, we must combine these four equations so as to eliminate the unwanted quantities, C_S, C_{Pr}, and C_{SPr}, and we must then solve the resulting equation for F_{bound}. (A simple guide to combining such equations is explained in Chapters 12 and 13.) The resulting equation is

$$F_{bound} = \frac{G \pm \sqrt{G^2 - 4C_{S.tot}C_{Pr.tot}}}{2C_{S.tot}}$$

where

$$G = C_{S.tot} + C_{Pr.tot} + (Keq)_{SPr}$$

An easy way to find out whether the radical is to be taken as positive or negative, and at the same time to check the derivation, is to assign arbitrary, but consistent, numerical values to the several variables in the manner discussed in Chapter 14. For example, if we arbitrarily let $C_S = 3$, $C_{Pr} = 4$, and $C_{SPr} = 5$, the other quantities *must* be $(Keq)_{SPr} = 2.4$, $C_{S.tot} = 8$, $C_{Pr.tot} = 9$, and $F_{bound} = 5/8$. Substituting these values into our equation for F_{bound}, we find that if the radical is taken as positive, $F_{bound} = 28.8/18$ which is preposterous. But if the radical be taken as negative, we find that $F_{bound} = 5/8$, which is correct. Hence, the sign must be negative.

For a concrete illustration of how each of the independent variables, $C_{S.tot}$, $C_{Pr.tot}$, and $(Keq)_{SPr}$ influences F_{bound}, see Figure 10-4.

CHAPTER 11

Exercise 1

For experiment 1, the values of $1/v$ corresponding to $v + 0.03 = 1.70$ and $v - 0.03 = 1.64$ are 0.588 and 0.610, giving a difference between a positive and a negative error of 0.022 for the transformed variable. But for experiment 6, the corresponding range of error (from $1/0.30 = 3.33$ to

$1/0.36 = 2.78$) is about 0.55 which is some 25 times as large as for experiment 1. Hence the datum from experiment 1 should be given $(25)^2 = 625$ *times as much weight* as the datum from experiment 6! Similar calculations will show that for a constant percentage error of 6 per cent, the ranges are $1/1.57$ to $1/1.77 = 0.072$ for experiment 1, and $1/0.31$ to $1/0.35 = 0.368$ for experiment 6, a 5-fold difference. Hence, the datum from experiment 1 should be given 25 times as much weight as the datum from experiment 6.

Exercise 2

The maximum response of which this muscle is capable seems to be a contraction of 40 mm. If we accordingly divide each response by 40 and multiply by 100 to convert it to per cent of maximum, we can look up the corresponding probits and plot the probit of per cent response against log dose. A straight line fitted to the points without tubocurarine crosses the ordinate for probit 5, corresponding to 50 per cent response, at about log 6.0. We may accordingly estimate $(Keq)_{DR}$ from Equation 11-25 as 6.0 μmoles per liter. If the data for the response in the presence of tubocurarine are similarly plotted, we find that they can also be fitted by a straight line which is parallel to the first, and about 1.30 log units to its right, crossing the probit 5 line at about log 120. Since a constant log difference of 1.30 units corresponds to an increase in the dose by a constant factor of 20, we conclude that for a given response, about 20 times as much acetylcholine was required in the presence of 10 μmoles per liter of tubocurarine as was required for the same response without tubocurarine, but with no change in the maximum response. The behavior of tubocurarine is thus consistent with its being a reversible competitive inhibitor of acetylcholine. (We would have been led to the same conclusion had we plotted the results in any of the ways illustrated by Figure 11-1, using the analogous symbols appropriate for concentration-effect relationships as shown in Table 11-1.) To calculate the dissociation constant for the receptor-inhibitor complex, we can employ Equation 11-16, and use the symbols D and R in place of S and E:

$$C_{DI} - C_{DI=0} = \frac{C_{DI=0}}{(Keq)_{IR}}(C_I) \qquad (F_{(Eff)_{max}} \text{ constant})$$

Taking the constant response as 50 per cent of maximum, this equation applied to our data becomes

$$120 - 6 = \frac{6.0}{(Keq)_{IR}}(10)$$

Solving for $(Keq)_{IR}$,

$$(Keq)_{IR} = 0.526 \ \mu M/L.$$

If we assume that the data can be explained by a log-normal distribution of tolerances in the population of muscle fibers being tested, the standard deviation, in log-dose units, would correspond to a one-probit interval. For our example, this is about 0.71 log-dose unit.

Exercise 3

Because the equation (Equation 11-9c) for the straight line is

$$1/v = 1/v_{max} + (k_M/v_{max})(1/C_S)$$

we need only set $1/v$ equal to zero and solve for $1/C_S$:

$$1/C_S = -1/k_M \text{ when } 1/v = 0$$

Hence, it is possible to estimate k_M by taking the reciprocal of this intercept with its sign changed.

Exercise 4

In Equation 11-16, if $(Keq)_{IE} = C_I$,

$$C_{S_I} - C_{S_{I=0}} = C_{S_{I=0}}$$

Hence,

$$C_{S_I} = 2C_{S_{I=0}}$$

Exercise 5

Solving Equation 11-9b for C_S in terms of the ratio (v/v_{max}),

$$(v/v_{max})k_M + (v/v_{max})C_S = C_S$$

$$C_S[1 - (v/v_{max})] = (v/v_{max})k_M$$

$$C_S = k_M(v/v_{max})/[1 - (v/v_{max})] \text{ or, } (v/v_{max})/[1 - (v/v_{max})] = (1/k_M)C_S$$

(Reiner actually suggests plotting C_S as ordinate so that the slope of the line will be k_M ; but this would be unorthodox since C_S is the independent variable.)

Exercise 6

Inverting Equation 11-9b and solving for $(v_{max}/v) - 1$,

$$(v_{max}/v) - 1 = k_M/C_S$$

Taking logarithms,

$$\log[(v_{max}/v) - 1] = \log k_M - \log C_S$$

Therefore, as is exemplified in Figure 11-1g, a plot of $\log [(v_{max}/v) - 1]$ against $\log C_S$ will give a straight line with an intercept of $\log k_M$ and a slope of -1.

Exercise 7

Whenever an exponent is to be evaluated from experimental data, it is wise to look for some logarithmic transformation. For the present problem, we could use the method discussed in Exercise 6 and illustrated by Figure 11-1e. The slope of the line will then be $-n$. We could equally well use a logarithmic plot of the equation suggested by Reiner (86) and discussed in Exercise 5:

$$\log \left(\frac{F_{(\text{Eff})\text{max}}}{1 - F_{(\text{Eff})\text{max}}} \right) = \log(1/K) + n \log C_D$$

where $F_{(\text{Eff})\text{max}} = (\text{Eff})/(\text{Eff})_{\text{max}}$, *i.e.*, the fraction of the maximum effect. This equation shows that a plot of the logarithmic function on the left against log C_D would give a straight line with an intercept of log $(1/K)$ and a slope of n.

Exercise 8

Multiplying Equation 11-12 by C_{S_I},

$$\frac{C_{S_I}}{v} = \frac{k_M}{v_{\text{max}}} \left(1 + \frac{C_I}{(K\text{eq})_{IE}} \right) + \left(\frac{1}{v_{\text{max}}} \right) C_{S_I}$$

which is analogous to Equation 11-9d.

For convenience, define $x = v/C_{S_I}$ so that $1/x = C_{S_I}/v$, and define

$$a = \frac{k_M}{v_{\text{max}}} \left(1 + \frac{C_I}{(K\text{eq})_{IE}} \right).$$

Then the equation just derived may be rewritten

$$1/x = a + C_{S_I}/v_{\text{max}} = a + \frac{C_{S_I}/v}{v_{\text{max}}/v} = a + \frac{1/x}{v_{\text{max}}/v}$$

Multiplying through by x,

$$1 = ax + v/v_{\text{max}}$$

Solving for v,

$$v = v_{\text{max}} - v_{\text{max}} a x$$

Restoring the original symbols,

$$v = v_{\text{max}} - k_M \left(1 + \frac{C_I}{(K\text{eq})_{IE}} \right) (v/C_{S_I})$$

which is analogous to Equation 11-9e.

Notice that both of these equations reduce to the analogous equations without inhibitor when $C_I = 0$.

Exercise 9

The *error* in C_S/v for a given *error* in v is equivalent to the *change* in C_S/v for a given *change* in v, i.e.,

$$d(C_S/v)/dv = -C_S v^{-2}$$

If this function be plotted against C_S, minimum error in C_S/v for a given error in v will be indicated by a minimum in the plotted curve (zero slope). In other words, we want to set

$$d(-C_S v^{-2})/dC_S = 0$$

But we cannot properly do this unless we note that v is itself a function of C_S (Equation 11-9b):

$$v = v_{max} C_S/(k_M + C_S)$$

Hence,

$$v^{-2} = \frac{k_M^2 + 2k_M C_S + C_S^2}{v_{max}^2 \, C_S^2}$$

Our differential equation now becomes

$$\frac{d}{dC_S}\left(-\frac{k_M^2 + 2k_M C_S + C_S^2}{v_{max}^2 \, C_S}\right) = \frac{d}{dC_S}\left(-\frac{k_M^2}{v_{max}^2}C_S^{-1} - \frac{2k_M}{v_{max}^2} - \frac{C_S}{v_{max}^2}\right) = 0$$

or,

$$\frac{k_M^2}{v_{max}^2}\cdot\frac{1}{C_S^2} - \frac{1}{v_{max}^2} = 0 \quad\text{or,}\quad k_M = C_S$$

which was to be proved.

CHAPTERS 12 AND 13

Exercise 1

A. The second sentence is in error. Whatever factors are important in determining the rate of approach to equilibrium during the *induction* of anesthesia are equally important in determining the rate of approach to equilibrium with zero concentration of anesthetic in the inspired air during *recovery* from anesthesia.

B. The second statement is just the opposite of the truth. In comparison with ether and chloroform, cyclopropane is relatively *insoluble* in plasma (S_B = about 0.5), and it is largely because of this *low* solubility that anesthesia occurs rapidly.

C. *Page 469.* The authors are evidently confusing "amount dissolved" which depends on partial pressure, and "solubility" which does not.

Page 471. Points 2 and 3 are grossly in error. Since ether is so highly soluble in blood, only a small proportion of the ether entering the lungs in the mixed venous blood becomes distributed into the alveolar air. The great majority remains in the arterial blood leaving the lungs.

D. "Volatility" has nothing whatsoever to do with the rate of induction of, or recovery from, anesthesia.

E. By "activity" is evidently meant potency as measured by the inverse of the concentration in blood needed for surgical anesthesia. But this has no influence upon rate of induction of anesthesia. For example, nitrous oxide is, in this sense, somewhat less potent than ether and very much less potent than chloroform. Yet if the speed of induction with the three agents were measured under comparable circumstances by using the "safe anesthetic concentration" in the inspired air (*i.e.*, the equilibrium concentration needed for maintenance of surgical anesthesia), it would be found that the speed of induction would be by far the fastest with nitrous oxide, very much slower with chloroform, and still slower with ether, this order being determined by their relative solubilities in blood.

Exercise 2

By using the tabular method, the derivations assigned in this exercise are perfectly straightforward and require no discussion.

Exercise 3

The exercise is too simple to need comment.

Exercise 4

Besides the two equations (13-14a and 13-15) of Table 13-3, which describe the exchange of gas at the lungs, we need only an equation which defines pulmonary clearance. In writing this equation, we must note that the net quantity of inert gas excreted by the lungs and hence cleared from the blood is the *difference* between the quantity of gas exhaled $\dot{Q}_E$ and the quantity inhaled, $\dot{Q}_I$. Hence,

$$(\dot{V}\text{cl}) = \frac{\dot{Q}_E - \dot{Q}_I}{C_{\text{ven}}} = \frac{\dot{V}_A C_A - \dot{V}_A C_I}{C_{\text{ven}}}$$

Now in combining this equation with the two mentioned above, which terms should we eliminate? If there may be inert gas in the inspired air, we cannot logically eliminate C_I or C_{ven} because the amount of gas excreted will depend in part upon the relative concentrations in the air and blood entering the lungs. On the other hand, we may quite logically eliminate the concentrations of gas in blood and air *leaving* the lungs, since these concentrations will be dependent upon all of the other factors. Accordingly, let us eliminate C_{art} and C_A, thereby obtaining

$$(\dot{V}\mathrm{cl}) = \frac{\dot{V}_A \dot{V}_B}{\dot{V}_A + S_B \dot{V}_B} + \frac{C_I \dot{V}_A}{C_{\mathrm{ven}}} \left(\frac{\dot{V}_A}{\dot{V}_A + S_B \dot{V}_B} - 1 \right)$$

which is the desired equation. Notice that *if C_I is zero*, the equation reduces to the very much simpler form:

$$(\dot{V}\mathrm{cl}) = \frac{\dot{V}_A \dot{V}_B}{\dot{V}_A + S_B \dot{V}_B}$$

Exercise 5

Given that

$$(\dot{V}\mathrm{cl}) = \frac{\dot{V}_A \dot{V}_B}{\dot{V}_A + S_B \dot{V}_B}$$

the partial derivative of $(\dot{V}\mathrm{cl})$ with respect to $\dot{V}_A$ is

$$\partial(\dot{V}\mathrm{cl})/\partial \dot{V}_A = S_B \dot{V}_B{}^2/(\dot{V}_A + S_B \dot{V}_B)^2$$

and the partial derivative of $(\dot{V}\mathrm{cl})$ with respect to $\dot{V}_B$ is

$$\partial(\dot{V}\mathrm{cl})/\partial \dot{V}_B = \dot{V}_A{}^2/(\dot{V}_A + S_B \dot{V}_B)^2$$

Setting these two partial derivatives equal to each other and solving for S,

$$S = (\dot{V}_A/\dot{V}_B)^2$$

This derivation shows that small increments of ventilation or perfusion are equally effective in increasing pulmonary clearance when the solubility of the gas being cleared is equal to the *square* of the "ventilation/perfusion" ratio.

Exercise 6

Your drawing of a model (really just the tonometer with its contained blood and air) of the system should make it clear that the amount of X originally present at a concentration of $C_{B.0}$ in the blood sample will distribute itself between the blood phase and the air phase according to its solubility, *i.e.*, according to the equilibrium ratio, $C_{B.\mathrm{eq}}/C_{\mathrm{air.eq}}$. To begin with, therefore, we have the equation of equilibrium:

$$C_{B.\mathrm{eq}}/C_{\mathrm{air.eq}} = S$$

Letting Q_{tot} be the total amount of X present in the system, conservation of X requires

$$Q_{\mathrm{tot}} = Q_{B.\mathrm{eq}} + Q_{\mathrm{air.eq}}$$

or, since $Q_{B.\mathrm{eq}} = C_{B.\mathrm{eq}} V_B$ and $Q_{\mathrm{air.eq}} = C_{\mathrm{air.eq}} V_{\mathrm{air}}$,

$$Q_{tot} = C_{B.eq}V_B + C_{air.eq}V_{air}$$

Likewise, conservation of volume requires that

$$V_{ton} = V_B + V_{air}$$

Finally, the concentration of X in the original blood sample is, by definition,

$$C_{B.0} = Q_{tot}/V_B$$

These equations contain eight symbols: The two *variables*, $C_{air.eq}$ and $C_{B.0}$; the three *parameters*, V_{ton}, V_B, and S; and the three *unwanted quantities*, $C_{B.eq}$, Q_{tot}, and V_{air}. By using the tabular arrangement described in the text, you should have no difficulty in eliminating the unwanted quantities and deriving the desired equation:

$$C_{B.0} = C_{air.eq}\left(\frac{V_{ton}}{V_B} + S - 1\right)$$

Exercise 7

At first glance, the data supplied seem too meager to allow a quantitative solution. But let us make a crude diagram of the situation and see what relationships can be discerned (Fig. A-2). Since X is unable to enter red cells, all of the symbols will refer to plasma, not whole blood.

Conservation of X (the Fick principle) requires

$$C_{art}\dot{V}_{art} = \dot{Q}_U + C_{ven}\dot{V}_{ven}$$

Conservation of water (*i.e.*, volume) requires

$$\dot{V}_{art} = \dot{V}_U + \dot{V}_{ven}$$

By definition of clearance

$$(\dot{V}cl) = C_U\dot{V}_U/C_{art}$$

(Note that clearance is always the volume of the *entering* plasma which is cleared of X per minute, not the volume of venous plasma.) *By definition*

$$\dot{Q}_U = C_U\dot{V}_U$$

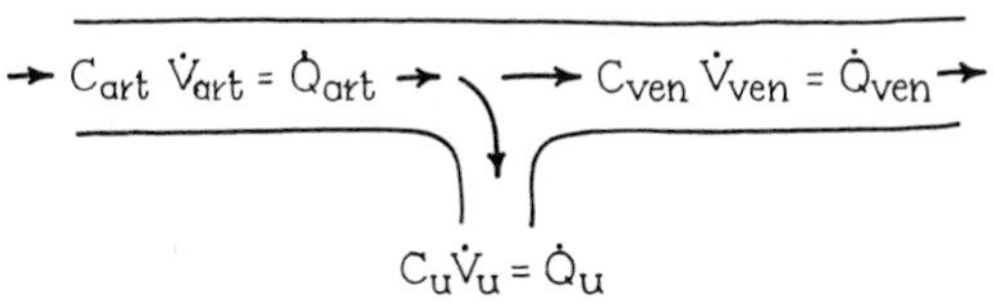

FIG. A-2. A Model of the System Analyzed in Exercise 13-7

TABLE A-1

Milligrams, Milliliters, Minutes

Equation No.	Equation	Find C_{art}	Find $\dot{V}_{art}$	Find $\dot{Q}_U$	$\dot{V}_{ven}$	C_{ven} (0.01)	$\dot{V}_U$ (4)	$(\dot{V}cl)$ (50)	C_U	$\dot{Q}_{in}$ (0.7)
1	$C_{art}\dot{V}_{art} = \dot{Q}_U + C_{ven}\dot{V}_{ven}$	✓	✓	✓	✓	✓				
2	$\dot{V}_{art} = \dot{V}_U + \dot{V}_{ven}$		✓		✓		✓			
3	$(\dot{V}cl) = C_U\dot{V}_U/C_{art}$	✓					✓	✓	✓	
4	$\dot{Q}_U = C_U\dot{V}_U$			✓			✓		✓	
5	$\dot{Q}_U = \dot{Q}_{in}$			✓						✓
6	$C_U\dot{V}_U\dot{V}_{art}/(\dot{V}cl) = \dot{Q}_U + C_{ven}\dot{V}_{ven}$	✗	✓	✓	✓	✓	✓	✓	✓	
7	$C_U\dot{V}_U\dot{V}_{art}/(\dot{V}cl) = \dot{Q}_{in} + C_{ven}\dot{V}_{ven}$	✗	✓	✗	✓	✓	✓	✓	✓	✓
8	$\dot{Q}_{in} = C_U\dot{V}_U$	✗		✗	✗		✓		✓	✓
9	$C_U\dot{V}_U\dot{V}_{art}/(\dot{V}cl) = \dot{Q}_{in} + C_{ven}\dot{V}_{art} - C_{ven}\dot{V}_U$	✗	✓	✗	✗	✓	✓	✓	✓	✓
10	$\dot{Q}_{in}\dot{V}_{art}/(\dot{V}cl) = \dot{Q}_{in} + C_{ven}\dot{V}_{art} - C_{ven}\dot{V}_U$	✗	✓	✗	✗	✓	✓	✓	✗	✓

Finally, since the concentration of X in arterial blood had become essentially constant, the rate of excretion of X must equal the rate of infusion of X:

$$\dot{Q}_U = \dot{Q}_{in}$$

Let us now make a table of these equations and the symbols used in them (Table A-1). We shall also write the known values above the appropriate symbols. The *units* (which must be consistently used) are also listed above the table. We proceed by eliminating the quantities whose values are unknown.

Solving Equation 10 in the table for $\dot{V}_{art}$,

$$\dot{V}_{art} = \frac{(\dot{V}cl)(\dot{Q}_{in} - C_{ven}\dot{V}_U)}{\dot{Q}_{in} - C_{ven}(\dot{V}cl)} = \frac{(50)(0.7 - 0.04)}{0.7 - 0.5} = \frac{33.0}{0.2} = 165 \text{ ml./min.}$$

If you made the mistake of neglecting the change in plasma volume due to urine formation as blood flows through the kidney, you overestimated plasma flow by 10 ml. per minute, even though the urine flow was itself only 4.0 ml. per minute. *Moral:* Volume of solvent as well as quantity of solute must be conserved.

Exercise 8

As is usually true, the key to the solution lies in making the appropriate assumptions. If we assume *instantaneous inspiration, uniform distribution of air flow and blood flow, and instantaneous diffusion* between alveolar air, lung tissue, and contained blood, we can treat the entire lung as a single compartment throughout which the inert gas entering the alveoli is distributed at time zero. (For convenience we shall also assume that the inspired mixture is at body temperature and saturated with water and that none of the inert gas returns to the lungs during the period of interest.)

With these assumptions in mind, let us draw a *model* (Fig. A-3) showing what happens when a breath of air containing the inert gas (indicated by stippling) is inspired. By depicting the inspired air as if it remained unmixed, the second diagram in Figure A-3 shows that the *quantity* of inert gas entering the alveoli, $Q_{in \to A}$ is

$$Q_{in \to A} = (V_I - V_D)C_I$$

whereas the *volume* of air in the alveoli after inspiration is

$$V_A = V_I + V_{FRC}$$

But according to our assumptions, the inert gas will be distributed (instantaneously) throughout the tissue and blood of the lung, as well as

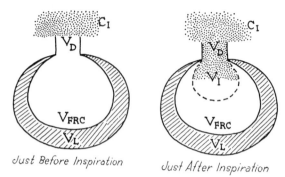

Just Before Inspiration   Just After Inspiration

Fig. A-3. A Model of the System Analyzed in Exercise 8, Chapter 13

throughout alveolar air. Its volume of distribution in the lung *referred to its concentration in alveolar air* will therefore be

$$(V\text{dist})_{L_A} = V_A + SV_L$$

You should now have no difficulty in calculating the *initial* concentrations in alveolar air, $C_{A.0}$, and in pulmonary venous blood, $C_{B.0}$.

To complete the analysis, we need only recognize that with the passage of time the pulmonary blood flow will "wash out" the inert gas from the lung compartment in the exponential manner which we have so often discussed before. The proportional rate at which washout will occur (the rate constant, k) will be the ratio of the blood flow, $\dot{V}_B$, to the volume of the compartment, *i.e.*, the volume of distribution in the lung but *now referred to blood concentration*:

$$(V\text{dist})_{L_B} = (V\text{dist})_{L_A}/S$$

so that k will be

$$k = \frac{\dot{V}_B}{(V\text{dist})_{L_B}} = \frac{\dot{V}_B S}{(V\text{dist})_{L_A}}$$

Completion of this exercise should now be straightforward. Your final equation should be

$$C_{B.t} = \left[\frac{SC_I(V_I - V_D)}{V_{\text{FRC}} + V_I + SV_L} \right] \exp\left[-\frac{\dot{V}_B St}{V_{\text{FRC}} + V_I + SV_L} \right]$$

CHAPTER 14

Exercise 1

It is at once evident that to have the proper units and to be dimensionally correct, u in the above equation should be per cent of dose per gram, not grams per second. In other words, the equation is actually for specific activity, defined elsewhere in the same article as a:

$$a = (100/M)e^{-kt}$$

(This error is even easier to identify in the original because the faulty equation is clearly stated to be an equation for specific activity.)

Exercise 2

This is a good example of the danger of trying to express a series of rather vague qualitative concepts in the guise of a superficially plausible equation. The equation is dimensionally correct, but becomes nonsense the moment we try to apply it to real concentrations. Let us calculate a value for K by substituting into the equation rough figures for the normal concentrations of the ions in plasma, expressed in millimoles per liter:

$$K = \frac{(140) + (5)}{(1.2) + (1) + (0.00004)} = \frac{145}{2.20004}$$

Now suppose that, without changing the concentrations of the other ions, we produce an exceedingly severe alkalosis, so that the pH of the plasma increases from 7.4 to 8.0 and tetany occurs. This change in hydrogen ion concentration will decrease the denominator to 2.20001 and will thereby increase the value of the "constant" K by about 0.00316 per cent. *The same change in K would be produced by increasing the concentration of sodium ion from 140 mM per liter to 140.002 mM per liter.* According to the passage quoted, one would therefore expect this trivial change in sodium concentration to produce the same change in excitability as increasing the pH from 7.4 to 8.0—an entirely preposterous conclusion.

We cannot "correct" the equation quoted since it is not the result of a mathematical derivation, and no correct equation corresponding to it exists.

Exercise 3

The statement written as an equation is

$$(V\text{dist}) = (\dot{V}\text{cl})/(\ln C_{P_1} - \ln C_{P_2})$$

The corresponding dimensional equation is

$$L^3 = L^3 T^{-1}$$

which is not correct. Since the equation is intended for calculation of a volume, it is presumably the term on the right which is in error. To make it dimensionally correct, we must multiply it by time. The only time involved in the passage quoted is the length of the clearance period. Let us try that:

$$(V\text{dist}) = (t_2 - t_1)(\dot{V}\text{cl})/(\ln C_{P_1} - \ln C_{P_2})$$

It is not obvious whether or not this equation makes sense. Let us therefore get back to primary factors by expressing the clearance in terms of the quantities from which it is usually calculated:

$$(V\text{dist}) = (t_2 - t_1)\dot{Q}_U/\bar{C}_P(\ln C_{P_1} - \ln C_{P_2})$$

where $\bar{C}_P$ is the logarithmic mean plasma concentration between t_1 and t_2, and is calculated as $\bar{C}_P = (C_{P_1} - C_{P_2})/(\ln C_{P_1} - \ln C_{P_2})$ (see Equation 6-56). Substituting this into our tentative equation,

$$(V\text{dist}) = (t_2 - t_1)\dot{Q}_U/(C_{P_1} - C_{P_2})$$

or, rearranging,

$$(V\text{dist})(C_{P_1} - C_{P_2}) = \dot{Q}_U(t_2 - t_1)$$

Now, at last, we have an equation which is obviously true, for the expression on the left is clearly equal to the decrease in the quantity of the substance within its volume of distribution between times 1 and 2, while the expression on the right is the total quantity of the substance excreted in the urine during the same period. Since we are told that there are no routes of loss other than the urine, these two quantities must be equal. We may conclude that our tentative revision of the original equation was indeed correct.

Exercise 4

Let us first rewrite the equation, with more familiar symbols:

$$\dot{Q}_{\text{tub}\to U} = C_U\dot{V}_U - \dot{V}_{\text{filt}}C_P F_{\text{free}}F_{\text{H}_2\text{O}/P} = \left[\frac{(\dot{V}_{\text{cl}})}{\dot{V}_{\text{filt}}} - F_{\text{free}}F_{\text{H}_2\text{O}/P}\right]C_P$$

It is now obvious that a quantity per unit time on the extreme left is being equated with a concentration on the extreme right, the coefficient of the concentration (in brackets) being dimensionless. This is impossible. But the dimensions of the expression in the middle are the same as those of $\dot{Q}_{\text{tub}\to U}$. So perhaps the error was made in deriving the expression on the extreme right from the expression in the middle. Suppose we tentatively accept

$$\dot{Q}_{\text{tub}\to U} = C_U\dot{V}_U - \dot{V}_{\text{filt}}C_P F_{\text{free}}F_{\text{H}_2\text{O}/P}$$

as correct, divide both sides by C_P, and solve for $\dot{Q}_{\text{tub}\to U}$:

$$\dot{Q}_{\text{tub}\to U} = C_P\left[(\dot{V}\text{cl}) - \dot{V}_{\text{filt}}F_{\text{free}}F_{\text{H}_2\text{O}/P}\right]$$

Now although the equation thus obtained is dimensionally correct, it fails to give a correct answer when a consistent set of numerical values is substituted into it. For example, suppose the total concentration of X in plasma, C_P, is 12 mg. per liter, with 40 per cent protein-bound and 60 per cent free. Then there would be 0.6 (12) $=$ 7.2 mg. of free X per liter of plasma. Suppose that the water content of plasma is 93 per cent by volume. Since all of the free X is dissolved in the water of plasma, the concentration of free X in plasma water will be 7.2 mg./0.93 L., or 7.74 mg. per liter. Since glomerular filtrate consists largely of plasma water, this may also be taken as the concentration of X in the glomerular filtrate. Let the glomerular filtration rate be 120 ml. per minute. Then the quantity of X filtered per minute is 7.74 (0.12) $=$ 0.929 mg. per minute. Suppose finally that the rate of excretion of X is 2.0 mg. per minute. Then the amount of X contributed by the tubules is 2.000 $-$ 0.929 $=$ 1.071 mg. per minute, and the renal plasma clearance of X is 2.000 mg. per minute per 12 mg. per liter $=$ 0.167 L. per minute or 167 ml. per minute. Substituting these consistent values into the revised equation, we have

$$1.071 = 12 \ [0.167 - 0.120 \ (0.6) \ (0.93)] = 1.200$$

which is incorrect.

If the source of the error is not evident to you, the safest procedure is to start from scratch and derive the equation yourself. Write down a *general* equation for each of the steps which were used in calculating the set of consistent numerical values used above. For example, the general equation corresponding to the first step would be

$$C_{\text{free}.P} = F_{\text{free}} C_P$$

By combining these general equations and solving for $\dot{Q}_{\text{tub} \to U}$ you should obtain

$$\dot{Q}_{\text{tub} \to U} = C_P \left[\dot{V}_{\text{cl}} - \dot{V}_{\text{filt}} \frac{F_{\text{free}}}{F_{\text{H}_2\text{O}/P}} \right]$$

Finally, check *your* derivation by substituting into your equation the consistent set of numerical values previously calculated:

$$1.071 = 12 \left(0.167 - 0.120 \frac{0.60}{0.93} \right) = 1.071$$

which is correct.

Exercise 5

We need go no further than simple analysis of dimensions and units to show that neither of these equations is correct. The first becomes

$$\text{ml./sec.} = \frac{(\text{in.})^2}{(\text{sec.}) \ (\text{cpm/ml.}) \ (\text{in./cpm})}$$

or

$$\text{ml./sec.} = (\text{in.})(\text{ml./sec.})$$

which is impossible.

The second becomes

$$\text{ml./sec.} = \frac{(\text{cpm})(\text{ml./cpm})(\text{in./cpm})}{(\text{in.})^2}$$

or

$$\text{ml./sec.} = \frac{(\text{ml.})}{(\text{cpm})(\text{in.})}$$

which is different, but equally impossible.

Since the method for obtaining data described in this paper is incredible, it would be pointless to try to find a "correct" equation for analyzing the results. However, careful study of the entire paper can be recommended to the reader as an extensive, though elementary, exercise in logical criticism.

Exercise 6

The equation is dimensionally correct. But if we were to check it by substitution of a consistent set of numerical values, we would find it invalid. However, we need not even go to the trouble of making up a set of consistent values. For if we allow each variable to assume (in turn) maximum and minimum values within the physiological range, we find by inspection that the equation becomes nonsense when the inspired P_{CO_2} approaches zero.

It will be safest to correct the error by deriving the equation yourself. Think of the expired air as composed of two parts: dead space air which has the same composition as inspired air (having exchanged no gas with the blood) and alveolar air. Using appropriate symbols for concentrations and volumes, write the corresponding conservation equations for 1) quantity of O_2, 2) quantity of CO_2, and 3) volume of gas. (*Assume* that the dead space is the same for CO_2 and O_2). Combine these equations to get rid of the volume terms, leaving only concentration terms. Now note that according to the ideal gas law, at constant temperature the partial pressure of any gas is directly proportional to its molar concentration,

$$P = RT(n/V)$$

Hence, by multiplying the molar concentrations by the constant, RT, we obtain the "alveolar air equation" in terms of partial pressures:

$$P_{O_2 \cdot A} = \frac{P_{CO_2 \cdot A}(P_{O_2 \cdot E} - P_{O_2 \cdot I}) + P_{CO_2 \cdot E}P_{O_2 \cdot I} - P_{O_2 \cdot E}P_{CO_2 \cdot I}}{P_{CO_2 \cdot E} - P_{CO_2 \cdot I}}$$

where the subscript I indicates inspired air; A, alveolar air; and E, expired air. Since it is difficult to get satisfactory samples of alveolar air for direct analysis, let us *assume* that CO_2 reaches distribution equilibrium between alveolar air and arterial (*i.e.*, pulmonary venous) blood. Then the partial pressure of CO_2 in arterial blood (which may be measured or calculated quite accurately) will equal the partial pressure in alveolar air:

$$P_{CO_2.art} = P_{CO_2.A}$$

Substituting this value for $P_{CO_2.A}$ in the above equation,

$$P_{O_2.A} = \frac{P_{CO_2.art}(P_{O_2.E} - P_{O_2.I}) + P_{CO_2.E}P_{O_2.I} - P_{O_2.E}P_{CO_2.I}}{P_{CO_2.E} - P_{CO_2.I}}$$

Since all of the terms on the right can be measured, the partial pressure of oxygen in alveolar air can be calculated. Notice that the above equation takes account of the possibility that CO_2 may be present in the inspired air. To obtain a correct form of the simpler equation published by Forster, we must *assume* that there is *no* CO_2 in the inspired air. The equation then reduces to

$$P_{O_2.A} = P_{O_2.I} - \frac{P_{CO_2.art}}{P_{CO_2.E}}(P_{O_2.I} - P_{O_2.E})$$

Exercise 7

Let us first rewrite the published equation with symbols which are easier to understand:

$$\log_{10}\left(\frac{1}{C_{osm}} - \frac{1}{154}\right) = a + b(\log_{10} \dot{Q}_{osm})$$

where
  C_{osm} is the concentration of the urine in milliosmoles per liter
  154 is the limiting concentration of the urine being approached (mOs./L.)
$\dot{Q}_{osm}$ is the rate of excretion of solute in the urine (mOs. per unit time)
Now let us apply this equation to each of the three situations described in the exercise.
  1. When $\dot{Q}_{osm}$ is small, C_{osm} is less than 154 mOs. per liter. Hence, $(1/C_{osm}) - (1/154)$ is *positive* but tends to *decrease* as Q_{osm} *increases*. The parameter b is therefore negative and should be so written in the equation.
  2. When $\dot{Q}_{osm}$ is small, C_{osm} is greater than 154 mOs. per liter, approaching 154 mOs. per liter as $\dot{Q}_{osm}$ increases. $(1/C_{osm}) - (1/154)$ is therefore *negative*. Since we cannot take logarithms of negative

numbers, the equation as written is inapplicable. Presumably the authors intended us to take the *absolute value* of the difference, *i.e.*, to regard it always as positive. If we do this, the equation can be applied, but again the parameter b is negative.

3. If $C_{osm} = 154$ mOs. per liter, the quantity $(1/C_{osm}) - (1/154)$ is zero. Since the logarithm of zero is indeterminate, the equation cannot be valid for this special case.

We are thus led to rewrite the equation as follows:

$$\log_{10} \left| \left(\frac{1}{C_{osm}} - \frac{1}{154} \right) \right| = a - b(\log_{10} \dot{Q}_{osm}); \quad C_{osm} \neq 154$$

Exercise 8

The equation is simple enough, and is dimensionally correct. Furthermore, it has a pleasingly symmetrical appearance (which, I presume, was why the author was stupid enough to write it in the first place!) But it fails to survive checking by numerical substitution of a set of values which are consistent with the model assumed. The fault lies in the tacit assumption that the iodide compartment has the same specific activity as the other two, which is not so. In fact, although $\dot{Q}_{I \to U_{bl}} - \dot{Q}_{I \to U}$ represents the rate of secretion of ordinary stable iodine from the blocked gland, it is easy to see that the corresponding rate for *radioactive iodine* must be the sum of $\dot{Q}^*_{T \to I_{bl}}$ and $\dot{Q}^*_{T \to F}$. But with the gland blocked, the only place for the radioactive iodine entering the iodide compartment to go is into the urine. Hence the above sum:

$$\dot{Q}^*_{T \to I_{bl}} + \dot{Q}^*_{T \to F} = \dot{Q}^*_{I \to U_{bl}} + \dot{Q}^*_{T \to F}$$

and the correct equation for Q_{thy} is therefore

$$Q_{thy} = \frac{Q^*_{thy}(\dot{Q}_{I \to U_{bl}} - \dot{Q}_{I \to U})}{(\dot{Q}^*_{I \to U_{bl}} + \dot{Q}^*_{T \to F})}$$

Exercise 9

Since the concentrations are presumably expressed in moles per liter (dimensions, L^{-3}), the several constants will have the following dimensions:

$$k_{-1}, k_2, k_3, k_{-4}, k'_{-1}, k_2', k_3', k'_{-4} \quad T^{-1}$$

$$k_1 \text{ and } k_1' \quad L^3 T^{-1}$$

$$k_4 \text{ and } k_4' \quad L^{3\alpha} T^{-1}$$

$$K_4 \text{ and } K_4' \quad L^{3\alpha}$$

Now let us examine the published equation dimensionally. The first parenthetical expression is a dimensionless ratio:

$$T^{-1}L^{-3}/T^{-1}L^{-3}$$

The subtrahend just before the equality sign is also a dimensionless ratio:

$$L^3T^{-1}L^{-3}/T^{-1} = T^{-1}/T^{-1}$$

Unity, in the longest parenthetical expression, is also dimensionless. Hence the remainder of that expression ought to be dimensionless, too. The dimensions of the numerator within square brackets are

$$L^3T^{-1}T^{-1}L^{-3}L^{-3} = L^{-3}T^{-2}$$

The dimensions of the denominator are

$$T^{-1}T^{-1}(1 + L^{3\alpha}T^{-1}L^{-3\alpha}) \cdots$$

There is no use going any further, for this is wrong! We simply cannot add unity, which is dimensionless, to reciprocal time.

Let us look over the original equation again. We find a symbol, K_4, which is duly defined below the main equation. But we do not find any similar symbol, K_4', even though its definition is also given below the original equation. Furthermore, though the symbol k_4' appears in the original equation, k_4 does not. This is surprising, for the whole problem otherwise has a nice air of symmetry. These considerations, coupled with our dimensional difficulty, strongly suggest that k_4' is simply a misprint for K_4'—a misprint which again emphasizes the great importance of being consistent in the use of symbols. And so it proves, for if we replace k_4' by K_4' in the demominator, all of our dimensional trouble disappears. $K_4'P^\alpha$ is dimensionless, and the entire denominator of the long square-bracketed ratio has the dimensions T^{-2}. Thus, the square bracketed expression itself has the dimensions L^{-3}. When this is raised to the power α and multiplied by K_4, with the dimensions $L^{3\alpha}$, we have a dimensionless expression which may properly be added to unity. Then the entire equation reduces to a dimensionless quantity equal to zero, as it should.

It is obvious that this simple dimensional analysis does not constitute a complete check of the equation. For example, if a symbol with the same dimensions as the correct one were misprinted, the error would not be revealed by dimensional checking.

Exercise 10

To begin with, the equation is needlessly complicated, symbols being used twice as often as required. It can easily be reduced to

$$C_{\mathrm{Na.reabs}.s} = C_{\mathrm{Na}.P} - C_{\mathrm{cr}.P}\left(\frac{C_{\mathrm{Na}.U.s} - C_{\mathrm{Na}.U.f}}{C_{\mathrm{cr}.U.s} - C_{\mathrm{cr}.U.f}}\right)$$

which is exactly the same equation. This equation is dimensionally correct, but if we start with the glomerular filtrate and make up reasonable numerical values, consistent with the author's assumptions, their equation does not, in general, yield the correct value for the concentration of sodium in the "proximal reabsorbate." The fault lies in the authors' trying to compare quantities which are not comparable. The concentration of creatinine in their stop-flow samples of urine was higher than in the free-flow samples, indicating that (because of additional water reabsorption during stoppage of flow) *1 ml. of stop-flow urine was derived from a larger initial volume of glomerular filtrate than 1 ml. of free-flow urine.* By calculating the quantity of sodium and of water reabsorbed *per milliliter of final urine*, the authors are thus comparing what happens to *different initial volumes of glomerular filtrate*. It should be obvious that such a comparison is invalid and that, even granting their basic assumptions, the equation,

$$Q_{\text{Na.during.}s} = Q_{\text{Na.tot.}s} - Q_{\text{Na.tot.}f}$$

is nonsense when reabsorption is calculated per milliliter of urine.

Let us accept the assumptions stated above, even though it seems very doubtful that sodium and water reabsorption in the more distal segments upon sudden restoration of flow is the same per milliliter of glomerular filtrate as during free flow. If we now apply the author's reasoning to the fate of *equal amounts of starting material*, and accordingly calculate sodium and water reabsorption *per milliliter of glomerular filtrate*, we obtain the following equation:

$$C_{\text{Na.reabs.}s} = \frac{C_{\text{cr.}U.s}C_{\text{Na.}U.f} - C_{\text{cr.}U.f}C_{\text{Na.}U.s}}{C_{\text{cr.}U.s} - C_{\text{cr.}U.f}}$$

In actual experiments, the authors found that $C_{\text{Na.}U.s}$ was usually about the same as $C_{\text{Na.}U.f}$. If these two quantities are equal, their erroneous equation reduces to

$$C_{\text{Na.reabs.}s} = C_{\text{Na.}P}$$

The authors were thus led to the conclusion that "the Na concentration in the proximal reabsorbate during stopped flow was plasma-like." From this conclusion, they infer that sodium and water are *passively* absorbed from the proximal convoluted tubules in accordance with an osmotic gradient. But the "correct" equation, based upon comparison of equal volumes of glomerular filtrate reduces to

$$C_{\text{Na.reabs.}s} = C_{\text{Na.}U.s} = C_{\text{Na.}U.f}$$

which (*if* the assumptions used in the calculation are accepted as valid) would require an entirely different interpretation.

APPENDIX B. LOGARITHMS

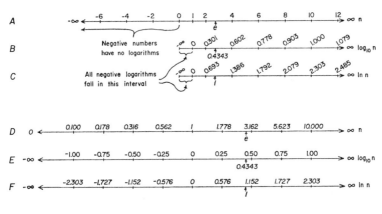

At the top of the figure, the real numbers are spaced arithmetically along a line (*line A*) which extends an infinite distance in both directions from zero. On an *arithmetic scale* of this sort, a constant finite distance measured anywhere along the line represents a *constant difference between two numbers*. For example, the distance between -6 and -4 is the same as the distance between -1 and $+1$, or -0.272 and $+1.728$, or, in general, n and $n \pm 2$ where n represents any real number. Furthermore, if we start at any point and lay off a series of adjacent constant distances, the corresponding numbers will form an *arithmetic progression*, for example, -6, -4, -2, 0, 2, 4, 6, etc. The corresponding logarithms, to the base 10 ($\log_{10} n$) and to the base e ($\ln n$), are placed along the two lines (B and C) below the arithmetic scale of numbers so that the logarithms corresponding to a given number appear immediately below that number. Notice that *negative numbers have no logarithms* and that all of the negative logarithms (lying between $-\infty$ and 0 and corresponding to the numbers between 0 and $+1$) are squeezed into a very small space on this kind of scale.

In the lower part of the figure, the logarithms (*lines E and F*) have again been placed vertically below the corresponding numbers (*line D*), but the spacing of the numbers is now logarithmic. On a *logarithmic scale*, a constant finite distance measured anywhere along the line represents a *constant difference between two logarithms*. For example, the distance between -0.75

and -0.50 along the $\log_{10}$ scale (a difference of 0.25 $\log_{10}$ unit) is the same as the distance between 0.25 and 0.50, or between -0.083 and 0.167, or, in general, between $(\log_{10} n)$ and $(\log_{10} n) \pm 0.25$.

Now with a constant log difference, the corresponding two *numbers* (antilogarithms) must bear a *constant ratio* to each other. In the above example, this ratio is the antilog of ± 0.25, *i.e.*, either 1.778 or 0.562, depending on whether the larger or the smaller number is placed in the numerator of the ratio. If we take equal distances successively along a logarithmic scale, so that the logarithms form an arithmetical progression, the corresponding numbers will form a *geometrical progression* since each successive term will differ from the previous one by a constant factor which is simply the constant ratio mentioned above. For example, the series of $\log_{10}$ values:

$$-0.50, \ -0.25, \ 0, \ 0.25, \ 0.50$$

or, if you prefer,

$$9.50 - 10, \ 9.75 - 10, \ 0, \ 0.25, \ 0.50$$

or,

$$\bar{1}.50, \quad \bar{1}.75, \quad 0, \quad 0.25, \quad 0.50$$

corresponds to the series of *numbers* (*i.e.*, antilogs):

$$0.316, \quad 0.562, \quad 1.000, \quad 1.778, \quad 3.162$$

where each number in the series can be obtained by multiplying the previous number by the factor 1.778.

Notice that logarithms to the base e (natural or Naperian logarithms) are directly proportional to logarithms to the base 10, but are *larger* by a factor of about 2.303:

$$\ln \ n = 2.303 \ (\log_{10} n)$$

$$\log_{10} n = 0.4343 \ (\ln n)$$

If you remember that the absolute value of the natural logarithm is always *larger* than the absolute value of the logarithm corresponding to the base 10, you will never be uncertain whether to multiply by 0.4343 or by 2.303 when changing from one base to another.

At the very bottom of the figure, part of *line D* has been reproduced as *line G*. *Line G*, consisting of two "log cycles," has been divided into whole tenths and whole units in order to illustrate more clearly what a logarithmic spacing of numbers looks like. Whenever we use such a scale, we are, in fact, dealing with the logarithms of the numbers, and not with the actual numbers marked upon the scale for our convenience. For example, a slide rule used for multiplication and division consists of two identical logarithmic scales so arranged that by sliding one scale along the other, we can easily add or subtract scale lengths. With logarithmic scales, this corresponds to adding or subtracting logarithms, and hence to multiplying or dividing numbers. For our convenience, the scales, though logarithmic, are marked in antilogarithms, *i.e.*, in the actual numbers which we want to multiply or divide.

With these points in mind, and with the figure as a guide, review the following facts about logarithms which are true for any base:

Let n and m represent any two real numbers which conform to the restriction that we cannot take logarithms of negative numbers.

$\log (n + m)$ cannot be related to $\log n$ and $\log m$ but must be handled as a single quantity

$\log nm = \log n + \log m$

$\log (n/m) = \log n - \log m = -(\log m - \log n) = -\log (m/n)$
(therefore, $\log (1/n) = -\log n$)

$\log n^m = m(\log n)$

$\log 1 = 0$

$\log 0$ and $\log \infty$ are indeterminate

Finally, it is essential to remember that the *mantissa*, or decimal portion of a logarithm to the base 10, is given in tables as a *positive* number. This positive mantissa tells us by what fraction of a log cycle the logarithm *exceeds* the whole-number logarithm (positive or negative), which marks the beginning of the cycle in which it lies. For example, $+0.750$, the mantissa for 0.562, is the same as the mantissa for 5.62 or 5,620 or 0.0000562. In each instance, the 0.750 tells us that the logarithm lies $75/100$ or $3/4$ of the way through its particular log cycle, always measuring from left to right along *scale E*. Therefore, to obtain the complete logarithm, we merely add the mantissa to the whole number, or *characteristic*, at the beginning of the log cycle. If the antilogarithm be expressed as in the second column of the following table, the characteristic will simply be the exponent of the 10:

ANTILOGARITHM		Characteristic plus Mantissa = LOGARITHM			
5,620.	or 5.62 (10^3)	3	$+$ 0.750	$=3.750$	
5.62	or 5.62 (10^0)	0	$+$ 0.750	$=0.750$	
0.562	or 5.62 (10^{-1})	-1	$+$ 0.750	$=\bar{1}.750$	$= -0.250$
0.000562	or 5.62 (10^{-4})	-4	$+$ 0.750	$=\bar{4}.750$	$= -3.250$

To facilitate conversion from logs to antilogs, negative logarithms are often written with a positive mantissa, *e.g.*, $\bar{1}.750$ or $9.750 - 10$. But whenever calculations are to be performed with the actual numerical value of a negative logarithm, it is almost always best to express it as a wholly negative number, *e.g.*, -0.25.

APPENDIX C. APPROXIMATIONS

True Value	Approximate Value	Fractional Error $\dfrac{\text{(True)} - \text{(Approx)}}{\text{(True)}}$	Approximate Value Is	Values for Which the Approximation Is in Error by about			
				0.01% of true value	0.1% of true value	1.0% of true value	10.0% of true value
$\dfrac{1}{1-x}$	$1+x$	x^2	Too small	$x = 0.01$	$x = 0.032$	$x = 0.1$	$x = 0.32$
$\dfrac{1}{1+x}$	$1-x$						
$(1+x)^2$	$1+2x$	$\left(\dfrac{x}{1+x}\right)^2$	Too small	$x = 0.01$	$x = 0.033$	$x = 0.11$	$x = 0.46$
$(1-x)^2$	$1-2x$	$\left(\dfrac{x}{1-x}\right)^2$	Too small	$x = 0.01$	$x = 0.031$	$x = 0.091$	$x = 0.24$
(A) $\sqrt{n} = \sqrt{xy}$ where $x \geq y$	$\dfrac{x+y}{2}$	$1 - \dfrac{1+(x/y)}{2\sqrt{x/y}}$	Too large	$\dfrac{x}{y} = 1.03$	$\dfrac{x}{y} = 1.09$	$\dfrac{x}{y} = 1.33$	$\dfrac{x}{y} = 2.43$
$\|\sqrt{x^2+y}\|$	$\left|x + \dfrac{y}{2x}\right|$	$1 - \dfrac{n+0.5}{\sqrt{n^2+n}}$	Too large	$n = 35$	$n = 11$	$n = 3$	$n = 0.7$
(B) $\|\sqrt{x^2-y}\|$	$\left|x - \dfrac{y}{2x}\right|$	$1 - \dfrac{n-0.5}{\sqrt{n^2-n}}$	Too large	$n = 36$	$n = 12$	$n = 4$	$n = 1.7$
		where $n = x^2/y$					

(A) This approximation provides a simple, and surprisingly accurate, method of obtaining square roots. Suppose you want the square root of n. Estimate its value and divide n by your estimate. Add the resulting quotient to your estimate and divide by 2. The result will usually be very close to the true value. If greater precision is needed, the procedure can be repeated.

Example: Estimate the square root of 7.78.

A little mental checking shows that 2.5 is too low, 3.0 is too high. We might, therefore, choose 2.7 as a provisional estimate.

$$7.78/2.7 = 2.8815$$

$$(2.8815 + 2.7)/2 = 2.791$$

This estimate is very close to the true value, 2.789.

(B) This approximation is often useful in solving quadratic equations. Any quadratic equation with a single variable z may be reduced to the form,

$$az^2 + bz + c = 0$$

whose general solution is

$$z = \frac{-b \pm \sqrt{b^2 - 4ac}}{2a}$$

If the absolute value of $4ac$ is small compared with b^2, we may be able to simplify the general solution very substantially by letting $b = x$ and $4ac = y$ in the approximation.

Example: Solve the equation $5z^2 - 20z + 3 = 0$

In this equation, $a = 5$, $b = -20$, and $c = 3$. Substituting these values into the general equation, we obtain the roots $z = 3.844$ and $z = 0.156$. But, if we let $b = x$, and $4ac = y$ in the approximation, we can reduce the general solution to

$$z = \frac{-b + \left(b - \dfrac{4ac}{2b}\right)}{2a} = -\frac{c}{b} = -\frac{3}{-20} = 0.150$$

and

$$z = \frac{-b - \left(b - \dfrac{4ac}{2b}\right)}{2a} = \frac{-2b + \dfrac{4ac}{2b}}{2a}$$

$$= -\frac{b}{a} + \frac{c}{b} = -\frac{-20}{5} + \frac{3}{-20} = 4.00 - 0.15 = 3.85$$

Exponential and logarithmic approximations. The following approximations represent one or more of the initial terms of certain infinite series. The infinite series themselves can be found in *The Handbook of Chemistry and Physics* (58). The tabulated values of x for errors corresponding to 0.1 per cent and 1.0 per cent of the true values are sufficiently accurate to serve as guides in deciding whether or not to use the indicated approximation. Better and better approximations may, of course, be obtained by using more and more terms, but this tends to spoil the simplicity which is the chief virtue of an approximation.

Quantity	Approximation	Values of x Which Will Give an Error of about	
		0.1% of true value	1% of true value
e^x	1	0.001	0.01
	$1 + x$	0.046	0.15
	$1 + x + (x^2/2)$	0.19	0.43
e^{-x}	1	0.001	0.01
	$1 - x$	0.044	0.13
	$1 - x + (x^2/2)$	0.17	0.36
$1 - e^{-x}$	x	0.002	0.02
	$x - (x^2/2)$	0.076	0.23
$\ln x$	$(x - 1)$	0.998 or 1.002	0.98 or 1.02
	$(x - 1) - [(x - 1)^2/2]$	0.946 or 1.054	0.83 or 1.17
$\ln (1 + x)$	x	0.002	0.02
	$x - (x^2/2)$	0.054	0.17
$\ln (1 - x)$	$-x$	0.002	0.02
	$-x - (x^2/2)$	0.054	0.17

Notice that both $1 - e^{-x}$ and $\ln (1 + x)$ have the same approximation in the above table. Hence, when x is sufficiently small,

$$(1 - e^{-x}) \simeq \ln (1 + x)$$

The arithmetic mean as an estimate of the logarithmic mean. Suppose that a quantity, y, declines exponentially toward an asymptote of zero. The *logarithmic* mean of y (see Section 6-15) between y_1 and y_2 is

$$\bar{y} = (y_1 - y_2)/\ln (y_1/y_2)$$

The *arithmetic* mean of y_1 and y_2 is simply $(y_1 + y_2)/2$. Let R be the ratio of y_1 and y_2 :

$$R = y_1/y_2$$

Then the *error* introduced by using the arithmetic mean instead of the logarithmic mean is

about 0.1 per cent when $R = 1.12$
1.0 per cent when $R = 1.41$
10.0 per cent when $R = 3.03$

APPENDIX D. SYMBOLS

Listed below are the major symbols employed throughout this book. In actual use, many of these symbols are made more specific by appending to them one or more subscripts. To list all of the specific variants thus obtained would require far too much space; the reader will usually find adequate definitions for each specific variant the first time it is used.

Symbols falling in the following five categories are in general not included in the list:

1. Symbols quoted from references for use in specific exercises or to illustrate points in the text.

2. Symbols used to designate dimensions (see Section 2-1).

3. Symbols used as *names* rather than as numerical quantities; for example, CO_2, carbon dioxide; S, a solute; P, plasma, etc.

4. Single-letter symbols (used very freely and often without specific definition) to denote general algebraic quantities, be they constants or variables. It is almost always clear from the context how these symbols are being used, so that there is little danger of their being confused with the symbols listed below.

5. Mathematical symbols such as $\sum$, "the sum of"; $f(x)$, "the function f of x"; $|x|$, "the absolute value (*i.e.*, sign ignored) of x." It is assumed that the reader is already familiar with standard mathematical notations such as these.

Symbol	Definition	Example of Units (Not Given for Dimensionless Quantities)	Defined or Discussed in
A	area	square meters	
a	acceleration	(feet per second) per second	
B	brightness	candles per square foot	
C	concentration	many	Table 2-2
c	velocity of light in a vacuum	2.998 (10^{10}) cm. per second	
(Cap)	capacitance	farads	
D	density	grams of x per cubic centimeter of x	Section 2-2
D	diffusivity or coefficient of diffusion	square centimeters per second	Equation 7-1a
E	energy	calories; ergs	
E	electromotive force or potential difference	volts	

427

e	base of natural logarithms	2.71828 ...	
(Eff)	effect of a drug	many	
F	force	dynes	
F	fraction		
G	gain (of an amplifier)		
g	acceleration due to gravity at earth's surface	980.665 (cm. per sec.) per second	
h	height	centimeters	
I	electric current	amperes	
I	input for amplifier	millivolts	
K	symbol used for various constants		
k	symbol used for various constants		
(K_{eq})	an equilibrium constant		Section 10-11
l	length	centimeters	
m	mass	grams	
N	number		
$\dot{N}$	number per unit time	per minute	
n	number		
"O"	output from amplifier	millivolts	
P	pressure	dynes per square centimeter	
P	probability		Section 6-8
pH	negative $\log_{10}$ of hydrogen ion concentration		Equation 10-32
pK_a	negative $\log_{10}$ of acid dissociation constant		Equation 10-33
pK_w	negative $\log_{10}$ of the "ion product" for water		Equation 10-38
Q	quantity	many	Section 2-2
$\dot{Q}$	quantity per unit time	many	
Q^*	quantity of an isotope used as a tracer		Section 9-7
q	electric charge (quantity of electricity)	coulombs	
(Q_{10})	temperature coefficient		Section 10-8
R	ratio		
R	ideal gas law constant	1.987 calories per °C-mole	
R	resistance to flow of electricity	ohms	
R	resistance to flow of fluid	mm. Hg/(liter per minute)	
r	radius	centimeters	
S	solubility of gas in liquid		Section 2-8
S	electric signal for amplifier	millivolts	
$s_{\bar{x}}$	standard error of mean of x	same units as x	
(Sp.ac)	specific activity. Ratio of isotope to ordinary atoms		Table 9-1
SS	sum of squares of deviations	same units as square of deviating variable	Section 3-10 and Exercise 3-4

T	temperature	degrees Kelvin or absolute	
t	time	minutes	
V	volume	liters	
$\dot{V}$	flow	liters per minute	
v	initial velocity of an enzymatic reaction	(millimoles per liter) per minute	Equation 11-7
$(\dot{V}\text{cl})$	clearance	milliliters per minute	Section 1–5 Section 8-4
$(V\text{dist})$	volume of distribution	liters	Section 8-5
W	weight	pound weight	Section 2-7
α	intrinsic activity of a drug		Section 1-4
α	Bunsen solubility coefficient or absorption coefficient	per atmosphere	Section 2-8
α	molar absorption coefficient	square centimeters	Exercise 6-4 (Discussion)
Δx	finite difference of, or change in x	same units as x	
η	viscosity	poise	
Π	osmotic pressure	dynes per square centimeter	Equation 2-18
π	ratio (circumference/diameter)	$3.14159\ldots$	

INDEX

Numbers in *italics* are not page numbers but give a Chapter number and an Exercise number. For example, *6-4* refers to Chapter 6, Exercise 4. For such entries, relevant material may be found both in the exercise at the end of the chapter and in the corresponding discussion in Appendix A.

Elogios para *Plataforma: cómo destacarte en un mundo ruidoso*

«He conocido a Michael Hyatt durante más de una década, y en todo ese tiempo lo he visto dominar casi todas las plataformas de medios sociales que aparecieron en escena. Ha usado el blog, Facebook, Twitter y otros para expandir su plataforma personal y pasar de ser un exitoso editor de libros a una marca nacional líder por sí misma. Créanme, este tipo sabe de qué está hablando... ¡así que presten atención!».

— **DAVE RAMSEY**
Autor de *best sellers del New York Times*
The Dave Ramsey Show

«Un libro generoso de un hombre que sabe de qué está hablando. Michael Hyatt ha construido una plataforma y tú también puedes hacerlo».

— **SETH GODIN**
Autor de *best sellers del New York Times*
Autor de *Somos todos raros*

«Una plataforma es absolutamente esencial para crear valor. Como gran fanático de Michael Hyatt, estoy entusiasmado de que comparta esto contigo. ¿Tu trabajo? Aprender esto e implementarlo. Tu éxito depende de ello».

— **CHRIS BROGAN**
Presidente de Human Business Works
Autor de *best sellers del New York Times*

«Michael Hyatt es una autoridad en el tema de crear una plataforma en nuestro mundo abarrotado, y *Plataforma* es la guía definitiva para construir paso a paso una plataforma... desde los cimientos. Si anhelas convertirte en un verdadero referente en este mundo, Hyatt puede enseñarte cómo. Con *Plataforma*, Hyatt ha logrado algo casi imposible: ¡crear una guía incluso *más* útil que su popularísimo blog!».

— **CLAIRE DÍAZ ORTIZ**
Innovación social
Twitter, Inc.

«He observado a Michael Hyatt construir su plataforma desde la base hasta convertirla en una de las más grandes del mundo. Y lo ha hecho con las estrategias y los consejos que presenta en este práctico libro. Cualquier autor, orador o pequeño empresario que quiera un plano para conseguir la atención y la visibilidad que desea, tienen que leer este libro».

— **JOHN C. MAXWELL**
Autor de *best sellers del New York Times*
y experto en liderazgo

«Cuando terminé la última página de *Plataforma*, me di cuenta de que tenía más notas útiles de este libro que de cualquier otro de negocios que hubiera leído en años. Esta es la guía definitiva para construir una presencia en línea. Michael Hyatt, uno de los pioneros de las redes sociales y el blog, comparte su exitoso plano para aumentar tu visibilidad. Aprende de su experiencia y ahorra tiempo, dinero y frustración siguiendo sus consejos paso a paso».

— **SKIP PRICHARD**
Presidente y gerente ejecutivo de
Ingram Content Group, Inc.

«En el mundo de hoy, tener una idea no es suficiente para lograr el contrato de un libro o de un disco, para hacer una película o conseguir fondos para comenzar una empresa. Se necesita una plataforma: seguidores conectados que estén dispuestos a probar, comprar y hacer correr la voz. Finalmente, en un solo libro, el plano para tu plataforma es revelado por el niño prodigio del blog Michael Hyatt. Terminarás este libro lentamente, porque sus consejos simples, pero poderosos harán que interrumpas la lectura con una urgencia imperiosa. Léelo y expande tu influencia».

— **TIM SANDERS**
Exdirector de soluciones de Yahoo
Autor de *Hoy en día somos ricos*